Statistical tests of the null hypothesis				
	Two-sample case		K-sample case	
One-sample case	Independent samples	Correlated samples	Independent samples	Correlated samples
Binomial test, p. 223 χ^2 one-way classification, p. 235	χ^2 test for two-way classification, p. 237 Fisher-Yates exact probability test, p. 243	McNemar test of change, p. 241	χ^2 test, p. 280	Cochran Q test, p. 286
	Median test, p. 246 Mann-Whitney U Test, p. 248	Sign test, p. 252 Wilcoxon matched pairs signed ranks test, p. 254	Median test, p. 282 Kruskal-Wallis one-way analysis of variance for ranks, p. 283	Friedman analysis of variance for ranks, p. 289
Student's t test, p. 216	Student's t test, p. 216	Student's t test, p. 221 Sandler's A test, p. 226	Analysis of variance, p. 262	

*Introduction to
statistics for the
behavioral sciences*

THE DORSEY SERIES IN PSYCHOLOGY

EDITOR   HOWARD F. HUNT   *Columbia University*

BARNETTE (ed.)   *Readings in Psychological Tests and Measurements*   rev. ed.

BARON & LIEBERT (eds.)   *Human Social Behavior: A Contemporary View of Experimental Research*

BENNIS, BERLEW, SCHEIN, & STEELE (eds.)   *Interpersonal Dynamics: Essays and Readings on Human Interaction*   3d ed.

COURTS   *Psychological Statistics: An Introduction*

DENNY & RATNER   *Comparative Psychology: Research in Animal Behavior*   rev. ed.

DESLAURIERS & CARLSON   *Your Child Is Asleep: Early Infantile Autism*

DEUTSCH & DEUTSCH   *Physiological Psychology*   rev. ed.

FISKE & MADDI   *Functions of Varied Experience*

FITZGERALD & McKINNEY   *Developmental Psychology: Studies in Human Development*

FLEISHMAN (ed.)   *Studies in Personnel and Industrial Psychology*   rev. ed.

FREEDMAN (ed.)   *The Neuropsychology of Spatially Oriented Behavior*

HAMMER & KAPLAN   *The Practice of Psychotherapy with Children*

HENDRY   *Conditioned Reinforcement*

KLEINMUNTZ   *Personality Measurement: An Introduction*

KOLSTOE   *Introduction to Statistics for the Behavioral Sciences*   rev. ed.

LIEBERT & SPIEGLER   *Personality: An Introduction to Theory and Research*

MADDI   *Personality Theories: A Comparative Analysis*   rev. ed.

MARKEL   *Psycholinguistics: An Introduction to the Study of Speech and Personality*

ROZEBOOM   *Foundations of the Theory of Prediction*

SALTZ   *The Cognitive Bases of Human Learning*

VON FIEANDT   *The World of Perception*

Introduction to statistics for the behavioral sciences

RALPH H. KOLSTOE
Professor of Psychology
University of North Dakota

1973 · Revised Edition
THE DORSEY PRESS *Homewood, Illinois 60430*
IRWIN-DORSEY INTERNATIONAL *London, England WC2H 9NJ*
IRWIN-DORSEY LIMITED *Georgetown, Ontario L7G 4B3*

© THE DORSEY PRESS, 1969 and 1973

Revised Edition

First Printing, May 1973

ISBN 0-256-01443-4
Library of Congress Catalog Card No. 72-96525
Printed in the United States of America

To SOREN OLAF KOLSTOE, Ph.D.,
Professor Emeritus of Psychology and Education,
State College, Valley City, North Dakota,
my father and first instructor of psychology.
It was he who first pointed out to me that,
if we were to advance our understanding of
human behavior, it would require greater use
of the methods of science including the use of
mathematics as a tool—especially that branch
of mathematics known as statistics.

Preface

THIS TEXTBOOK was written for a first course in applied statistics in that broad area called the behavioral sciences. I set a goal for the first edition that the book should be conceptually honest, should emphasize statistics as a tool to use in understanding behavior, but at the same time should be "readable." That purpose has not changed in this revision.

During the writing stages, multiple-choice test items were used to determine student comprehension of each portion of the book. If the failure rate on any given item was more than 40 percent, that section was rewritten and tested again the next semester. Most material was tested at least twice to insure that student comprehension was high.

The response to the first edition has been gratifying. This Revised Edition has profited by the thoughtful comments of instructors and students throughout the United States and Canada. This feedback has led me to delete, to add, and to modify material; but this Revised Edition still reflects my basic premises concerning statistics as they are applied to the behavioral sciences.

Intelligent use of statistics seems to involve three separate but related tasks: (1) decisions as to which statistical model is most appropriate for a given situation; (2) the computational facility to obtain the correct statistics numerically; and (3) the interpretation of the meaning of statistical answers.

The first of these, the appropriateness of a statistical model, is emphasized in the body of the textbook. Central to student understanding is some structure or framework into which the various concepts and techniques can be fitted. The measurement approach of the late S. S. Stevens (nominal, ordinal, interval, and ratio scaling) worked well in the first edition and has been retained in this Revised Edition.

The second task, computational facility, is approached in two ways.

Examples in the textbook are clearly separated from the on-going text information and the workbook has been revised to retain the close co-ordination with the text. The workbook materials have proved useful with a great variety of computational devices varying from hand computation in self-instructional situations to the use of remote terminals tied to large computers.

The third task, interpretation of statistical results, is emphasized throughout the text and new material was written for chapter 13.

Changes were made in all chapters. The notation for the mean has been changed to the use of a bar over the variable (e.g., $\overline{X}$). Additional tests were added, including the Mann-Whitney U test, the Wilcoxon signed-ranks test, the Cochran Q test, the Friedman analysis of variance for ranks, the Fisher-Yates exact probability test, Sandler's A test, and the contingency coefficient. The material in chapter 13 of the first edition on reducing the probability of the Type II error has been moved to chapter 10 and the entire treatment of hypothesis testing has been expanded. The material on matched samples which appeared in chapter 13 of the first edition was also moved to chapter 10. New material on matched samples was included in each of chapters 10, 11, and 12. Summary charts were added to chapters 8, 10, 11, and 13. The references at the end of each chapter were reviewed and changed as appropriate to direct students to sources for more advanced information.

Some instructors suggested that in the Revised Edition, chapter 2, "Organizing Data," and chapter 3, "Using Numbers in Measurement," should be reversed in order. I prefer introducing the students to numbers as quickly as possible and then presenting the scaling characteristics later. However, interchanging these two chapters has worked well for others with no apparent difficulty in student learning.

The additional statistical tests and other materials that have been added to the Revised Edition have made the text more flexible. The important task, to have students achieve a high degree of understanding of the fundamental concepts in statistics remains unchanged. By using the basic material included in chapters 1–8 and 10 and then selecting material from chapters 9, 11, 12, and 13, each instructor can build a course which best fits a particular situation.

Many individuals and agencies have contributed materially to the development of this book. The National Science Foundation provided funding for two summers to allow me to pursue advanced study in statistics at North Carolina State University and at the University of Florida. The Social Science Research Council provided financial support to allow me to participate in a seminar in data analysis and measurement theory conducted at Stanford University by Professor Lincoln Moses.

To all my students these past years, I owe a large debt. It is not possible to mention everyone who has made helpful comments for the Revised Edition, but a few must be mentioned. Invaluable and detailed criticism of the manuscript of this edition was provided by Professor Hilda Wing, University of North Dakota (Psychology); Professor Howard Hunt, Columbia University (Psychology); Professor J. D. Robson, University of Utah (Sociology); and Professor George Wolford, Dartmouth College (Psychology). A special debt is owed Ms. Ruth Smith, departmental secretary extraordinaire, who managed the myriad of details that surround the writing of a book—not the least of which was protecting my time for writing.

I am indebted to the Literary Executor of the late Sir Ronald A. Fisher, F.R.S., to Dr. Frank Yates, F.R.S., and to Oliver and Boyd, Ltd., Edinburgh, for permission to reprint Tables G, J, N, and a portion of Table D from their book, *Statistical Tables for Biological, Agricultural, and Medical Research.*

Last but certainly not least, I am indebted to my wife Carolyn for her constant support, encouragement, and understanding beginning with our undergraduate days and continuing to the present.

April 1973   RALPH H. KOLSTOE

Symbols used in text

X whole score in X

x deviation score in X

Σ summation sign—Greek letter sigma (upper case)

$\bar{X}$ sample mean

μ population mean—Greek letter mu

S standard deviation (sample value)

s standard deviation (estimate of population value)

σ standard deviation of population—Greek letter sigma (lower case)

S^2 variance (sample value)

σ^2 variance (population value)

$x!$ x factorial, product of x times all integers smaller than x

Z Z score

$s_{\bar{X}}$ standard error of the mean (estimate of population value)

$\sigma_{\bar{X}}$ standard error of the mean (population value)

S_p standard error of a proportion

C.I.   confidence interval

r Pearson product-moment correlation coefficient

Y' Y score predicted from regression equation

$S_{y \cdot x}$ standard error of estimate

S_r standard error of a correlation coefficient

ρ true correlation—Greek letter rho

r_s Spearman rank-difference correlation coefficient

C contingency coefficient

r_{xx} reliability coefficient

S_e standard error of measurement

$s_{D\bar{X}}$ standard error of the difference in means

C.R.   critical ratio

"t"   student's t test

A Sandler's A test

α probability of occurrence of a Type I error—Greek letter alpha

β probability of occurrence of a Type II error—Greek letter beta

df degrees of freedom

χ^2 chi-square test—Greek letter chi
U Mann-Whitney U test
T Wilcoxon T test
F F ratio
SS sum of squared deviations
MS mean squared deviations
H Kruskal-Wallis one-way analysis of variance for ranks
Q Cochran Q test
χ_r^2 Friedman analysis of variance for ranks

Contents

1. Introduction   **1**

Science and the study of behavior. Statistics and the study of behavior: *Descriptive statistics. Inferential statistics.* Measurement and the use of statistics. Need for statistics in the behavioral sciences.

2. Organizing data   **7**

Grouping scores: *Rearranging scores. Managing a large number of scores. Frequency distributions.* Real and apparent limits of numbers: *Continuous numbers. Discrete numbers.* Graphic representation: *Cartesian coordinates.* Description of frequency distributions: *Skewness. Kurtosis.* Percentiles and percentile ranks: *Cumulative frequency distribution. Computation of percentiles. Percentile ranks.*

3. Using numbers in measurement   **39**

Scaling: *Nominal scale. The ordinal scale. The interval scale. The ratio scale. Summary of the characteristics of nominal, ordinal, interval, and ratio scales. Scaling characteristics in the behavioral sciences.* Working with numbers and symbols: *Order of arithmetic operations. Fractions. Signed numbers. Simple algebraic equations. Algebraic summation.*

4. Measures of central tendency   **62**

Mean (arithmetic average): *Definition of the mean. Computation of the mean. Computation of the mean from grouped data. Coding scores for computation. Combining means.* The median. The mode. The geometric and the harmonic mean. Summary of the use of the mean, median, and mode: *Nominal data. Ordinal data. Interval or ratio data.* Central tendency in skewed distributions. Statistical inference: *Bias. Efficient. Consistent.*

5. Measures of variability   **79**

Determining variability for nominal data. Determining variability for ordinal data: *The semi-interquartile range.* Determining variability for interval and ratio data: *Average deviation. Standard deviation. Computation of the standard deviation. Coding scores for computation. Tchebysheff's inequality.* Summary of use of range, semi-interquartile range, average deviation, and standard deviation: *Nominal data. Ordinal data. Interval or ratio data.* Statistical inference: *Bias. Efficient. Consistent.* The Z score: *Comparing measurements from different distributions. Mean of Z scores. Standard deviation of Z scores.*

6.   **The normal curve and probability**   **104**

The normal curve: *The normal curve and human characteristics. Area of the normal curve. Uses of the table of the normal curve.* Probability: *Definition of probability. The additive theorem of probability. The multiplicative theorem of probability. Contingent probability. The binominal expansion and probability. The binomial expansion and the normal curve.*

7.   **Sampling theory**   **131**

Selecting a sample from a population: *Defining the population. Stratifying the population. Random selection. Multi-stage sampling. Sampling from a finite population. Use of random numbers.* Distribution of sample means: *Standard error of the mean. Confidence intervals. Central limit theorem.* Establishing confidence interval with nominal data: *Standard error of a proportion. Standard error and sample size.*

8.   **Correlation and regression**   **153**

Index of co-relationship between two variables. The use of r in prediction. Accuracy of prediction. Testing the significance of an r. Computation of the Pearson product-moment correlation coefficient: *Z score method of computing* r. *Deviation score method of computing* r. *Whole-score method of computing* r. Assumptions involved in the interpretation of r. Indices of association with noninterval data: *Spearman rank-difference correlation coefficient. Contingency coefficient.*

9.   **Use of statistics in tests and measurements**   **182**

Reporting test scores: *Reporting scores for nominal data. Reporting scores for ordinal data. Reporting scores for interval or ratio data. The Z score. The derived score.* The use of correlation in the evaluation of tests: *Reliability of tests. Validity of tests. Effects of unreliability on validity.*

10. **Significance of differences between two groups with interval or ratio scaling**   199

Establishing a confidence interval around the difference in means: *The standard error of the difference in means.* Testing specific hypotheses about differences in means: *The null hypothesis. Making decisions concerning the null hypothesis. Reducing the probability of the Type II error.* Student's *t* statistic. Matched samples: *Standard error of the difference in means. Degrees of freedom. Computation of* t. *Difference method of computing* t. *Sandler's* A *statistic.* Robustness of the *t* statistic.

11. **Significance of differences between groups with nominal or ordinal scaling (nonparametric statistics)**   232

Testing for significance of differences with nominal data: *The binomial test. The chi-square test. The McNemar test of change. The Fisher-Yates exact probability test.* Testing for the significance of differences with ordinal data: *The median test (independent samples). The Mann-Whitney* U *test (independent samples). The sign test (correlated samples). The Wilcoxon matched-pairs signed-ranks test (correlated groups).*

12. **Introduction to the analysis of variance and multi-group nonparametric statistics**   259

Testing for differences between variances. The F test for testing differences among means: *Analyzing variance. The F ratio for testing differences among means.* Computation of the analysis of variance: *Testing for significant differences between pairs of means.* Complex analysis of variance. Comparisons of several groups with nominal or ordinal scaling: *Testing for significance of differences among independent groups with nominal data (the chi-square test). Testing for significance of differences among K independent groups with ordinal data (the median test for more than two groups). Kruskal-Wallis one-way analysis of variance for ranks. Testing for significance of differences among K correlated groups with nominal data (The Cochran Q test). Testing for significance among K correlated groups with ordinal data (the Friedman analysis of variance for ranks).*

13. **Planning and interpreting experiments**   295

Characteristics of experiments: *Independent and dependent variables. Extraneous variables. Noncomparable groups.* Use of statistics in planning experiments: *Estimating sample size. Using subjects as their own control. The experiment as a "fair" test of the hypothesis.* Characteristics of the dependent variable: *Reliability of the dependent variable. Scale characteristics of the dependent variable.* Interpreting research results: *Random sampling. Proportion of variance accounted for. Behavioral sciences and human welfare.*

Appendix: Part A—Tables   318

A: Table of squares and square roots of the numbers from 1 to 1,000, 318

B: Percent of total area under the normal curve between mean ordinate and ordinate at any given Z score distance from the mean, 332

C: Table of binomial coefficients, 333

D: Values of r at the .05 and .01 levels of significance, 334

E: Values of r_s (rank-order correlation coefficient) at the .05 and .01 levels of significance, 335

F: Values of F at the .05 and .01 significance levels, 336

G: Distribution of t, 338

H: Critical values of Sandler's A, 339

I: Table of probabilities associated with values as small as observed values of x in the binomial test, 340

J: Critical values of chi-square, 341

K: Critical values in the Fisher-Yates test, 342

L: Probabilities associated with values as small as observed values of U in the Mann-Whitney test (small samples), 357

M: Critical values of U in the Mann-Whitney test, 360

N: Table of random numbers, 364

O: Critical values of T in the Wilcoxon matched-pairs signed-ranks test, 367

P: Critical values of χ^2_r in the Friedman ANV for ranks, 368

Appendix: Part B—List of computing formulas   370

Index   377

1

Introduction

EVER SINCE man first wondered about the nature of the world about him, he has speculated about man himself. The understanding of the behavior of animals had immediate survival value—to obtain food and to avoid becoming food. But the understanding of fellowman was also necessary for survival. We have developed many different approaches to the study of man. Certainly great literature and the arts are important approaches to understanding man. Within the past 100 years the study of man has taken a new and different approach. As the methods of science increased our understanding of the physical world, these methods began to be applied to the understanding of man himself. The very term *behavioral science* would seem to imply that the methods and techniques of science can be applied to the study of behavior.

SCIENCE AND THE STUDY OF BEHAVIOR

The use of science as a method commits the student of behavior to objectivity in his observation and requires that what has been observed must be capable of being verified by other trained and qualified observers. These requirements of objectivity and verifiability have led scientists toward mathematics as a tool in their research. Among the many advantages of mathematics for science is that of accuracy of communication. If the observer can express his observations in terms of numerals, other observers can test and check (verify) his observations more accurately and objectively.

Any attempt to pinpoint the first application of mathematics to the study of man would certainly create more controversy than the issue deserves, but the name of Fechner must stand out in any such account. Gustav Theodor Fechner (1801–87) was a German physicist. In his *Elements of Psychophysics,* Fechner attempted to relate changes in the physical world to changes in man's perception of the world. It was known during Fechner's time that the amount of change in a physical stimulus that is necessary to be perceived as a change is related to the magnitude of the original physical stimulus (Weber's law). Fechner extended these ideas to the area of psychophysical parallelism.

Wilhelm Max Wundt (1832–1920), at Leipzig, used mathematics in attempting to determine the elements and the structure of consciousness; and it was in Wundt's laboratory that the young American psychologist, James McKeen Cattel (1860–1944) began his studies on individual differences. In England, Sir Francis Galton (1822–1911) turned toward that branch of mathematics called statistics to aid in his studies of human differences. Émile Durkheim (1858–1917), the French sociologist, insisted that the social sciences must become empirical. His famous work *Suicide* was a statistical rather than moral or philosophical study of this phenomenon.

STATISTICS AND THE STUDY OF BEHAVIOR

Today the behavioral scientist uses any branch of mathematics which proves to be useful to him, but one branch of mathematics is used much more than others. This is statistics. The word *statistics* has the same root as our modern term *state,* referring to a political unit. Statistics was a necessary tool of the state. If taxes were to be levied and armies raised, it was necessary to enumerate many characteristics of the population. Although other uses of statistics have in recent years become relatively more important than simple enumeration, *descriptive* statistics are still essential to our complex society. Even in that small part of the world, the college or university, the administrator must know the average class size, the range of class sizes, how many classes have 5–10 students enrolled, how many have 35–45 students enrolled, how many have 300–1,000 enrolled, etc.

The field of statistics is often divided into two branches: (1) *descriptive* statistics, which is concerned with the enumeration and description of an existing situation; and (2) *inferential* statistics, which is concerned with conclusions regarding groups (population) arrived at inductively from the study of samples from the larger group.

DESCRIPTIVE STATISTICS.   Every ten years the U.S. Bureau of the Census is required by the Constitution of the United States to take a census of the population. This is a complete enumeration, and the data generated from this census would be considered descriptive. That is, the data describe the situation that existed at the point in time when the census was taken. The important characteristic of descriptive statistics is that *all* members of the group (population) are included. The size of the population is not a consideration. The population could be as small as the number of students in a specific course or, like the U.S. census, could include all the people in the United States.

INFERENTIAL STATISTICS.   During election years in the United States, we are bombarded with polls which attempt to predict the outcome of the election. These polls consist of a small sample of individuals who are "measured" with respect to their voting behavior, and from this sample an inference is made concerning the probable voting behavior of the entire population of voters. The important characteristic of inferential statistics is that a *sample* from a group (population) is used to make inferences about the population from which it is obtained. The question of the degree to which the sample is representative of the population becomes critical.

The prediction of the action of a population from the examination of a sample is important in all areas of human endeavor. Statistics, as a branch of applied mathematics, cuts across all branches of knowledge. Techniques developed in one area usually have application in others. One area which has both utilized and contributed to the development of inferential statistics is agriculture. During the past 20 or 30 years the United States has seen one of the most surprising reversals in history. We have been concerned with an *over*production of farm products that has led to many different programs designed to reduce the fantastic productivity of our farms at the same time that the percentage of the population engaged in agriculture has *dropped* drastically. This change in agricultural productivity can be attributed to a variety of factors, most important of which have been the application of science to agriculture and the implementation of scientific findings through automation and improved management techniques.

Let us explore the kind of study that is fairly typical in agriculture. The agricultural experiment station of one of our state universities has developed a new variety of wheat. This was developed by careful cross-breeding and selection. The new variety should have many desirable characteristics, including a higher yield per acre. The critical question now becomes: Will this new variety of wheat, which we shall call Nodak, give a higher yield of high-protein wheat than those varieties that are currently available? To test this question, seed is sent to many different experimental farms in the area, and Nodak wheat is raised along with the older varieties of wheat. After the grains have been harvested, it is possible to compare the average yield per acre of the different varieties of wheat seed. Let us assume that Nodak wins by approximately two bushels to the acre more than the other varieties. Is this sufficient evidence to warrant large-scale production? This is the kind of question that modern inferential statistics can answer. Statistics can give us an excellent estimate of what would happen if the experiment were repeated or if farmers were to switch to Nodak for large-scale planting. It is this type of application of science (the controlled experiment) and statistics (the planning and analysis of the experiment) that has done much to revolutionize agriculture in the past generation.

These same techniques can be and are being applied to problems of the behavioral sciences. From the research of sociologists, psychologists, cultural anthropologists, and educators, it would appear that the early experiences of children are an important factor in their later performance in the formal school situation. Is it possible to provide these early experiences to preschool children and determine the impact on later learning in the classroom? Children from "culturally deprived" environments could be selected and exposed to varying "enrichment" programs. Later the school performance of these children could be compared with that of other groups who did not receive these enrichment experiences. By using modern inferential statistics, it would be possible to determine if these early experiences are having the desired effect and if these types of programs should be instituted on a large-scale basis. Controlled experiments can test many of these ideas, and the experiments can be planned and analyzed by using modern statistics.

The use of statistics within the behavioral sciences varies a great deal. Psychology utilizes the techniques and methods of statistics in many of its endeavors. Political science has just recently increased reliance in this field.

MEASUREMENT AND THE USE OF STATISTICS

As was mentioned earlier, the scientist, whenever possible, attempts to express his observations in the form of numbers. Many areas of measurement are well established with standard units, as in the physical sciences with feet, pounds, etc. In many areas of the behavioral sciences, measurements are not nearly so sophisticated. But let us imagine ourselves back many hundreds of years before the development of our present measures of length. Suppose that it was important to measure the height of people. We might divide height into three major categories: tall people, middle-sized people, and short people. Then we could count the number of individuals in each of these classes. If these three categories of measurement were not fine enough for our purposes, we could break each category into finer classes such as tall-tall people, middle-tall people, and so on. Unfortunately, this is a rather loose form of measurement, in that the terms *tall, middle,* and *short* could be easily misinterpreted by others. Just how tall is middle? Short? It would be far better if we could find a more specific measurement unit of height than the categories of tall-tall, middle-tall, and so on.

We could use the unit that our English forefathers provided for us, the length of the king's foot. We shall now measure people as to how many king's-foot lengths they are in height. We would place a piece of wood alongside the king's foot and carefully make a mark on the wood to produce a portable measuring stick. We could put this stick alongside people and be able to state that this tall-tall person is 6½ king's feet tall. Unfortunately, our king is mortal; upon his death we have a new king, who happens to have bigger feet, and we discover that our tall-tall person is no longer 6½ king's feet tall but has shrunk to 6 feet. Yet we know that the person's height has not changed. Perhaps the measuring unit is a little more flexible than we desire. We then agree among ourselves that all kings shall have the same foot size, which shall be a fixed length marked on a piece of wood or stone. As kings come and go, we shall not measure their feet but rather keep this standard length, so that, as in the present case, where England's king is a queen (who I presume, has a smaller foot than her predecessor), it will not change the basic measuring unit.

In the behavioral sciences, in many cases, we have not arrived at the place where we have the equivalent of a king's foot which may be used as a standard. The sociologist describes differences in the social-economic sphere of human endeavor as upper class, middle class, and lower class, with finer gradations of upper-upper, middle-upper, etc., very similar to our old tall, middle, and short measurement of height. Is an IQ of 110 on the Stanford-Binet test the same as an IQ of 110 on the Wechsler-Bellevue scales? This lack of clarity of measurement does not mean that the behavioral scientist should despair of being able to use numbers and statistics in his studies. Fortunately, statistics have been developed which allow us to work with even the crudest forms of measurement.

Need for statistics in the behavioral sciences

The day is past when the serious student of any of the behavioral sciences can pursue his interest in depth without at least an elementary knowledge of statistics. More and more, the latest writings are using statistics to analyze and transmit new knowledge. It is virtually impossible to read a journal article or even an intermediate-level textbook without finding that much of the critical content of the article or book is expressed in the language of statistics. For the curious student, a sound knowledge of statistical techniques is essential for the research to explore those aspects of behavior which are as yet untapped.

The subject matter of the behavioral sciences is the behavior of man and lower animal forms. Statistics is not in and of itself a part of that subject matter; statistics is for the behavioral scientist what the microscope is for the biological scientist. Understanding the uses of a microscope does not mean that a person is a biologist, but the biologist would be severely handicapped in his work if he did not know how to use the microscope. So it is for the behavioral scientist. Rare is the present-day behavioral scientist who can hope to make important contributions to knowledge if he does not have an understanding of the uses of statistics.

If an introductory textbook is to be useful as a learning device, not all topics of interest or potential usefulness can be included. The first requirement of such a book is clarity of communication of the basic principles of the area. On the other hand, textbooks are used later as reference sources by students and should be written with that purpose in mind. In this book, some topics which may prove useful to the student in more advanced work have been omitted, but a list of carefully selected references has been placed at the end of each of the following chapters. These references were selected on the basis of three criteria: (1) the probable need of the student for the information at some future time, (2) the amount of mathematical background necessary to comprehend the advanced material, and (3) the availability of the references in most libraries.

The references included at the end of each chapter, except where noted, do not require mathematics beyond algebra. A student who has successfully mastered the material in this textbook should have a minimum of difficulty in understanding the topics as presented in the selected references.

2

Organizing data

Grouping scores

AN EXPERIMENT was reported[1] dealing with the persistence of an old,
no longer adaptive response in white rats. Certain procedures were followed which produced a strong persistent response to one arm of a Y
maze. After this training, food was shifted to the other arm of the maze,
and the number of trials before the animals shifted to the new food
side of the maze was recorded for each animal. For our purposes, we
shall examine the data from only two of the groups used in the larger
study. One group was allowed to continue in the maze with no change
in procedure except that the location of food had been shifted to the
arm of the maze opposite the arm that had been used in the preliminary
training. This group was labeled the zero-second-delay or control group.
The second group was delayed in a small chamber located just before
the choice point in the maze. The rats were delayed in this chamber
for five seconds and were labeled the five-second-delay or experimental
group. The data for these subjects are presented in table 2–1 in the order
that they were obtained from the experiment.

These are the data from the experiment. How do we extract meaning
from the numbers? How can we organize these numbers to make interpre-

---

[1] R. H. Kolstoe, M. Kleban, and A. Utecht, "Effects of Delay on the Maintenance
of Fixated Behavior in Albino Rats," *Genetic Psychology,* vol. 105 (1964), pp.
275–82.

TABLE 2–1

Zero-second-delay (control) group		Five-second-delay (experimental) group	
Subject	Number of trials before shift	Subject	Number of trials before shift
C1	42	E1	8
C2	40	E2	34
C3	43	E3	38
C4	16	E4	16
C5	42	E5	32
C6	44	E6	48
C7	28	E7	8
C8	22	E8	28
		E9	47

tation easier? By glancing over the numbers, we note that animal E6 took the greatest number of trials to switch to the new response, and animals E1 and E7 took the least number of trials. All three of these subjects were in the five-second-delay group. What about the zero-second-delay group? What is the range of scores in this group? Note that the highest score was 44, obtained by animal C6, and the lowest score 16, from animal C4. The range of scores for the zero-second-delay group appears to be 42 − 16 = 26; whereas the range in scores for the five-second-delay group appears to be 48 − 8 = 40. From this, it appears that there is more variation in the behavior of the animals in the five-second-delay group than in the zero-second-delay group.

REARRANGING SCORES

Perhaps if we rearranged these scores within each group from highest to lowest, we could obtain a clearer idea of what is happening when delay is introduced in the experiment (see table 2–2). From this arrangement, it becomes relatively simple to obtain a rough indication of the middle part of each of the two sets of data. Since there are eight subjects in the control group, the middle part of this distribution would be a point above which there are four subjects and below which there are four subjects. One half of the subjects in the control group scored 42 or above, and one half of the subjects scored 40 and below. The experimental group contains nine subjects. The middle part of this group can be found by locating the score of the subject that ranks fifth from the

TABLE 2–2

Zero-second-delay (control) group		Five-second-delay (experimental) group	
Subject	Number of trials before shift	Subject	Number of trials before shift
C6	44	E6	48
C3	43	E9	47
C1	42	E3	38
C5	42	E2	34
C2	40	E5	32
C7	23	E8	28
C8	22	E4	14
C4	16	E1	8
		E7	8

bottom of the group (or the subject that ranks fifth from the top of the group). Subject E5 is the middle subject by score, and its score was 32. From simple inspection of the data, we have been able to determine that the experimental group appears to be more variable in its behavior (wider spread or range of scores within the group), but the control group seems to have a higher middle part of the distribution, with the midscore in the interval between 40 and 42, whereas the midscore for the experimental group was 32.

There are several other ways that we could rearrange these numbers that might make it more convenient for us to determine similarities and differences between the two groups. One technique would be to order all 17 scores in one grouping from highest to lowest but identify each score as to control and experimental groups (see table 2–3).

From this arrangement of the scores, we can easily determine the midscore of all subjects (the score of the ninth subject, ignoring the control or experimental identification), which is 34 trials, the score earned by subject E2. We can also determine how many of the subjects in each of the two groups scored above this point or below this point. Note that three subjects from the experimental group (E6, E9, and E3) scored above this point, while five of the control subjects were above this middle score (C6, C3, C1, C5, and C2).

The data are thus arranged and organized for only one major purpose: the convenience of the investigator, to help him extract the maximum amount of information from the data. With just a few scores, as in this case, it is a relatively simple matter to rearrange the data for an adequate

TABLE 2–3
All subjects grouped together

Subject identification	Number of trials to switchover
E6	48
E9	47
C6	44
C3	43
C1	42
C5	42
C2	40
E3	38
E2	34
E5	32
C7	28
E8	28
C8	22
C4	16
E4	16
E1	8
E7	8

description of the results of the experiment. If we were to go a step further and ask what would happen if this study were repeated with another group of subjects, we would need to use *inferential* statistics. The question concerning the observed differences in apparent variability of the two groups and also the observed differences in the middle scores can be answered by inferential statistics. An examination of some of these inferential procedures will be postponed until chapters 10, 11, and 12 of the text.

MANAGING A LARGE NUMBER OF SCORES

As the range of scores and the number of scores increase, the simple rearrangement of scores that was used above is not usually adequate for the research worker to understand his data. Table 2–4 shows 80 scores obtained from a sample of freshman students. These 80 scores are arranged in the alphabetical order of the names of the students. What does this set of numbers tell us about the students? What is the highest score? The lowest score? With a little effort, we can determine that 96 is the highest score and the lowest score is 30. Where do most of the numbers fall? Are they evenly distributed between 30 and 96? Do they tend to "bunch" near 30? Near 96? These are questions that are difficult to answer when the numbers are arranged in this particular order.

TABLE 2–4
Test scores obtained from 80 freshman students

56	92	75	85	81
87	48	36	42	77
38	45	65	47	72
66	63	96	50	67
33	41	38	90	60
59	44	46	85	48
61	80	49	81	52
62	69	44	55	57
84	51	72	59	52
43	54	30	55	78
46	54	95	60	48
73	47	64	93	63
75	39	56	43	54
53	71	41	49	69
50	77	52	53	58
88	40	52	51	74

FREQUENCY DISTRIBUTIONS

These 80 scores could be reordered into a frequency distribution extending from 30 through 96 which will reduce the 80 different scores into 67 possible score values. This simple frequency distribution is presented in table 2–5.

GROUPED FREQUENCY DISTRIBUTIONS. Although the frequency distribution of table 2–5 is definitely an improvement over the arrangement of the numbers in table 2–4, it is still rather difficult to determine the characteristics of the data. What is needed is some method of categorizing or grouping these numbers so that there will be fewer than 67 groups of scores to examine. We could group the scores by units of 20: 30–49; 50–69; 70–89; 90–109. Certainly four categories of groups are much easier to examine than 67.

Note in table 2–6 that when this large a grouping is made, we have lost a great deal of information about the numbers. We are treating all scores from 30 through 49 as if they were identical. The easiest grouping we could make would be one category: 30–96. Now all scores are in one group, but we have lost all the advantages of measurement—we can no longer differentiate among the individuals. We are caught in a dilemma—on the one hand, it is much too difficult (even impossible) to "read" the meaning in a very large set of scores without some kind of categorizing or grouping of these scores; but on the other hand, group-

TABLE 2–5
Simple frequency distribution of test scores from 80 freshman students

Score	Frequency distribution	Score	Frequency distribution	Score	Frequency distribution	Score	Frequency distribution
96	1	79	0	62	1	45	1
95	1	78	1	61	1	44	2
94	0	77	2	60	2	43	2
93	1	76	0	59	2	42	1
92	1	75	2	58	1	41	2
91	0	74	1	57	1	40	1
90	1	73	1	56	2	39	1
89	0	72	2	55	2	38	2
88	1	71	1	54	3	37	0
87	1	70	0	53	2	36	1
86	0	69	2	52	4	35	0
85	2	68	0	51	2	34	0
84	1	67	1	50	2	33	1
83	0	66	1	49	2	32	0
82	0	65	1	48	3	31	0
81	2	64	1	47	2	30	1
80	1	63	2	46	2	$N = 80$	

ing leads to a loss of information concerning the differences in individual scores which are placed in the same category. Too few groups lead to too great a loss of information from the data; too many groups lead to difficulty in analyzing and understanding the data. Over the years, it has been found that when the number of categories is 20 or more, interpretation of the data becomes more difficult than necessary. But when the number of categories becomes less than 10 or 12, too much information is lost. An easy way out of this problem is to aim for 15 groups or categories of scores. This makes a nice compromise between 10 and

TABLE 2–6
Frequency distribution of the test scores from 80 freshman students (interval width = 20)

Score interval	Frequency distribution
90–109	5
70– 89	18
50– 69	33
30– 49	24
	$N = 80$

20. If we are to have about 15 categories, how wide will each be? What should the width of the interval (i) be?

SELECTING THE INTERVAL SIZE. If we divide the range of the scores by 15, we have the interval which will provide 15 categories. What is the range of our numbers? The highest is 96, the lowest 30. Is the range then 66? Not quite. When we subtract 30 from 96, we are actually excluding the 30 from the range. Since our data include the number 30, we want the range of scores starting with 30 through 96. Perhaps this can be illustrated by using smaller numbers. What is the range of scores from 5 through 10 (both 5 and 10 included)? Let us write the numbers and count them: 5, 6, 7, 8, 9, 10. Notice that there are six digits counting 5 through 10. Thus, $10 - 5 = 5$ because we are excluding 5 from the set of numbers under consideration. We can determine the range, then, by two equivalent methods: (1) Find the highest number and subtract from it one less than the lowest number in the set ($96 - 29 = 67$); or (2) find the highest number, subtract from it the lowest number, and add 1 ($96 - 30 + 1 = 67$).

Going back to the original problem of determining the interval size (i), we find that the range is 67, and we desire about 15 categories. If we divide 67 by 15, this should yield the correct interval size: $67 \div 15 = 4.467$, which we may round off to 4.5. Using the interval size of 4.5, we can begin to set up the groupings. Let us begin with 30 at the bottom of the first category. The next category would begin at 34.5, the next at 39, the next at 43.5, the next at 48, etc. This would become messy. Why are we grouping the numbers in the first place? We are grouping them for convenience, and decimals are not convenient for most people. Remember that there is nothing sacred about 15 categories. That is a convenient starting place for 12 and 20 categories. Instead of using $i = 4.5$, we have the option of using $i = 4$ or $i = 5$. Which is more convenient? Five is, for two reasons. First, man was developed with 5 digits on each hand, and from this we have developed a number system based on the digits of the two hands—10. Working with intervals of 5 or 10 seems to be relatively easy for most people. Secondly, after the numbers have been grouped, there are many situations which will require us to determine and work with the midpoint of the interval. All intervals that are even (2, 4, 6, 8, 10, 12, etc.) will have a fraction as the midpoint. What is the midpoint of the interval 4–7? The answer, of course, is 5.5. What is the midpoint of the interval 5–9? The answer is 7. Intervals that consist of an odd number have an integer (whole number) for the midpoint. Intervals that consist of an even number have a fraction for the midpoint. Suppose that in dividing the range by 15 to obtain a first approximation to a desirable interval size, we obtain $i = 11$. Should we use 11 as the interval size? Probably not. The midpoint of the interval would be an integer, but it is more convenient for most of us to work with multiples of 10 (10, 20, 30, etc.) than with multiples of 11 (11, 22, 33, etc), so in this case the convenience of the 10s system outweighs the convenience of the integer midpoint. As was pointed out previously, the basic reason behind all grouping is convenience of interpreting and understanding data. There is nothing difficult or magic in the decisions in grouping numbers. It is done for one basic reason and one reason only: convenience. If the grouping is made too "convenient," we may lose too much information. We then must strike a balance between convenience and the loss of information.

TABLE 2–7
Three frequency distributions of the test scores from 80 freshman students

$i = 3$		$i = 5$		$i = 20$	
Score interval	Frequency distribution	Score interval	Frequency distribution	Score interval	Frequency distribution
96–98	1	95–99	2	90–109	5
93–95	2	90–94	3	70– 89	18
90–92	2	85–89	4	50– 69	33
87–89	2	80–84	4	30– 49	24
84–86	3	75–79	5		
81–83	2	70–74	5		
78–80	2	65–69	5		
75–77	4	60–64	7		
72–74	4	55–59	8		
69–71	3	50–54	13		
66–68	2	45–49	10		
63–65	4	40–44	8		
60–62	4	35–39	4		
57–59	4	30–34	2		
54–56	7				
51–53	8				
48–50	7				
45–47	5				
42–44	5				
39–41	4				
36–38	3				
33–35	1				
30–32	1				
	$N = \overline{80}$		$N = \overline{80}$		$N = \overline{80}$

Returning once more to the original problem of establishing a grouping for the data in table 2–4, it would appear that an interval of 5 will best suit the needs of more than 12 categories but less than 20 and will provide an interval that is convenient. In table 2–7, the data presented in table 2–4 have been grouped using three different interval widths. Notice that with too many groups ($i = 3$) not much pattern is apparent in the data. Using $i = 5$, we can see a marked pattern in the numbers; but where we use just a few groups ($i = 20$), this pattern disappears again.

Real and apparent limits of numbers

CONTINUOUS NUMBERS

Suppose that we were to measure a man's weight and find it to be 185 pounds. Does this mean that the person weighs exactly 185 pounds? Is it possible that he weighs 185.1 pounds? Or 184.9 pounds? In this case, we are only interested in measuring to the nearest pound. We have no need for a finer discrimination in weight. Suppose we obtain an extremely accurate scale and determine that our person does in fact weigh 184.93 pounds. Is this exact? No. What is meant is that our person weighs at least 184.925 pounds but no more than 184.935 pounds. When we state that a person weighs 185 pounds, we are stating that of the three permissible numbers, 184, 185, and 186, this individual is closer to 185 than to the other two numbers. The *real* limits on the weight 185 are 184.5 to 185.5. Whenever we measure a *continuous* variable, the obtained number can never be exact. It indicates a band of possible numbers extending one-half unit below the recorded number and one-half unit above. This extension one-half unit above and one-half unit below is referred to as the *real limits* of the number. Perhaps a picture of the measurement scale will help clarify the relationship between the real and apparent limits, as in grouped data. Using $i = 5$, figure 2–1 shows these relationships for the interval 45–49.

Most continuous variables are measured in such a way that the midpoint of the measurement interval is recorded to stand for the entire interval. Thus, 185 pounds is the midpoint for the range of weights extending from 184.5 to 185.5. If we elected to record weight in 10-pound intervals, a weight of 160 pounds would refer to the range of 155 to 165 pounds. There are a few special cases in common usage where the stated measure-

FIGURE 2–1
Illustration of the relationship between the
real and apparent limits of an interval

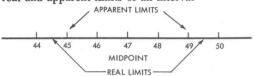

ment is not the midpoint of the interval under consideration. Age is an example of this. If people are asked how old they are, they normally report their age in years. Thus, if an individual states that he is 23 years old, what are the real limits of his age? Does this mean that he is at least 22 years and 6 months old, but not more than 23 years and 6 months? Not usually. In measuring age, most individuals report their age to their *last* birthday. Thus the person who states that he is 23 years old is at least 23 years of age but has not yet reached his 24th birthday.

DISCRETE NUMBERS

Suppose that we count the number of people in a classroom and find that there are 52 people present. What are the real limits of this number? What do we mean when we state that there are 52 people in the room? Does this mean 51.5 or 52.5? Of course not. Since we count people in units which are *discrete,* the numbers assigned to people or other discrete units are *discrete numbers.* The limit on the number assigned to the 52 people in the room is 52. The limit does not extend one-half unit below and one-half unit above the number because we are counting discrete units, not measuring a continuous variable.

Graphic representation

Often, we can get a much better idea of what the data are trying to tell us by drawing a picture of the scores. Perhaps it is too strong to say that "one picture is worth ten thousand words," but it is no exaggeration to say that a picture can usually make it much easier for the analyst to understand a set of data as well as to communicate the meaning to someone who is not so familiar with the basic data.

Graphing of data is simply representing the findings in the form of a picture—the better to convey the meaning of the numbers. There are no limits as to the kinds of graphs or pictures that could be used to represent a set of data; but since the purpose of graphing data is ease of communication, a number of rules or conventions have developed over time. These rules are not different in principle from many others we use. If you are driving an automobile in the United States, it is wise to drive on the right-hand side of the road. There is no particular reason why we couldn't drive on the left-hand side of the road, as do the British. What is necessary is that there be some agreement as to the "proper" side of the road. Where lack of agreement can lead to extremely serious consequences, laws are passed to enforce conventions. No one is apt to be seriously injured or killed because of neglecting to observe the conventions in graphing data, but the consequence is usually failure to communicate, and the very purpose of drawing the picture is defeated.

CARTESIAN COORDINATES

In graphing data, the Cartesian coordinate system is generally used—an X-axis and a Y-axis at right angles to each other. By convention, the horizontal axis (abscissa or X-axis) is used for the independent variable of the study, and the vertical axis (ordinate or Y-axis) is used to represent the dependent variable. Some years ago the writer was asked to review some data and the accompanying graph which had been prepared by a colleague. He had drawn a graph of his data, but it didn't "look right" to him. It was found that he had used the ordinate of the Cartesian coordinates for his independent variable and the abscissa for the dependent variable. Violation of the convention had confused the very person who had drawn the graph!

THE POLYGON. If the frequency distribution of a set of scores is to be presented graphically, the scores should be placed along the abscissa or X-axis, and the frequency of scores in each category should be plotted along the ordinate or Y-axis. Figure 2–2 shows the data from table 2–7, with $i = 5$, plotted in the form of a frequency polygon. Along the horizontal or X-axis, we place the score intervals, starting with 25–29, 30–34, etc. Along the vertical or Y-axis, we place the frequencies 1, 2, 3, etc.

Note that there are two individuals who scored in the interval 30–34. Since grouping the data has lost the identification of the individual scores, the midpoint of the interval is used in plotting these grouped data. The midpoint of the interval 30–34 is 32; thus a point, 32 on the X-axis and 2 on the Y-axis, is plotted to represent the frequency of scores in

FIGURE 2–2
A frequency polygon of the test scores from 80 freshman students

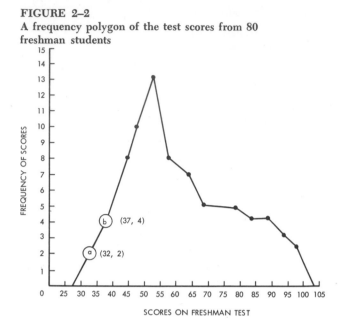

SCORES ON FRESHMAN TEST

this interval (see fig. 2–2). Notice that the midpoint of the interval using the apparent limits 30–34 is the same as the midpoint using the real limits 29.5–34.5. In the interval 35–39, there are four scores, so for this interval the plotted point is the midpoint of the interval on the X-axis (37) and the point 4 on the Y-axis. This procedure is continued until a point has been plotted depicting the frequency of scores in each category. These points are then connected with straight lines drawn between adjacent points.

In constructing a polygon, the "tails" of the graph are generally brought to zero to obtain a "closed" graph. This is done by using two additional categories: one below and one above the obtained scores. In Figure 2–2, this was done by adding the interval 25–29 at the low end of the X-axis and the interval 100–104 at the high end. The midpoint of the lower interval is 27, and the frequency of scores in this interval is 0; thus a point is marked at the coordinate points 27 and 0, and connected with a straight line to the coordinate point 32 and 2. A similar procedure is followed for the interval 100–104.

THE HISTOGRAM. Using the same Cartesian coordinate system and the same data that were used for figure 2–2, a different type of picture or graphic representation could be presented. For the interval 30–34 along the X-axis and at the point 2 along the Y-axis, we can draw a straight line from the point 29.5 and 2 to the point 34.5 and 2. We then draw a line from each end of this short line down to the X-axis. This produces a "bar" which represents the frequency in this interval. In fact, the term *bar graph* is probably used more often for this type of graphing than the more formal term *histogram*. One interesting feature of the histogram is that in each interval the area within the bar is directly proportional to the frequency within the interval. We could construct a small rectangle at frequency 1 across the entire interval. This would represent one score in the interval. The second "block" could then be placed on top of the first. The interval 34.5–39.5 would have four of these blocks. A histogram of the grouped data for the 80 freshmen is shown in figure 2–3. Should

FIGURE 2–3
A histogram of the test scores from 80 freshman students

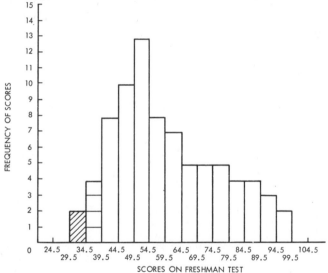

there be a gap between adjacent bars, or should they touch each other? Since a continuous variable is being graphed, the upper limit of the one bar should be the lower limit of the next higher bar.

Notice that in figure 2–3, the real limits are used rather than apparent limits. In actual practice most writers use the apparent limits rather than the real limits. This is done for convenience. Notice that the real limit of 29.5 requires four units of space on the graph. The apparent limit of 30 uses only two units of space. If nothing further is to be done with the graph, than using the apparent limits will not cause any difficulty. If, however, the graph is going to be used to estimate other characteristics of the data, then the use of the apparent limits will introduce an "error" of one half a unit.

If we were to graph the number of males and the number of females in a classroom, we would place the categories male and female along the horizontal or X-axis and the frequency along the vertical or Y-axis. A histogram could then be constructed to depict these data. Should the edges of the categories meet, or should a space be left? Since the categories male and female represent discrete categories rather than a continuous variable, a space should be left between these categories. Whenever a histogram is used to represent a set of data, the adjacent edges of the bars should touch if the data are of a continuous nature, and a space should be left between the bars if the categories represent a discrete variable. (See fig. 2–6 for a histogram based on discrete data.)

To show better the relationship of the histogram and polygon with the same data, these two graphs are superimposed on one another and presented in figure 2–4. The apparent limits have been used in drawing figure 2–4 since no further computations are to be based on this picture. As was mentioned earlier, the area under each bar of the histogram is proportional to the frequency of the scores in that category. The same

FIGURE 2–4
A frequency polygon and a histogram of the test scores from 80 freshman students

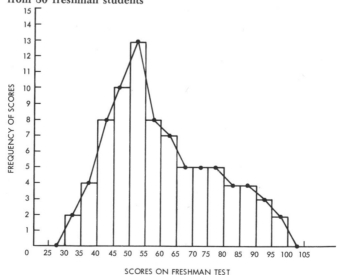

is not true for the polygon. Notice, however, that if the ends of the polygon are brought to zero on the X-axis, then the *total area* under the polygon is equal to the total area enclosed by the histogram. In either case, with a continuous variable the area under the histogram and the area encompassed by the polygon are proportional to the total frequency of scores.

The data presented have concerned test scores, and these have been graphed as if the data were continuous. Should test scores be treated as continuous or discrete? Most tests are scored by assigning an integer for a correct answer and a zero for an incorrect answer. In a 50-item multiple-choice test a student may have 40 correct answers and 10 incorrect, and obtain a score of 40. In the manner of scoring these tests, it is not possible for a student to obtain a score of 39.6. He can only obtain scores in discrete units. Either the item is correct, or it is incorrect. He either earns a point or does not earn a point. With a variable such as weight, it is always possible to use a finer scale and obtain as precise a measurement as needed. We treat tests scores as continuous variables even if at first glance it appears that they should be discrete. It is assumed that the underlying variable being measured is continuous and the integer nature of the scores is an artifact of the technique used in the measurement. In the typical test there are some items that are, in fact, not "test" items at all. The answer is so obvious that it is clear to almost all that it is the correct answer. There are also items which for a given student seem to have all correct answers. In this latter case the student may utter a little prayer and pick an answer. If lucky, he receives a point for his "knowledge." If unlucky, it is scored zero. For a group of students the one item may range from extremely easy for one student to extremely difficult for another. Yet all students who do select the correct answer are measured as if they possessed exactly the same amount of knowledge concerning that item. But this is not different from measuring weight. Many individuals may obtain a weight "score" of 185 pounds, but this reflects the crudeness of the measurement or the fact that a more precise measure is not needed. If weight is measured in whole pounds, this does not mean that the variable being measured is discrete. A decision as to the discrete or continuous nature of a variable should be based on the conceptualization of the variable, not on the particular technique of measurement.

OTHER CONVENTIONS IN GRAPHIC REPRESENTATION. In figure 2–5 the data on freshman test scores have been graphed in three different polygons. In one sense, all of these graphs are correct. The correct frequency has been plotted with the appropriate score category. But the graphs certainly do not look the same. Figure 2–5*a* seems to indicate

FIGURE 2–5
Three polygons of exactly the same data, illustrating the differences in the ratio of the Y-axis to the X-axis (figure 2–5*a* has a ratio of 3/1; figure 2–5*b* has a ratio of 3/4; and figure 2–5*c* has a ratio of 1/3)

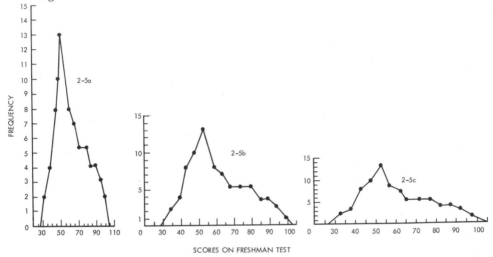

SCORES ON FRESHMAN TEST

that the data are extremely "bunched" in the middle part of the distribution. Figure 2–5*b* has the same general appearance as figure 2–2, and figure 2–5*c* seems to indicate a relatively flat distribution of the data. The difference in these three figures is the ratio of the X-axis to the Y-axis. Over the years, it has become standard practice to have the Y-axis approximately two thirds to three fourths the length of the X-axis. This convention has been followed so much that we almost automatically assume that this approximate ratio is being maintained when we examine a graph. Notice that in figure 2–5*a* the Y-axis is about three times the length of the X-axis, producing the appearance of a tightly bunched distribution. In figure 2–5*b* the ratio of the Y-axis to the X-axis is approximately ¾, and thus it looks like the other figures where these data were presented before. In figure 2–5*c* the ratio of the Y- to the X-axis is about ⅓, giving the appearance of a flat distribution of a large range. Since the purpose in graphing is ease of communication, it is rather imperative that this convention of the relationship between the X- and Y-axis be followed. Where the convention is violated seriously, communication undoubtedly occurs, but it may be the communication of an erroneous impression.

By convention, the scores along the abscissa or X-axis usually start with a number slightly smaller than the lowest obtained score or score interval. In the graphs of the data obtained from the 80 freshman, the smallest number on the X-axis has been 25. Although it is not necessary to start the X-axis at zero, this is necessary for the Y-axis. The Y-axis should have zero at the point where the two axes join. This convention can often create a problem. Figure 2–6 shows some data collected from

FIGURE 2–6
Two different graphs of the same data to illustrate a discontinuous graph representation

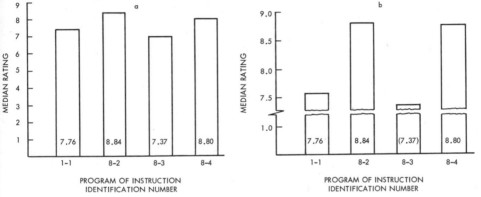

electronics maintenance specialists in the United States Army.[2] Notice that the ratio of the ordinate to the abscissa is $3/4$ and the ordinate starts at zero. The data are presented as a histogram because the categories along the base line (X-axis) do not represent a continuum in the usual sense. From this graph, it would appear that there was little or no difference in these groups. Statistical analysis of these data indicated that there were statistically significant differences between the groups. The graph does not communicate these differences because the scale units used on the ordinate are too large relative to the measurements used. To present finer scale units, it would be necessary to violate either the convention of the ratio of the two axes or the convention that the Y-axis must start

[2] R. H. Kolstoe, R. S. Czeh, and G. B. Rozran, *Ordnance IFC Electronics Maintenance Personnel: Analysis of Activities with Implications for Training,* Technical Report 37 (Washington, D.C.: Human Resources Research Office, George Washington University, March, 1957), Part II, T–38.

at zero. Figure 2–6*b* conveys the differences much better than figure 2–6*a*. Notice that the ordinate begins at zero but the straight line is broken to indicate that a segment of the graph has been removed. This is a warning to the reader to pay careful attention, in that something is not standard. Not all graphs using Cartesian coordinates maintain the convention of the Y-axis starting at zero. In the section on correlation, most of the graphs (scatterplots) of bivariate distributions (relationship between two variables) will not follow this convention. In the case of graphic representation of independent and dependent variables or frequency distributions, it is wise to follow the above convention to insure that the reader of the graph will interpret the data correctly.

LABELING OF GRAPHS. It is perhaps a waste of time and energy even to discuss the question of labeling of data—especially graphs. Anytime one picks up a book, magazine, or journal, the graphs are labeled adequately. They give the reader all the relevant information that is needed to interpret the graph. There is a title to convey the general idea of what is being presented. The two axes are labeled so that the reader can easily determine what is being plotted on the X-axis and the units being used, and the same is true of the Y-axis. Any special labeling of different lines is included. This is done, of course, because the author is concerned about the communication process and the material must pass through the hands of at least one editor who will be alert to the problem of adequate communication.

Graphs are generally labeled very clearly when the intention is to publish the material. But most graphs are first drawn so that the investigator can better understand his own data. Since he knows what is being graphed, he often does not bother to label completely. The graph is then placed in a file, and sometime later the author cannot understand his own picture. Without proper labeling, the graph may soon lose its power to communicate—even to the person who drew the picture.

Description of frequency distributions

When data are plotted in a histogram or polygon and there are a reasonably large number of scores, a distinctive pattern begins to emerge. With a great deal of measurement of humans and human activities, the distribution of measures tends to have a few low scores, many scores "bunched" in the middle of the distribution, and a few "high" scores. Such a distribution is shown in figure 2–7. If a line were drawn through

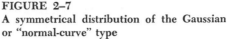

FIGURE 2–7
**A symmetrical distribution of the Gaussian
or "normal-curve" type**

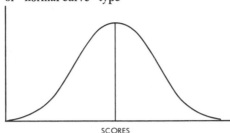

SCORES

the middle of the distribution and the one half superimposed on the other, we could see that the curve is symmetrical; that is, each half is the mirror image of the other half. As was stated above, many data from living matter tend to distribute themselves in a symmetrical "bell-shaped" manner. One special symmetrical curve that is bell-shaped is called the *Gaussian*[3] *distribution* or the *normal curve*. This particular symmetrical distribution has great importance in the field of statistics, and we shall examine some of the properties of this normal curve in chapter 6.

The term *normal curve* should not be interpreted to mean that we expect all distributions to fall into this particular symmetrical bell-shaped form—or that only behavior which shows this symmetrical type of distribution is normal behavior. Much activity of interest to the behavioral scientist follows distributions quite different from the symmetrical bell-shaped type, and it is still "normal" in the sense of usual or to be expected.

SKEWNESS

The distribution of family income in the United States is definitely nonsymmetrical. A few families have annual incomes in the million-dollar range. The distribution of family income has an extremely long "tail" in the direction of high incomes. A nonsymmetrical distribution is called a *skewed distribution*. Techniques are readily available in advanced texts to compute the degree of skewness. When this is done, it is found that a symmetrical distribution has zero skewness. A nonsymmetrical distribution in which the "long tail" is toward the high scores will have an index of skewness which is positive in sign, and we refer to such a dis-

---

[3] Named for the German mathematician C. F. Gauss (1777–1855).

tribution as being *positively skewed*. The distribution of annual family income in the United States is positively skewed. If the long tail is in the direction of the low scores in the distribution, the index of skewness will be negative in sign, and we refer to the distribution as being *negatively skewed*. When the index of skewness is computed, we can determine the exact degree of skewness from the size of the index (large index indicates a marked nonsymmetry or skewness; small number indicates small degree of skewness) and the direction of skewness from the sign of the index (a positive sign indicates that the long tail is in the direction of high scores). In general, it is not necessary to know the exact degree of skewness of a set of data, and a visual examination of the frequency distribution is usually adequate to determine negative, zero, or positive skewness. (See fig. 2–8.)

FIGURE 2–8
Illustration of two skewed frequency distributions

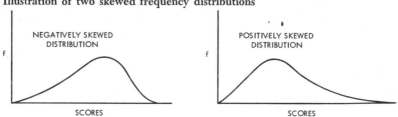

KURTOSIS

Figure 2–9 shows three symmetrical distributions which are quite different from each other. All three have zero degree of skewness, but the distribution of figure 2–9a appears very narrow and "peaked" in the center. The middle distribution has the familiar bell shape, while the distribution pictured in figure 2–9c is relatively flat. The degree of flatness or peakedness of a frequency distribution is referred to as *kurtosis*. A peaked distribution such as pictured in figure 2–9a is called *leptokurtic*. The Gaus-

FIGURE 2–9
Illustration of three degrees of kurtosis

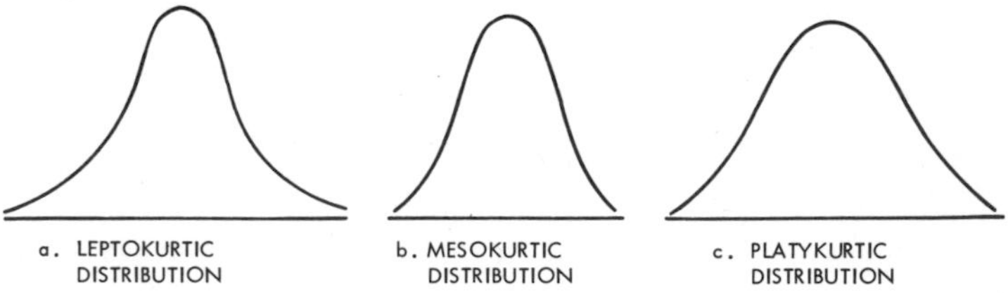

a. LEPTOKURTIC   b. MESOKURTIC   c. PLATYKURTIC
   DISTRIBUTION   DISTRIBUTION   DISTRIBUTION

sian or normal distribution is described as *mesokurtic;* and the wide, flat type of distribution is called *platykurtic.* As in the case of skewness, it is relatively easy to compute the degree of kurtosis of a distribution, and the computing formulas may be found in most advanced textbooks. For most purposes the general descriptive terms *leptokurtic, mesokurtic,* or *platykurtic* are sufficient to describe the general shape of the distribution. This can be determined accurately enough by visual inspection of the frequency distribution.

The two general descriptive terms of skewness for departure from symmetry and kurtosis as to degree of peakedness or flatness are usually sufficient to communicate the shape of a frequency distribution.

Percentiles and percentile ranks

After a set of scores has been obtained, we are usually interested in the relative position or meaning of these scores. A student who receives a score of 63 on an examination is not usually satisfied with just knowing the score. What was the highest score on the examination? The lowest? Does the 63 indicate a *C*? One way to describe the relative position of a score is in terms of the percentage scoring above or below that particular point. In the beginning of this chapter, some data from a fixation study with white rats were presented. In those data it was determined that the midscore for the group of 17 subjects was a persistence score of 34 trials. This tells us that about 50 percent of the subjects had scores above 34 and 50 percent had scores below 34. Determining relative standing is reasonably straightforward in working with a small set of scores. With a large number of scores, it is usually more convenient to work with grouped distributions.

CUMULATIVE FREQUENCY DISTRIBUTION

To facilitate the determination of the relative ranking of scores, it is helpful to add another element to the frequency distribution. This is done by adding a column to the distribution labeled *cumulative frequency.* In this column we enter the frequency of cases in and below each specific category. In table 2–8, we have used the data obtained from the 80 freshman students with two additional columns for cumulative frequency and cumulative percentage. Notice that in the columns labeled *cumulative frequency* the bottom category has the frequency of two. The

TABLE 2–8
Cumulative distribution of the test scores from
80 freshman students

Score interval	Frequency distribution	Cumulative frequency	Cumulative percentage
95–99	2	80	100
90–94	3	78	98
85–89	4	75	94
80–84	4	71	89
75–79	5	67	84
70–74	5	62	78
65–69	5	57	71
60–64	7	52	65
55–59	8	45	56
50–54	13	37	46
45–49	10	24	30
40–44	8	14	18
35–39	4	6	8
30–34	2	2	2

category 35–39 has the entry in the lower category (two) plus the frequency in its own category (four) for a total cumulative frequency of six. The same procedure is followed for each subsequent category. In the interval 80–84 the cumulative frequency is 71, which is the sum of the cumulative frequency below this category (67) plus the frequency of 4, which is the frequency in the category. For the last (top) category of 95–99, the cumulative frequency is 80, which is the total number of scores. The cumulative frequency column shows the number of scores in or below the category in which we are interested.

After the cumulative frequency column has been computed, it becomes a simple matter to compute a cumulative percentage column for the data. We simply divide each entry in the cumulative frequency column by the total number of cases in the entire distribution. Referring to table 2–8, if we divide the cumulative frequency for category 30–34 (two) by the total frequency of scores (80), we obtain 2.5 percent. We do not record percentages to decimals in most situations, so this figure would be rounded to 2 percent and entered in the cumulative percentage column. In the category 80–84, we divide the cumulative frequency of 71 by the total frequency of 80 and obtain the cumulative percentage of 89 percent. Note that the entry in the last (top) category will always be 100 percent. The cumulative frequency distribution and the cumulative percentage dis-

FIGURE 2–10

Cumulative frequency distribution and cumulative percentage distribution of scores on freshman test

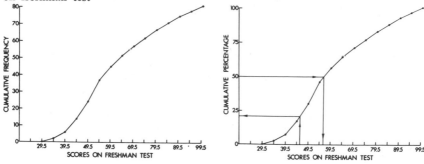

tribution are shown in figure 2–10. Since the cumulative percentage graph is to be used to compute some characteristics of this data, care was used to label the curve with the *real* limits rather than the apparent limits (see p. 19).

The cumulative percentage curve allows us to answer such questions as: What is the middle score in the distribution? What is the dividing point for the top 10 percent of the group? Any such question concerning percentage standing can be read directly from a graph such as that shown in figure 2–10. This type of cumulative curve is referred to as an *ogive*.

Let us turn to the problem of determining the 50-percent point for the data from the 80 freshmen. We are looking for that point in the distribution above which are located 50 percent of the cases and below which are located 50 percent of the cases. We can estimate this middle point by using the graph of cumulative percentages. In figure 2–10 a line has been drawn horizontally from the 50-percent point on the Y-axis to the point where it intersects the cumulative percentage curve. At this point a second line has been drawn down to the X-axis. Using this graphic approach, we would estimate that the middle score is about 57. The midpoint in a distribution is often called the *50th percentile,* since it is that point below which 50 percent of the scores fall and above which we find 50 percent of the scores. Another common name for the 50th percentile is the *median*—it is the middle of the distribution.

COMPUTATION OF PERCENTILES

If more accuracy is required for the median, it can be computed from the grouped distribution. In table 2–8 the cumulative percentage column has an entry of 46 percent for the interval 50–54 and an entry of 56 percent for the interval 55–59. This means that the median must lie within the interval 55–59. The median cannot be as high as 59.5 (upper real limit of the interval), since we know that 56 percent of the scores are at or below this point. The median cannot be as small as 54.5, since 46 percent of the scores are at or below that point. In order to compute the median, we must first multiply the total N by 50 percent to determine the number of cases above and below the median. In this case $N = 80$, so we shall need to identify the 40th case. In the cumulative frequency column we note that there are 37 cases cumulated through the category 50–54. It should be noted that when the scores have been accumulated through the category, we are referring to the upper real limit of the category. In this case, 37 scores are at or below the point 54.5. In order to find the point at or below which we have 40 cases, we shall have to move three scores into the next higher interval. When we grouped the data, we lost the identity of the individual scores within the category, so we cannot pinpoint the lowest three scores of the eight scores in the interval 55–59. The best assumption that we can make is that these eight scores are spread evenly over the entire interval; thus, we need to find three eighths of the distance within this interval. Since the score distance within the interval is five units, we must multiply the total score distance (five) by three eighths to obtain 1.9, which is the number of scores needed. The next step is to add this distance of 1.9 scores to the lower real limit of the interval ($1.9 + 54.5 = 56.4$). The median of the distribution is 56.4; that is, 50 percent of the scores are 56.4 or lower, and 50 percent of the scores are 56.4 or higher. Notice that this is quite close to the estimate made from visual inspection of the cumulative percentage graph.

Figure 2–11 further illustrates the process of shifting from order of observations to the score which represents the midpoint of the order. The data from the interval 55–59 of table 2–8 are used for this illustration.

In determining percentiles, we are looking for a score such that a given percentage of individuals scored at or below this point. Many schools organize the freshman English class into groupings depending upon the background of the students. It might be determined that 15 percent of the students are ready to profit from an accelerated or "enriched" program and 15 percent need remediation in basic English, while the middle 70 percent of the new freshmen should be enrolled in the regular college English course. What score on the English placement examination is such that 15 percent of the students will score at or below this point? What

FIGURE 2–11
Illustration of computation of the median from grouped data

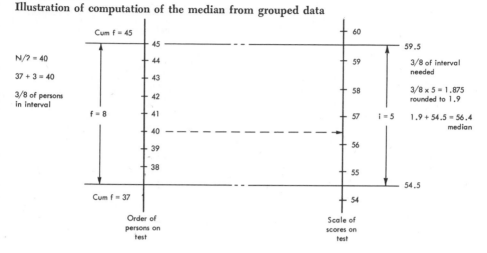

score is such that 85 percent of the students will score at or below this point?

An inspection of the cumulative percentage column of table 2–8 reveals that the 15th percentile must be in the interval 40–44. The interval 35–39 shows a cumulative percentage of 8 percent, whereas the interval 40–44 shows a cumulative percentage of 18 percent. The next question is how many of the scores must be at or below the 15th percentile point. This can be found by multiplying the total number of scores (N) by the given percentage desired (15 percent). Thus, we are seeking the score of the 12th person from the bottom of the distribution. The full computation of the 15th percentile is given in example 2–1.

It is rather difficult to express the steps in the computation of percentiles in a neat, concise manner. A summary of the steps is presented below:

$$\text{percentile} = \left[\frac{(N)(\%) - c.f. \text{ in interval below}}{f}\right] \times i + l.r.l. \qquad (2\text{–}1)$$

where:

N is the total number of cases
%   is the desired percentage
$c.f.$ is the cumulative frequency in the interval below
f is the frequency of cases in the interval
i is the width of the interval
$l.r.l.$ is the lower real limit of the interval

Example 2-1   Computation of percentiles from grouped frequency distribution

Determination of any percentile	*Determination of the 15th percentile*
I. Multiply the total frequency (*N*) by the percentage to determine the number of scores at or below that point	I. $80 \times 15\% = 12$ scores
II. Find the score corresponding to the *n*th individual:	II. Find the score corresponding to the 12th individual:
A. Locate the class interval in which the *n*th case is recorded in the cumulative frequency column	A. Class interval 40–44 has the cumulative frequency of 14
B. Subtract from the *n*th case the cumulative frequency of the interval below the one in which the *n*th case is located	B. $12 - 6 = 6$
C. Divide the value from step B by the frequency distribution in the interval in which the *n*th case is located	C. $6/8 = .75$
D. Multiply the result of step C by the interval width (*i*)	D. $.75 \times 5 = 3.75$; round to 3.8
E. Add the result of step D to the upper real limit of the category just below the one in which the *n*th case is located	E. $3.8 + 39.5 = 43.3$

The 15th percentile is equal to the score of 43.3

The following numerical expression of formula 2–1 is taken from the data of example 2–1:

$$\left[\frac{(80)(.15) - 6}{8} \right] \times 5 + 39.5 = 43.3$$

Although the computation of percentiles is straightforward, their interpretation is often incorrect. The percentile is a statement of the point at or below which a given percentage of scores fall. The percentiles indicate a ranking but not distance between scores; thus the *distance* between the 55th percentile and 60th percentile is *not* necessarily the same as the *distance* between the 90th percentile and 95th percentile. Percentiles show the order of individuals, not the distance between individuals. From the above statements, we could say that 5 percent of the scores are between the 55th percentile and the 60th percentile, and also that 5 percent of the scores are between the 90th percentile and the 95th percentile.

The median was computed for these data and was found to be 56.4. In example 2–1, we found the 15th percentile to be 43.3. From table 2–8, it is a simple matter to compute the 85th percentile, which is 80.75, rounded to 80.8. Suppose that, quite incorrectly, someone found the "average" of the 15th percentile and 85th percentile—$(.15 + .85)/2 = .50$ and $(43.3 + 80.8)/2 = 62.05$. The conclusion might be that the median (50th percentile) is equal to 62.05. This is incorrect, since we have already computed the median for these data and found it to be 56.4. Percentiles indicate the *relative ranking* of scores; they do not indicate the distance between scores. Since percentiles involve ranking, it may be very misleading to "average" percentiles.

PERCENTILE RANKS

A percentile is always a score. The 15th percentile of the data presented in table 2–8 was the score of 43.3. Often, we are interested in the reverse question. John Jones received a score of 47 on the test administered to the freshmen. How does his score compare to the other 79? We are interested in determining the *percentile rank* of the score of 47. Since we are asking the reverse question, the determination of the percentile rank of a score is the reverse procedure for finding the score which corresponds to a given percentile. Referring to table 2–8, let us examine the problem of determining the percentile rank of the score of 47. Note that the score of 47 is in the interval from 45 to 49; also note that 30 percent of the scores are at or below 49.5 (the upper real limit of the interval) and 18 percent of the scores are at or below the score of 44.5 (the lower real limit of the interval). The score of 47 must have the percentile rank somewhere between 18 percent and 30 percent.

One method of determining the percentile rank of the score of 47 is to use the cumulative percentage ogive. This procedure is illustrated in figure 2–10. It is first necessary to locate the score of 47 along the abscissa. The next step is to draw a line directly up to the ogive and then draw a line from the ogive to the ordinate. As shown in figure 2–10, this graphic solution yields a percentile rank of 23 for the score of 47.

To determine the specific percentile rank arithmetically, we shall have to determine the number of scores which are at or below the score of 47.

We must first determine how far the score of 47 is into the interval. This can be found by subtracting: $47 - 44.5$ (the lower real limit of the interval) $= 2.5$ units into the interval. What proportion is this of the total interval? This may be found by dividing 2.5 by the interval width $(i = 5)$, or $2.5/5 = .50$. This tells us that our score of 47 is one half the distance into the interval 45–49, so we shall need to find one half the people who scored in this interval. Examination of table 2–8 indicates that 10 scores are in the interval 45–49, and we need to find one half of them, which is $5(10 \times .50 = 5)$. We have now determined that five individuals scored 47 or lower, but higher than 44.5. We need to find the total number of individuals who scored at or below 47. This can be determined by adding the cumulative frequency up to 44.5 (14 individuals) to the five individuals in the range of 44.5–47 $(14 + 5 = 19)$. But we wish to determine the percentile rank of the score of 47, so we must divide 19 by the total number of scores (80). Thus the percentile rank of the score of 47 is $19/80 = 23.7$ percent, rounded to 24 percent. This tells us that 24 percent of the scores in the distribution are located at or below the score of 47. Notice that this is quite close to the estimate made from visual inspection of the cumulative percentage ogive.

Figure 2–12 further illustrates the process of shifting from the score

FIGURE 2–12
Illustration of computation of percentile ranks from grouped data

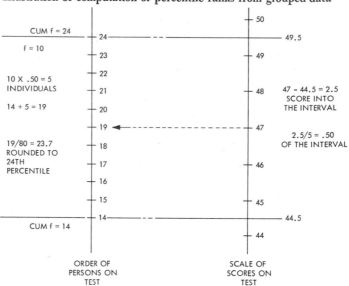

of the individual to the order of observations. The data from the interval 45–49 of table 2–8 are used for this illustration. The specific steps for computing the percentile rank of a given score and an example are given in example 2–2.

Example 2-2 Using the data in table 2-8, determine the percentile rank of the score 83

I. Determine the number of cases falling at or below the specific score
   A. Subtract from the score the lower real limit of the interval in which the score is located
   B. Divide the result of step A by the interval width (i)
   C. Multiply the result of step B by the number of cases in the interval (f)
   D. Add the result of step C to the *cumulative frequency* of the interval *below* the one in which the score is located

II. Express the *number* of cases at or below the score as the *percentage* of cases, dividing the result of step D by the total number of cases (N) and multiplying by 100

I. Determine the number of cases falling at or below the score of 83
   A. $83 - 79.5 = 3.5$
   B. $3.5/5. = .70$
   C. $(.70)(4) = 2.8$
   D. $2.8 + 67 = 69.8$

II. $(69.8/80) \times 100 = .8725 \times 100 = 87.25\%$, rounded to 87%

The percentile rank of the score of 83 is 87 percent

A summary of the steps in computing percentile ranks is presented in formula 2–2 below:

$$\text{Percentile rank} = \frac{\left[\dfrac{(X - l.r.l.)}{i} \times f + c.f. \text{ in interval below}\right]}{N} \times 100 \qquad (2\text{--}2)$$

where:

X is the score
$l.r.l.$ is the lower real limit of the interval
i is the width of the interval
f is the frequency of cases in the interval
$c.f.$ is cumulative frequency in interval below
N is the total number of cases

The following numerical expression of formula 2–2 is taken from the data of example 2–2:

$$\frac{\left[\dfrac{(83 - 79.5)}{5} \times 4 + 67\right]}{80} \times 100 = 87\%$$

Summary

Perhaps the most important consideration in organizing data is convenience in determining the meaning of the scores. With just a few scores, simple rearrangement will usually be sufficient for adequate communication; but with a large number of scores, it becomes necessary to group them in some meaningful way. If the number of groups used is small (less than 10 or 12), too much information is lost in the grouping. Too many groups (more than about 20) make it difficult for most people to extract the information from the data easily and conveniently.

A simple approach to obtain between 10 and 20 groups is to divide the total range of scores (highest score minus the lowest plus one) by 15. This will produce an interval size that will yield about 15 groups. The final determination of the interval size to be used should be based on convenience. Three conveniences in selecting an interval size are: (1) whole numbers are more convenient to use than decimals; (2) odd numbers will have an integer for the midpoint of the interval, whereas even numbers will have a decimal; and (3) multiples of five are more convenient for most people than other multiples (e.g., seven).

When numbers are assigned to continuous variables, these numbers represent an interval of possible values. The number 14 when assigned to weight measurements indicates that the weight may be as low as 13.5 pounds or as high as 14.5 pounds. It is the midpoint of the interval. The number 63.4 when assigned to a height measurement would indicate the interval extending from 63.35 to 63.45—it is the midpoint of the interval which extends one-half unit below and one-half unit above the recorded number.

Numbers assigned to discrete objects do indicate the exact number of objects. We do not count people in parts. The number 14 when assigned to a count of people indicates exactly 14 people, not 13.5 or 14.5. The meaning of a number is determined by the nature of the objects or things being measured or counted.

The Cartesian coordinate system (X-axis and Y-axis at right angles to each other), is the most often-used method of presenting data graphically. If a frequency distribution is being graphed, the score values are plotted along the abscissa (X-axis), and the frequency in each category is plotted along the ordinate (Y-axis). If the results of a study with a dependent variable and independent variable are being presented, the independent variable is plotted along the abscissa and the dependent variable along the ordinate.

The *polygon* is a line graph which is plotted by placing a point at the midpoint of each interval along the abscissa and connecting these points by a series of straight lines to form the polygon.

The *histogram* or bar graph is constructed by drawing a bar or enclosed rectangle for each of the categories along the abscissa. The width of the bar should cover the entire range of the category (for continuous numbers), and the height will indicate the value along the ordinate. For continuous data, each bar of the graph should touch the adjacent bars. For discrete data, a space should be left between adjacent bars.

For accuracy and convenience in communication, the ordinate should begin at zero, and the ratio of the height of the ordinate to the width of the abscissa should be about $3/4$. When it is necessary to violate the conventions in graphing, some indication of the violation should be placed on the graph.

Many of the data of the behavioral scientist distribute themselves into a symmetrical bell-shaped curve called the *Gaussian distribution* or *normal curve*. A distribution which is nonsymmetrical is called a *skewed* distribution. A distribution is *positively skewed* if the long tail is toward the large scores. The distribution is described as *negatively skewed* if the long tail is in the direction of the small scores.

The term *kurtosis* is used to indicate the degree of peakedness or flatness of a distribution. The normal curve can be described as *mesokurtic*. A distribution which is more peaked than the normal curve is referred to as *leptokurtic,* and a *platykurtic* distribution is flatter than the normal curve.

One way of organizing and describing a set of scores is to determine the percentage of the groups who score at or below a point. This type of scaling indicates the *relative order* in the scores and not the distance between scores. A *percentile* is always a score which indicates a percentage at or below that point; thus the 50th percentile would be that score at or below 50 percent of the group scores. A *percentile rank* is the percentage ranking on a given score. The answer to the question "What is the percentile rank of a score of 27?" will be a percentage figure and will indicate the percentage of the group who scored at or below the score of 27. Certain percentiles are used a great deal and have been given special names. The 50th percentile is also called the *median.* The 10th percentile is called the *first decile,* and each multiple of 10 is referred to as a *decile*. The 80th percentile is called the *eighth decile*. Other especially named percentiles are the 25th percentile or first quartile, and the 75th percentile or third quartile. (The median is sometimes referred to as the *second quartile*.)

References

PERCENTILES AND PERCENTILE RANKS

PAUL BLOMMERS and E. F. LINDQUIST, *Elementary Statistical Methods in Psychology and Education* (2d ed.; Boston: Houghton Mifflin Co., 1960), chap. iv.

GRAPHING

QUINN McNEMAR, *Psychological Statistics* (4th ed.; New York: John Wiley & Sons, Inc., 1969), chap. ii.

SKEWNESS AND KURTOSIS

QUINN McNEMAR, *Psychological Statistics* (4th ed.; New York: John Wiley & Sons, Inc., 1969), pp. 25–30.

3

Using numbers in measurement

Scaling

MOST PEOPLE who work with numbers and statistics are not interested in the numbers themselves, but in the things they represent. Since it is more convenient to use numbers (symbols) to represent the objects or behavior in which we are interested, it becomes necessary to understand the different ways in which people use numbers. At first glance, this last statement seems a little silly. We all know that numbers are simply numbers and the way people use numbers doesn't vary—or does it?

Suppose that a person were to take an automobile trip across the United States in an east-west direction. As you know, the United States federal highways that run in a general east-west direction have been given even numbers, and federal highways which run in the general north-south direction are designated by odd numbers. Since our hypothetical friend is to drive in an east-west direction, let us suppose that he drives half of the distance on U.S. 20 and the rest of the distance on U.S. 30. Now a question that could be asked would be: What would be the average highway on which this person drove? Of course, the question itself is ridiculous. But let us pursue it strictly from the standpoint of numbers. If we take the symbol for one highway (20) and add this to the symbol for the other (30), we would achieve a number of 50. Since only two highways are involved, the average of the two would be U.S. Highway 25. Since this is an odd-numbered federal highway, we would be in the position of saying that on this east-west trip, on the average, our friend drove on a highway which runs north and south! Obviously, we cannot, or should not, average highway numbers together. This example merely serves to illustrate that our use of numbers is not always as straightforward as it initially appears.

The assignment of numbers to objects is often referred to as *scaling*. Although the numbers themselves are merely symbols, we need not be too concerned with the properties of the symbols; but when we let these numbers represent objects, we must be concerned with the scaling characteristics of the objects. We may be able to find the average of the numbers 20 and 30, but this does not mean that we can find the "average" of two highways. In order to avoid confusion, we should perform only those manipulations of the numbers (symbols) which could also be performed on the objects themselves. Most people, without thinking about these differing properties of objects, do understand the procedure of adding, subtracting, and averaging the objects. People do not try to average highways. The confusion arises when we assign numbers to the objects and then try to manipulate the numbers as if they no longer represented the objects.

S. S. Stevens[1] has suggested that we classify the scalable properties of objects into four general classes. He suggests that we refer to these four classes as *nominal, ordinal, interval,* and *ratio* scales. Thus, in a sense, we can say that we have four different sets of rules which we must observe when we start to manipulate numbers (symbols) which represent objects. These rules are relatively simple to understand, and such understanding certainly makes it easier to manipulate symbols and have the resulting "answers" relate to the objects in which we are interested. The rest of this section will be devoted to the rules for the construction and use of the nominal, ordinal, interval, and ratio scales.

NOMINAL SCALE

The first and most basic operation involved in scaling is that of observing differences. We can differentiate people according to sex—call those of one kind men, the other kind women. We might classify male college students according to their affiliation or nonaffiliation with fraternities. If we have a clear and unambiguous way of discriminating among objects

---

[1] S. S. Stevens, "Mathematics, Measurement and Psychophysics," in S. S. Stevens, ed., *Handbook of Experimental Psychology* (New York: John Wiley & Sons, 1951), pp. 1–49.

or people, we have the simplest scale of measurement, the nominal scale. For convenience, we generally wish to assign symbols to the various cate gories or classes of the nominal scale. The word *nominal* implies that the symbols are merely names. The names could be words (*male, female*), or maybe letters of the alphabet (*a* or *b*), or numbers.

There are only two basic rules by which symbols are assigned to objects to form a nominal scale:

1.   Two classes which are *different* with respect to the quality being measured may not have the same symbol assigned to them. Using highways as an illustration, it would be extremely confusing if several separate highways were given the same name or number.

2.   Two objects which are the *same* with respect to the quality being measured may not have different symbols assigned to them. Again, referring to highways, think how confusing it would be if the highway number on the same road were changed every few miles.

If we have decided to classify voters according to party affiliation and called the groups *Democrats, Republicans, Independents,* and *Other,* we may substitute the symbols *Dem., GOP, Ind.,* and *Oth.* Or we may use *A, B, C,* and *D* or 1, 2, 3, and 4. If we have chosen numbers for symbols, we must remember that the formal rules of arithmetic which apply to numbers do not necessarily apply to the objects.

Suppose we ask voters how they plan to vote in the next election and the expected answers were: Republican, Democratic, and Vegetarian. Suppose we decide to call a Republican answer 1, a Vegetarian answer 2, and a Democrat answer 3. Suppose the sample of voters answered as follows: 49 percent Republican, 2 percent Vegetarian, and 49 percent Democratic. Now we "average" the numbers assigned and find that the average number is two. A conclusion then will be that on the average the voters intended to vote for the Vegetarians. Such a conclusion is certainly not warranted by the data. The numbers can be added and subtracted, but the categories and responses cannot. In obtaining an average, we have ignored the fact that the rules of arithmetic which apply to numbers do not apply for the categories for which these numbers stand.

In our everyday affairs we are not usually bothered by the misuse of numbers assigned to objects on a nominal scale because we don't usually assign numbers as names. We generally use some convenient word other than a number to label members of a group or to identify individuals. Some of the exceptions to this have been mentioned above (e.g., highways). Other uses of numbers for names can be found in the identification of individual participants in sporting events. The number on the shirt of a baseball player is simply a convenient label and does not tell us anything about his position on the team or his proficiency.

In recent years the possibility of misuse of numbers assigned to nominal scales has become greater. Most of us are acquainted with the punched cards used on many modern data processing machines. The columns of these cards are numbered, but these numbers may be used to designate different categories. The positions within the columns (rows) are also numbered, and these punches may again refer to classes. For example, column 18 of a card may be used to designate sex of an individual; and within this column, row 1 may be used for male and row 2 for female. Although this is not a serious problem, there does exist the possibility of incorrect manipulation of the numbers reported from a data run from one of these types of machines.

Agnew and Pyke[2] report a case of inappropriate averaging of numbers which is related to computer usage. In order to facilitate record keeping of medical problems in Canada, The Dominion Bureau of Statistics has developed a different number code to be used for the diagnosis of hospital patients. A young man, age 24 years, became a patient in a hospital. One physician diagnosed the problem as suspected brain tumor. A second physician diagnosed the problem as schizophrenia. A stenographer in the hospital had the responsibility of coding the diagnosis and reporting it to the Dominion Bureau of Statistics. The code number for suspected brain tumor was 308; for schizophrenia, 300. The stenographer hit upon the solution of averaging the two numbers, thus the diagnosis on this patient became 304—senility! A rather rare diagnosis for a 24-year old.

THE ORDINAL SCALE

We can often do more than simply discriminate differences among objects. We may be able to go a step further and specify that one object has more or less of some given quality than the other objects. When we can make the finer discrimination of "more than" or "less than," we may be able to establish an ordinal scale. To do this, we must show that the objects so ordered conform to three properties. If this is the case, then an ordinal scale has been constructed.

We shall illustrate these principles and the establishment of an ordinal scale by using three categories of items: buildings, people, and pencils.

[2] Neil McK. Agnew and Sandra W. Pyke, *The Science Game An Introduction to Research in the Behavioral Sciences* (Englewood Cliffs, N.J.: Prentice–Hall, 1969), pp. 95–97.

If we are to order these objects, they must possess something in common—a quality which can be ordered. In this case we shall use the common property of height. These classes of objects have the property of connectedness or communality—they are related to each other in some specific way (height). But we must go further than this to be able to scale these groups of objects. We must be able to say that if buildings and people are not of the same height, then either the buildings must be taller than the people, or the people must be taller than the buildings. Symbolically, this could be stated as follows: If $A \neq B$, then either $A > B$ or $A < B$. This statement should be read: If A is not equal to B, then either A is greater than B, or A is less than B. As you can see, the symbol $\neq$ means "is not equal to"; $<$ means "is less than"; $>$ means "is greater than."

Once we have determined a common property existing among our categories, we have to make sure that this is a stable relationship. Thus, if we find that buildings possess more height than people and we wish to have an ordinal scale, we cannot have a situation in which people sometimes have more height than buildings. Symbolically, this may be stated: If $A > B$, then $B \not> A$, and should be read as: If A is greater than B, then B cannot be greater than A. The symbol $\not>$ means "is not greater than." This characteristic of stability of the relationship between objects is called the *property of asymmetry*. Asymmetry means "not symmetric"; if two objects stand in some definite relationship to each other (nonequal relationship), then this relationship cannot be reversed. If a building has more height than a person, the person cannot have more height than the building if we are to develop an ordinal scale for buildings, people, and pencils.

The third property of an ordinal scale deals with the relationship between all the objects that are to be scaled. If we find that buildings do have more height than people and people have more height than pencils, then we must be able to say that buildings have more height than pencils. Using symbols once more: If $A > B$ and $B > C$, then $A > C$, and should be read as: If A is greater than B and B is greater than C, then A is greater than C. This is called the *property of transitivity*. This property can be summarized by saying that if a relationship exists between two things (A and B) and a second relationship exists between one of these things and a third thing (B and C), then the same type of relationship must hold between the first and the third thing (A and C).

For some years, behavioral scientists have been interested in the social relationships that exist among animals of the same species. It has been observed that chickens seem to observe a definite order of feeding. This has been referred to as a *pecking order*. A series of studies by F. D. Klopfer[3] of Washington State University involving this social dominance in pigs will illustrate the problem of establishing an ordinal scale. Klopfer used a situation in which a hungry pig was trained to press a lever to obtain a pellet of food. The pig was required to press the lever and *then walk across the pen* to secure the food. In one small experiment, three pigs were used. We can call them A, B, and C. Each pig was trained separately on the lever-pressing response, and then the pigs were placed in the pen two at a time. If pig A presses the lever, the other pig will be closer to the food than pig A will be. What happens in this type of situation? Will the pigs cooperate, fight, press the lever alternately, or do something else? Klopfer found that a stable relationship develops. One pig comes to dominate the other, so that the one pig will press the lever, and then go and eat while the other pig waits. Thus a form of social dominance is established. Now let us see if we can establish an ordinal scale of social dominance for these three pigs. For an ordinal scale to be formed, we must show that the behavior of the three pigs meets the three properties discussed above.

1.   *The property of connectedness.* Do we have a common characteristic which can be shown to vary systematically with the pigs? This was established rather rapidly. In all pairings, one pig was dominant with respect to the other member of the pair in regard to the behavior under consideration.

2.   *The property of asymmetry.* This refers to the stability of the difference. In this case, pig A was dominant with respect to pig B ($A > B$), and this relationship was maintained from one testing period to the next. This same stable dominance was true of all pairs of pigs tested.

3.   *The property of transitivity.* Since the first two properties were found to hold for the three pigs, all that is needed is to show that as we shift from one pair of animals to a different pair, the same order of dominance is maintained. It was found that A was dominant with respect to B and that this relationship was stable ($A > B$ and $B \not> A$); also, B was dominant with respect to C, and this relationship was also stable ($B > C$ and $C \not> B$). The final test would be to place pigs A and C together in the pen and observe the resulting social behavior. Once again, dominance was

---

[3] Personal communication, 1952.

shown, and this was stable over testing periods. Pig C was dominant with respect to A! For the property of transitivity to hold, the following relationship was necessary: $A > B, B > C,$ *and* $A > C$. Since it was found that C was dominant with respect to pig A $(C > A)$, the property of transitivity did not hold, and an ordinal scale of social dominance for these pigs could not be established.

In the above examples we have labeled the groups with words (*buildings, people,* and *pencils*) or letters (A, B, or C), and we could have used numbers as our symbols. In that case the numbers could be used as names. But if we have established an ordinal scale (the properties of connectedness, asymmetry, and transitivity hold for the objects) and we decide to number the groups, then we should assign our numbers in such a manner as to indicate the order in our groups. Most of the time, we use numbers in a systematic way. One sequence might run 1, 2, 3, 4, 5, etc., or we might use numbers in the order 2, 4, 6, 8, etc. We call this the *conventional order* of numbers. When we assign numbers to objects on an ordinal scale, we must assign these numbers so that they will in fact indicate order in the objects. Returning for the moment to our ordinal scale of buildings, people, and pencils, we can say (with respect to height) that buildings > people > pencils. If we assign numbers to the groups, the numbers themselves should convey the order of height of the objects. Thus, we might assign the number 1 to pencils, 2 to people, and 3 to buildings. Symbolically, we could say $3 > 2 > 1$. It would be extremely confusing to assign 2 to buildings, 1 to people, and 3 to pencils. In this latter situation the symbolic statement of the relationship would be $2 > 1 > 3$. The relationship $1 > 3$ might prove to be very confusing to us. If we use numbers to identify our groups on an ordinal scale, the conventional order of the numbers must correspond to the order in which the groups have been placed.

In addition to using the number system simply to identify different objects (nominal scale), we can also use the number system to indicate the order of objects (ordinal scale).

In chapter 2, we determined the 50th percentile, the 15th percentile, and the 85th percentile for the scores obtained from the 80 freshman students. We found that the median (50th percentile) was equal to 56.4. The 15th percentile was 43.3, and the 85th percentile was 80.8. The "average" of the 15th percentile and 85th percentile was found to be 62.05, which was *not* equal to the median. Percentile ranks constitute an ordinal scale. They show the order of the individual scores, but not the distance between scores. The "average" of the numbers assigned to an ordinal scale may be misleading. When numbers are assigned to an ordinal scale, the order of the numbers is important, but their absolute values are not.

Behavioral scientists are often defensive about the sophistication of measurement in the physical sciences as compared to the behavioral sciences. When the earth scientist wishes to measure the hardness of different substances, he is forced to utilize an ordinal scale. How do we measure the hardness of a rock? The geologist does this by determining which substances scratch which other substances. A diamond will scratch glass, but glass will not scratch a diamond. The scale of hardness is called Mohs' scale and is literally an ordering of which minerals will scratch the others.

THE INTERVAL SCALE

In addition to discriminating between objects and ordering them in terms of more or less of a given quality, we may be able to *quantify* these comparisons. Sometimes it may be possible to say not only that Jane is taller than Mary, but also that Jane is three inches taller. John's temperature is two degrees higher than normal. To make these quantitative comparisons, we must have a unit of measurement. If a unit of measurement has been established, an interval scale may have been achieved. When we have achieved the scaling precision for an interval scale, the objects being scaled must have met all the requirements for an ordinal scale and also the requirements of additivity. Most of us have never stopped to consider just what is meant by adding. As we progress through school, we learn to manipulate numbers by adding, subtracting, multiplying, and dividing, but we seldom consider the meaning of these operations. We learn that highway numbers, when added, have no particular meaning; but we do not attempt to understand the general process of addition.

If we wish to add the length of two objects together, we can lay them end to end. The new length extends from the outer end of one object to the outer end of the other. If we wish to add weights, we put both weights together in the same pan of a beam balance.

Before we can examine the characteristics of additivity, we must first define equality. What do we mean when we say two things are equal? A simple answer is that they are the same. In the formal system of mathematics, equality means no difference. For our purposes, equality may be defined as follows: A and B are equal $(A = B)$ if A is not greater

than B $(A \not> B)$ and A is not less than B $(A \not< B)$. This definition
works very well when we stay within the system of formal mathematics.
However, when we move into the real world and begin to measure objects,
people, or the behavior of people, we are in the empirical world where
equality is defined as "no difference that can be detected by a specified
measuring instrument." Suppose that two youngsters have the same weight
when weighed on a bathroom scale. We would say that they have equal
weight—no difference was found with our measuring instrument. If we
use an extremely precise scale weighing to one millionth of an ounce,
we might have great difficulty in discovering two people with the "same"
weight. Equality, then, in the empirical system, does not mean *same;*
it means *no difference that can be detected with a given measuring device.*

Having defined equality, we may examine some of the characteristics
of additivity. When we add objects together, the order of addition should
have no effect on the results. Symbolically, this may be stated as: If
$A + B = C$, then $B + A = C$. In the number system we can state that
$2 + 3 = 5$ and $3 + 2 = 5$. This characteristic is called the *commutative
property*—order of addition will not affect the results. This principle
can be extended to a series of additions. The order in which a series
of additions is performed will not affect the results. Symbolically,
$A + B + C = B + A + C$. In numerical terms, $2 + 3 + 4 = 3 + 2 + 4$.

When an equality exists $(A = B)$ and we add an equal amount to
both A and B, the result must still be equal. Symbolically, this can be
stated as: If $A = B$ and $C = D$, then $A + C = B + D$. In numerical
terms, this could be $2 + 3 = 5$ and $2 = 1 + 1$; then $(2 + 3)$
$+ 2 = 5 + (1 + 1)$, or $7 = 7$. If equals $(C = D)$ are added to equals
$(A = B)$, the results $(A + C = B + D)$ must be equal. This relationship
is referred to as the *axiom of equals.*

One other characteristic of additivity should also be considered. If
two things are equal and some quantity other than zero is added to one
of these equal objects, then the equality must be destroyed. Symbolically,
this can be stated as: If $A = B$ and $C > 0$, then $A + C > B$. This can
be called the *addition of discriminable amounts.*

It should be pointed out that these characteristics are an integral part
of our number system as we use it in everday living. We can manipulate
numbers just as we always have. The problem comes when we assign
the numbers to the objects. Unless we can show that the objects can
be manipulated in the various ways discussed above, the results obtained
by adding and subtracting the numbers will not necessarily tell us anything
about the objects in which we are interested.

One question that often arises concerns the operations of subtraction, multiplication, and division. It may be well to point out at this time that these other operations are special cases of addition. This can be illustrated most easily with multiplication. Four times three is a "shortcut" way of adding $3 + 3 + 3 + 3$.

Behavioral scientists are frequently interested in measuring qualities for which no convenient equal-interval scale exists. Attitudes are one example. We may be able to establish a scale on attitude toward school integration, and reliably order individuals on these attitudes. If individual A checks five items on the scale and individual B checks ten items, we may be able to say that person B has a more favorable attitude toward integration than does person A. We cannot say anything about the distance between these two individuals unless we can establish that the scale of attitudes possesses the characteristics of additivity. This is not to say that interval scales cannot be established for attitudes but rather than the authors of such scales must present adequate evidence of the interval nature of the scales before we are willing to infer anything other than order from the obtained measurements. L. A. Stone and R. James[4] published an article dealing with the prestige of different teaching specialties in the high schools. They were not only able to report an order of teaching specialties from least prestigeful to most prestigeful, but they also established the distance between specialties—they met the requirements of interval scaling with their particular sample of subjects. Traits such as liberal-conservative and introversion-extroversion usually yield ordinal scales when measured. The units should not be added or used to infer distance unless interval scaling has been achieved. The numbers represent order but do not imply an amount of difference.

One problem which has engrossed some psychologists for over one hundred years is the relationship between changes in the physical world (often measured by a scale with equal units) and a human's perception of those changes. If the light of one candle is added to the light of one other candle, humans will notice a difference in brightness. If the light of one candle is added to the light of one thousand candles no difference will be noticed. The physical characteristics may be additive but the human *perception* may not be additive.

The area of psychophysics is in part concerned with the problem of the relationship between changes in the physical world and the human's

---

[4] L. A. Stone and R. James, "Interval Scaling of the Prestige of Selected Secondary Education Teacher-Specialties," *Perceptual and Motor Skills,* vol. 20 (1965), pp. 859–60.

perception of those changes. But psychophysics is also concerned with the problem of establishing measurement scales for variables which do not have an obviously direct tie to physical scales (e.g., empathy). The problem of developing *adequate* measures for many phenomena in the behavioral sciences has been with us for many years and will remain one of the most basic and pressing problems for a long time to come.

THE RATIO SCALE

The ratio scales we use are generally those of the physical sciences, such as weight, height, length, etc. If we have an absolute zero point in addition to the unit of measurement, we have a ratio scale.

Since we generally use numbers in terms of ratio scaling, perhaps it would be best to point out some specific differences between interval and ratio scales. In our monetary system, $10 is twice as much money as $5. Zero dollars is no money—an absolute zero point—absence of money. One dollar is worth as much as any other dollar. Our monetary system represents a ratio scale.

The manner in which heat is generally measured will serve to illustrate the difference between an interval scale and a ratio scale. One degree centigrade is equal to any other degree centigrade. The units are equal, but the zero point is arbitrary. The temperature of the freezing point of water (under specified conditions) is taken as the arbitrary zero point; the boiling point (again under specified conditions) is given the value of 100. The heat distance in between may be divided into 100 equal units, but the reference points are arbitrary. Any other set of equal units could be substituted, and the Fahrenheit scale represents such a linear transformation. Thirty-two degrees Fahrenheit may be substituted for zero degrees centigrade, and 212 degrees Fahrenheit for 100 degrees centigrade. The temperature distance between the reference points remains constant. Only the size of the units changes when we change from one scale to another. Within each scale the interval between degrees remains constant. These interval scales permit statements such as "John's temperature is 2 degrees higher than normal." Such scales—because they do not utilize an absolute zero point—do not permit us to say that 10 degrees centigrade is twice as hot as 5 degrees centigrade. The zero we are using is not absolute. Zero degrees centigrade or Fahrenheit does not indicate the complete absence of heat. It is an arbitrary reference point—not an absolute zero.

The interval and ratio scales are similar, and the rules by which numbers are assigned are the same with one exception. For a ratio scale, we apply our operations and our numbers to the total amount measured from an *absolute* zero point. For an interval scale, we apply the operations to differences from some arbitrary zero point.

SUMMARY OF THE CHARACTERISTICS OF NOMINAL, ORDINAL, INTERVAL, AND RATIO SCALES

To aid the student in differentiating between these four levels of measurement, the main characteristics of each type of scaling are summarized below.

NOMINAL SCALES.   Nominal scaling must meet the following requirements.

1.   Two classes which are *different* with respect to the quality being measured may not have the same symbol assigned to them.

2.   Two objects which are the *same* with respect to the quality being measured may not have different symbols assigned to them.

ORDINAL SCALES.   Ordinal scaling must meet the following requirements:

1.   The requirements of nominal scaling.

2.   The property of connectedness, which means that there must be some common element or property relating the objects to be scaled.

3.   The property of asymmetry, which means that there must be a stability of any observed differences to be scaled.

4.   The property of transitivity, which means that an ordered relationship which holds in one part of the scale must also hold for all other parts of the scale.

INTERVAL SCALES.   Interval scaling must meet the following requirements:

1.   The requirements of ordinal scaling.

2.   The establishment of a unit of measurement which permits additivity. Additivity includes:

a)   The commutative property, so that the order of addition will not affect the results.

b)   The axiom of equals, which means that when equal quantities are added to other equal quantities, the results must be equal.

c)   The addition of a discriminable amount to one portion of an equality must result in an inequality.

RATIO SCALES.   Ratio scaling must meet the following requirements:

1.   The requirements of interval scaling.

2.   The establishment of a true zero point.

SCALING CHARACTERISTICS IN THE BEHAVIORAL SCIENCES

Many of the variables used in the behavioral sciences have the characteristics of interval or ratio scales. The number of responses, magnitude of response, time per response, etc., are all measurements which may be interval or ratio in nature. Many other measurements do not meet the requirements of additivity, and thus the numbers assigned indicate order but not distance. It is incorrect to think that there is any basic difference in measurement between the behavioral sciences and the physical sciences. In any situation the scientist uses those measurements which can best aid in understanding the phenomenon under study. In all cases the scientist should be aware of the measurement or scaling characteristics of the measures being employed.

It has been said that the measurement of intelligence as expressed in the IQ is different from physical measurement because of the absence of an absolute zero point. There is no question that we have been unable to define zero intelligence. Zero on an intelligence test, and consequently zero IQ, does not indicate zero intelligence. Some years ago, L. L. Thurstone applied psychophysical scaling techniques to the measurement of intelligence and attempted to determine an absolute zero point. In the case of the IQ scale, before we can attempt to assign a zero point meaningfully, we must first establish a unit of measurement which will meet the requirements of additivity. Notice that a person with an IQ of 100 is not twice as intelligent as a person with an IQ of 50. Such a statement would require an absolute zero point. What about three individuals with IQ's of 100, 110, and 120? Is the person with an IQ of 120 as much different from the person with an IQ of 110 as the person with an IQ of 110 is different from the person with an IQ of 100? The answer is, we don't know. We can say that 15 degrees centigrade is as much warmer than 10 degrees centigrade as 10 degrees centigrade is warmer than 5 degrees centigrade. An interval scale exists for the measurement of heat. We do not have an interval scale for the measurement of intelligence. Does this, then, imply that all that IQ scores can tell us is the order of people and not the distance between? The answer is, both yes and no. IQ scores show order and a very rough indication of distance.

Throughout this section, we have discussed the problem of scaling as if there were four separate and distinct types of scales: nominal, ordinal, interval, and ratio. In actuality, the scaling characteristics of different objects vary in a continuous way from nominal to ratio. Certain types of measurements (such as test scores) probably do not meet all the requirements for an interval scale, but the measurement is better than ordinal. We could name some of these in-between points, but it would not add much to our understanding of the basic problems of measurement and how we assign numbers to our measured objects.

The last portion of this chapter deals with some of the rules of formal arithmetic which are used when we combine numbers. These are the ways in which numbers can be manipulated meaningfully if the measures are interval or ratio in nature.

Working with numbers and symbols

Although we must always be alert to the manner in which we treat numbers assigned to data (nominal, ordinal, interval, or ratio scaling), the establishment of equal units of measurement permits us to use numbers in conformity with most of the rules of arithmetic. Most of these were learned in the first eight grades, and many are used habitually and seem self-evident. The purpose of this material is to serve as a brief review for those students who may have forgotten some of the rules for the manipulation of numbers. Many students will find this section unnecessary.

ORDER OF ARITHMETIC OPERATIONS

ADDITION AND SUBTRACTION.   The order in which numbers are added or subtracted does *not* affect the results.

Example:

$$3 + 4 + 5 = 12; \quad 4 + 5 + 3 = 12$$

or

$$a + b + c = d; \quad b + c + a = d$$

MULTIPLICATION AND DIVISION.   The order in which numbers are multiplied or divided does *not* affect the results.

Example:

$$3 \times 4 \times 5 = 60; \quad 4 \times 5 \times 3 = 60$$

or

$$a \times b \times c = d; \quad b \times c \times a = d$$

or

$$abc = d; \qquad bca = d$$

USE OF PARENTHESES.   Any expression enclosed in parentheses is to be treated as a single number.

Example:

$$3(4 + 5) = 3(9) = 27$$

In some cases—as in the above example—it is easier first to combine the expression within the parentheses—$(4 + 5)$ to (9) and then multiply. In other cases we may multiply each expression in the parentheses by the term preceding—

$$3(4 + 5) = 3(4) + 3(5) = 12 + 15 = 27$$

USE OF THE RADICAL SIGN ($\sqrt{}$).   Any expression under a radical sign is to be treated as a single number.

Example:

$$\sqrt{4 + 9} \neq 2 + 3; \quad \sqrt{4 + 9} = \sqrt{13} = 3.606$$

MIXED ARITHMETIC OPERATIONS.   If an expression involves the operations of addition and/or subtraction as well as multiplication and/or division, the multiplication and division *must* be performed first, then the addition and subtraction.

Example:

$$12 + 5 \times 4 - 18 \div 6 = 12 + 20 - 3 = 29$$
$$12 + 5 \times 4 - 18 \div 6 \neq 17 \times 4 - 18 \div 6$$

or

$$a + b \times c - d \div e = a + bc - d/e$$

FRACTIONS

ADDITION OR SUBTRACTION.   Before we add or subtract, fractions must be changed so that they have a common denominator.

Example:

$$\tfrac{3}{4} + \tfrac{4}{5} = (\tfrac{3}{4})(\tfrac{5}{5}) + (\tfrac{4}{5})(\tfrac{4}{4}) = \tfrac{15}{20} + \tfrac{16}{20} = \tfrac{31}{20} = 1\tfrac{11}{20}$$
$$\tfrac{3}{4} + \tfrac{4}{5} \neq \tfrac{7}{9} \text{ (incorrect)}$$

or

$$\frac{a}{b} + \frac{c}{d} = \frac{ad}{bd} + \frac{cb}{bd} = \frac{ad + cb}{bd}$$

MULTIPLICATION.   To multiply fractions, multiply the numerators and multiply the denominators.

Example:

$$\frac{1}{2} \times \frac{2}{3} \times \frac{3}{4} = \frac{1 \times 2 \times 3}{2 \times 3 \times 4} = \frac{6}{24} = \frac{1}{4}$$

or

$$\frac{a}{b} \times \frac{c}{d} \times \frac{e}{f} = \frac{ace}{bdf}$$

MULTIPLICATION OR DIVISION OF NUMERATOR AND DENOMINATOR BY SAME EXPRESSION.   As long as both the numerator and the denominator are multiplied or divided by the same number or symbol, the value of the fraction does not change.

Example:

$$\frac{1}{2}(\frac{3}{3}) = \frac{3}{6} = \frac{1}{2}; \quad \frac{5}{6}(\frac{4}{4}) = \frac{20}{24} = \frac{5}{6}$$

or

$$a/b(c/c) = ac/bc = a/b$$

DIVISION.   To divide a fraction, invert the divisor and multiply.

Example:

$$\frac{1}{2} \div \frac{3}{4} = \frac{1}{2} \times \frac{4}{3} = \frac{1 \times 4}{2 \times 3} = \frac{4}{6} = \frac{2}{3}$$

SIGNED NUMBERS

People who live in the northern states are no strangers to negative numbers. Even the thought of -20 degrees Fahrenheit is apt to cause shivers. Signs prefixed to a number can be used to show deviations above or below any reference or zero point. Thus, if the average test score of a group of students is 85, a score of 80 is five points below the average, whereas a score of 90 is five points above. These could be written as -5 and 5, respectively. In general, any number without a sign is considered to be a positive number, but any negative number is *always* written with the negative sign prefixed to the number.

ADDITION. *Two or more numbers with the same sign:* Add the numbers and prefix the common sign.

Example:

$$2 + 3 + 4 + 5 = +14 \text{ (more commonly, 14)}$$
$$(-2) + (-3) + (-4) + (-5) = -14$$

Two numbers with unlike signs: Find the numerical difference between the two numbers and prefix the sign of the larger.

Example:

$$(+4) + (-2) = +2$$
$$(-4) + (+2) = -2$$

A group of numbers with unlike signs: Add the positive numbers and add the negative numbers; find the difference between the two sums and prefix the sign of the larger sum.

Example:

$$(+2) + (-3) + (+4) + (-5) + (+6) = (+12) + (-8) = +4$$
$$(-2) + (+3) + (-4) + (+5) + (-6) = (+8) + (-12) = -4$$

SUBTRACTION. To subtract one signed number from another, change the sign of the subtrahend (number to be subtracted) and proceed as in addition.

Example:

$$(+5) - (-3) = (+5) + (+3) = 8$$
$$(+5) - (+3) = (+5) + (-3) = 2$$

MULTIPLICATION. *Like signs:* Multiplication of two numbers with like signs gives a positive product.

Example:

$$(+3) \times (+4) = +12; \quad (+a)(+b) = +ab$$
$$(-3) \times (-4) = +12; \quad (-a)(-b) = +ab$$

Unlike signs: Multiplication of two numbers with unlike signs gives a negative product.

Example:

$$(+3) \times (-4) = -12; \quad (+a)(-b) = -ab$$
$$(-3) \times (+4) = -12; \quad (-a)(+b) = -ab$$

When three or more numbers are to be multiplied together, they are multiplied in pairs, and the final operation involves the multiplication of two numbers following the rules given above.

Example:

$$(+3) \times (+4) \times (-5) = (+12) \times (-5) = -60$$

or

$$(+3) \times (+4) \times (-5) = (+3) \times (-20) = -60$$
$$(-3) \times (-4) \times (-5) = (+12) \times (-5) = -60$$

or

$$(-3) \times (-4) \times (-5) = (-3) \times (+20) = -60$$

DIVISION.   *Like signs:* Division of numbers with like signs gives a positive quotient.

Example:

$$+4/+2 = +2; \quad -4/-2 = +2$$

Unlike signs: Division of numbers with unlike signs gives a negative quotient.

Example:

$$+4/-2 = -2; \quad -4/+2 = -2$$

SIMPLE ALGEBRAIC EQUATIONS

Only one rule is necessary to remember in working with equations, and that is: *We may perform any operation (addition, subtraction, multiplication, etc.) to one side of an equation as long as we perform the same operation to the other side of the equation.*

ADDITION.

Example:

$$2 + 3 = 5$$

Add 2 to each side:

$$(2) + 2 + 3 = (2) + 5$$

or

$$x = X - M$$

Add M to both sides of the equation:

$$x + M = X - M + M$$
$$x + M = X$$

SUBTRACTION.

Example:

$$2 + 3 = 5$$

Subtract 3 from each side:

$$2 + 3 - 3 = 5 - 3$$
$$2 = 2$$

or

$$X = x + M$$

Subtract x from both sides of the equation:

$$X - x = x + M - x$$
$$X - x = M$$

MULTIPLICATION.

Example:

$$2 + 3 = 5$$

Multiply both sides by 2:

$$(2)(2 + 3) = (2)(5)$$
$$10 = 10$$

or

$$\frac{\Sigma X}{N} = M$$

Multiply both sides by N:

$$(N)\frac{\Sigma X}{N} = M(N)$$
$$\Sigma X = MN$$

DIVISION.

Example:

$$\Sigma x^2 = NS^2$$

Divide both sides by N:

$$\frac{\Sigma x^2}{N} = \frac{NS^2}{N}$$

$$\frac{\Sigma x^2}{N} = S^2$$

ALGEBRAIC SUMMATION

To summate means simply to add. In working with numbers, there is little problem in summing a series of different numbers. We simply add them and write the total. When we attempt to sum algebraic terms, the resulting expression often becomes long and difficult to handle. For convenience in overcoming the awkwardness, symbols can be substituted for the *operations* we perform with numbers. When addition is desired, the capital Greek letter sigma (Σ) is used to denote this. The symbol Σ should be read as "the summation of." When we have a large number of scores from the same group which vary from each other, it is often convenient to let a letter (such as X) stand for any one of these scores. If we have five scores that we wish to add (sum), this could be written as: $X_1 + X_2 + X_3 + X_4 + X_5$. However, if we have 100 scores, this procedure becomes awkward. The summation sign (Σ) is used to indicate the addition of a series of values. If the symbol X is used to stand for a score, and a subscript is used to refer to a specific score (X_2), a general symbol such as i may be used for any specific score (X_i).

Where all scores in a set are to be summed, this is often indicated by $\sum\limits^{N} X_i$, where it is understood that i will begin with the first score. In most beginning treatments of statistics the expression ΣX is used to indicate that the summation includes all N scores in the set, and should be read as "sum all the scores."

In working with summations, it is well to remember that Σ is merely a symbolic method of representing the operation of addition and the usual rules and shortcuts of addition can be used.

SUMMING A CONSTANT.   When a constant (unchanging value) is to be summed, we may obtain the same result by multiplying the constant by N

(the number of times the constant occurs). Let k be the symbol for a constant.

Example:

$$\Sigma k = 2 + 2 + 2 + 2 + 2 = (5)(2) = 10$$
$$\Sigma k = k + k + k + k + k = Nk$$

SUMMING TWO OR MORE TERMS. The summation of two or more terms is the same as the sum of the sums, taken separately.

Example:

$$\Sigma(X + Y + Z) = \Sigma X + \Sigma Y + \Sigma Z$$

If

$X_1 = 1$	$Y_1 = \ \ 2$	$Z_1 = -2$
$X_2 = 2$	$Y_2 = -2$	$Z_2 = \ \ 3$
$X_3 = 3$	$Y_3 = \ \ 1$	$Z_3 = \ \ 4$

then

$$\Sigma(X + Y + Z) = (X_1 + Y_1 + Z_1) + (X_2 + Y_2 + Z_2) + (X_3 + Y_3 + Z_3)$$
$$\Sigma(X + Y + Z) = (1 + 2 - 2) + (2 - 2 + 3) + (3 + 1 + 4)$$
$$(1) + (3) + (8) = 12$$

or

$$\Sigma X + \Sigma Y + \Sigma Z = (X_1 + X_2 + X_3) + (Y_1 + Y_2 + Y_3) + (Z_1 + Z_2 + Z_3)$$
$$\Sigma X + \Sigma Y + \Sigma Z = (1 + 2 + 3) + (2 - 2 + 1) + (-2 + 3 + 4)$$
$$(6) + (1) + (5) = 12$$

thus

$$\Sigma(X + Y + Z) = \Sigma X + \Sigma Y + \Sigma Z$$

SUMMATION OF A VARIABLE TIMES A CONSTANT. This is equal to the constant times the summation of the variable. Let $k = 2$ and $\Sigma X = X_1 + X_2 + X_3$.

Example:

$$\Sigma(kX) = (2)(2) + (2)(3) + (2)(4) = 4 + 6 + 8 = 18$$

or

$$(2)(2) + (2)(3) + (2)(4) = 2(2 + 3 + 4) = 2(9) = 18$$

thus

$$\Sigma(kX) = k\Sigma X \ (k \text{ times the sum of } X)$$

or

$$\Sigma aX = aX_1 + aX_2 + aX_3 + aX_4 = a\Sigma X \text{ (}a \text{ times the sum of } X\text{)}$$

SUMMATION OF A VARIABLE DIVIDED BY A CONSTANT. This is equal to summation of the variable and the total divided by the constant.

Example:

$$\sum \left(\frac{X}{k}\right) = \tfrac{2}{2} + \tfrac{3}{2} + \tfrac{4}{2} + \tfrac{5}{2} = 1 + 1\tfrac{1}{2} + 2 + 2\tfrac{1}{2} = 7$$

$$\frac{\Sigma X}{k} = \frac{2+3+4+5}{2} = \frac{14}{2} = 7$$

thus

$$\sum \left(\frac{X}{k}\right) = \frac{\Sigma X}{k}$$

or

$$\sum \left(\frac{X}{a}\right) = \frac{X_1}{a} + \frac{X_2}{a} + \frac{X_3}{a} + \frac{X_4}{a} = \frac{\Sigma X}{a}$$

There is nothing mysterious about these various rules in working with numbers. Most of them are well known in practice, but most students have never tried to verbalize the rules or put them in symbolic form.

Summary

The use of numbers allows the behavioral scientist to utilize both the convenience of the number system and the power of the logic within formal mathematics. The major limitation lies in the different ways in which we assign numbers to objects and the meaning that is contained in this numbering system. S. S. Stevens has proposed that we differentiate between four types of scales and that the numbers applied to objects scaled in these four different ways convey different amounts of information.

If we can only place objects in different categories, then we are working with a *nominal scale,* and numbers assigned to these categories only represent difference.

When objects can be ordered along some continuum, we have achieved an *ordinal scale.* Numbers assigned to such a scale show differences in categories and also rank or order along the continuum.

The establishment of an *interval scale* allows for most of the usual operations of arithmetic (addition, subtraction, and so on). Numbers assigned to an interval scale indicate the *distance* between objects.

With an interval scale and a zero point which indicates the absence of the quantity being measured, we have obtained a *ratio scale*. Numbers assigned to objects which form a ratio scale convey both distance between objects and the ratios between measurements. We can say that one object is twice as heavy as another object.

In working with numbers, two questions should be kept in mind:

1. What are the characteristics of the data to which we are applying numbers (nominal, ordinal, interval, and ratio)?

2. Are the operations we perform on the data justified?

With these questions before you, much of the mystery of statistics—as it applies to data—will disappear.

References

LEVELS OF MEASUREMENT

VIRGINIA SENDERS, *Measurement and Statistics* (New York: Oxford University Press, 1958), chap ii.

SIDNEY SIEGEL, *Nonparametric Statistics for the Behavioral Sciences* (New York: McGraw-Hill Book Co., Inc., 1956), pp. 21–30.

S. S. STEVENS, "Mathematics, Measurement, and Psychophysics," chap. i in S. S. STEVENS (ed.), *Handbook of Experimental Psychology* (New York: John Wiley & Sons, Inc., 1951).

ARITHMETIC OPERATIONS

ANDREW R. BAGGALEY, *Mathematics for Introductory Statistics, A Programmed Review* (New York: John Wiley & Sons, Inc. 1969).

W. L. BASHAW, *Mathematics for Statistics* (New York: John Wiley & Sons, Inc. 1969). This is a much more extensive treatment than the other references listed and contains some matrix algebra, set algebra, and basic probability.

VIRGINIA SENDERS, *Measurement and Statistics* (New York: Oxford University Press, 1958), chap. 0.

H. M. WALKER, *Mathematics Essential for Elementary Statistics* (rev. ed.; New York: Henry Holt & Co., Inc., 1951). This entire book is an excellent review source of basic mathematical operations.

4

Measures of central tendency

IN CHAPTER 2, it was noted that rearranging data allows a reader to make a general description of a set of numbers by inspection of the frequency distribution. There are four different indices that are used to describe a frequency distribution: (1) skewness, (2) kurtosis, (3) central tendency, and (4) variability. Skewness refers to the departure from symmetry. Kurtosis refers to the degree of peakedness or flatness of a distribution. A precise numerical specification of the degree of skewness or kurtosis is not necessary at a beginning level in statistics, and this text will not deal with these two measures in any further detail.

The other two measures used to describe a frequency distribution are of great importance in statistics. These two measures have been hinted at in chapter 2, but this chapter and chapter 5 will concentrate on the determination of the location (central tendency) and the scale (variability) of a distribution of numbers.

Mean (*arithmetic average*)

If a student is presented with a set of numbers and is asked to determine the average of this set, he can usually determine the average with no difficulty. Most college students have been computing averages since grade school. If, on the other hand, the statement is made that "the mean of a set of numbers is equal to the sum of the individual scores divided by the number of scores in the set," and that this can be represented by the equation

$$\bar{X} = \frac{\Sigma X}{N},$$

(4-1)

these same students will throw up their hands in horror and say that they never were any good in math and they simply don't understand how to compute a mean. The mean is simply the name for the average that the students have been able to compute for years. Within statistics, the term *average* has several different meanings, so that it is advisable to use the more precise term *mean* to describe this measure of location or central tendency of a distribution. As was pointed out in the last section of the previous chapter, certain symbols in mathematics refer to operations to be performed on numbers. We are all familiar with the plus sign $(+)$ as an operator that states: Add together the two numbers on either side of the symbol. The uppercase Greek letter sigma (Σ) is used in statistics to indicate an operation to be performed on numbers. This symbol indicates that one is to add all numbers of the type that follow the symbol. In the equation for computation of the mean, ΣX states that one should add all the X's. If we let X stand for any score in the set, then the ΣX reads: "Add all X's" or "Summate all X's." The symbol N is used to refer to the number of scores. In this case it would mean that there are N X's in the distribution of scores. The equation for the mean of the distribution

$$\bar{X} = \frac{\Sigma X}{N}$$

should be read as "Summate all the scores" (ΣX) and "Divide by the number of scores" (N).

DEFINITION OF THE MEAN

Notice that we defined the mean of the distribution as being the "average." This is not accurate. As was mentioned, "average" can refer to several different measures of location or central tendency. The mean of a distribution is defined as that point in a distribution about which the sum of the deviations is equal to zero. If we examine this statement, we can see that a new term, *deviations*, has been introduced. By deviation, we mean the difference between a score and the mean of the set of scores. The symbol used to indicate a deviation about the mean is the lowercase letter for the symbol used to indicate the obtained score. Thus, if X refers to any score in a distribution, x refers to the deviation of that score from the mean. In equation form, this would be: $x = X - \bar{X}$. The deviation is the difference between the obtained score (X) and the mean $(\bar{X})$. The definition of the mean was that point in the distribution about which the sum of the deviations is equal to zero. Putting this into equation form, we would write $\Sigma x = 0$. Let us now substitute for the deviation the method of obtaining it and proceed with a small amount of algebraic manipulation, to determine the value of the mean.

$$x = X - \bar{X}$$
$$\Sigma x = \Sigma(X - \bar{X}) = 0$$

In chapter 3, we found that the sum of the difference in a series of numbers is equal to the difference in the sums of these numbers. Thus

$$\Sigma(X - \bar{X}) = \Sigma X - \Sigma\bar{X}$$
$$\Sigma x = \Sigma X - \Sigma\bar{X} = 0$$

But the mean is a constant for all X's in the set, and the sum over a constant is equal to the constant times the number of times it occurs. Thus

$$\Sigma\bar{X} = N\bar{X} \text{ which may be substituted in the equation}$$

$$\Sigma X - N\bar{X} = 0$$

Adding $N\bar{X}$ to both sides of the equation yields

$$\Sigma X = N\bar{X}$$

Dividing both sides of the equation by N yields

$$\frac{\Sigma X}{N} = \bar{X}$$

$$\bar{X} = \frac{\Sigma X}{N}$$

We arrive at the usual formula for the mean. Notice that this is not the definition of the mean. The mean is defined as that point in a distribution about which the sum of the deviations is equal to zero. This formula tells us how to find that point about which the sum of the deviations is equal to zero.

With five scores (1, 3, 6, 7, and 8), the mean would be equal to five;

that is, the sum of the deviations about the mean will be equal to zero. The mean is the point of balance. This is shown in figure 4–1.

FIGURE 4–1
The mean as the point of balance in a set of numbers

COMPUTATION OF THE MEAN

If there are only a few scores in a distribution, computing the mean by adding up all the scores and dividing by N involves very little labor; but if there are many scores, the labor involved can be reduced considerably by taking advantage of some of the rules for combining numbers that were reviewed in chapter 3. In example 4–1, the score 24 occurs six times. Each time we come to the score 24, we could add it into the sum and proceed in the usual way. But if many of the scores are recurring, we can shorten the labor by multiplying that score by the number of times it occurs and then adding this total into the sum of scores. Symbolically, we could write an equation for the mean as:

$$\bar{X} = \frac{\Sigma f X}{N}. \tag{4-2}$$

Notice that a new symbol has been placed in the formula for the mean—f. This stands for the frequency with which a given score occurs. In the case of the example below, the score of 24 occurs with a frequency of 6; thus, this equation tells us to multiply 6 by 24 to obtain 144 and add (Σ) with all other combinations of scores times their frequency of occurrence. This procedure is illustrated in example 4–1.

COMPUTATION OF THE MEAN FROM GROUPED DATA

If the mean were to be computed from data that has been grouped in a frequency distribution, it would be impossible to determine the value of each individual score. Every score in a given category becomes the same as every other score in that category. Since we cannot identify the individual scores, we must use one value to stand for all scores in the interval. The value that is used is the midpoint of the interval. (Note the advantage of using i as an odd number so that the midpoint is an

---

Example 4–1   Illustration of the computation of the mean with several recurring scores

The following 30 scores were obtained in the following order: 25, 27, 20, 23, 29, 28, 21, 26, 23, 23, 24, 25, 24, 27, 22, 26, 22, 25, 24, 28, 23, 24, 27, 25, 26, 22, 25, 24, 24, 21. These can be rearranged conveniently into a frequency distribution.

Score	f	fX
29	1	29
28	2	56
27	3	81
26	3	78
25	5	125
24	6	144
23	4	92
22	3	66
21	2	42
20	1	20
$\Sigma f = 30$		$\Sigma fX = 733$

The $\Sigma f = 30$ and is all the observations, or N

The $\Sigma fX = 733$, which is the sum of all the scores

$$\bar{X} = \frac{\Sigma fX}{N} = \frac{733}{30} = 24.4$$

---

integer.) This midpoint becomes our best estimate of these scores. To determine the mean of the distribution of grouped data, we would multiply the midpoint of each interval (Mid) by the frequency of scores within the interval (f) and add this to all other scores. Symbolically, this can be stated:

$$\bar{X} = \frac{\Sigma f \, \text{Mid}}{N}. \tag{4–3}$$

This procedure is illustrated in example 4–2.

Example 4–2   Illustration of the computation of the mean from grouped data

Score interval	f	Midpoint	f Mid
35–39	2	37	74
30–34	1	32	32
25–29	4	27	108
20–24	9	22	198
15–19	8	17	136
10–14	3	12	36
5–9	2	7	14
0–4	1	2	2
	$N = 30$		$\Sigma f \, \text{Mid} = 600$

$$\bar{X} = \frac{\Sigma f \, \text{Mid}}{N}$$

$$\bar{X} = \frac{600}{30} = 20.0$$

The student will note that these three techniques for computing the mean are equivalent. In the first formula encountered

$$\bar{X} = \frac{\Sigma X}{N}$$

the score stands for the midpoint of an interval. The interval is only one unit wide; thus a score of 24 has the real limits of 23.5 and 24.5. Where is the "true" value of the score? We are in effect taking the midpoint of the interval to stand for the best estimate of our measurement within this interval. The absence of the symbol f for frequency within an interval can be taken as a frequency of one, and multiplying a number by one does not change its value, so in this case the symbol f is not needed.

There is one difference in the three procedures mentioned above. In grouping data, measured differences in scores are lost, and these scores are treated as if they were the same. As the interval becomes wider and the number of groups decreases, this loss of information can become serious. Earlier, it was mentioned that the number of categories or groups should not be less than 10 to 12. This is because of the inaccuracies that will be introduced by grouping. Advanced textbooks give a correction that may be applied to adjust for the inaccuracy introduced by grouping; however, if the number of groups is more than about 12, the amount of correction for grouping becomes so small as to be of little consequence.

CODING SCORES FOR COMPUTATION

If desk calculators are not available, certain arithmetic shortcuts can be used to ease the computational labors involved in computing the mean. These shortcuts involve some form of coding the data. The simplest form of coding involves adding or subtracting a constant from each score in the distribution. This same constant is then subtracted or added (reverse of the original operation) after the mean of the coded data has been determined. Example 4–3 shows the computation of the mean using this

---

Example 4-3 Illustration of the computation of the mean with coded data

A constant of 10 has been subtracted from each of the 30 scores in example 4–1. These have been arranged into a frequency distribution.

Score	f	fX_c
19	1	19
18	2	36
17	3	51
16	3	48
15	5	75
14	6	84
13	4	52
12	3	36
11	2	22
10	1	10
	$\Sigma f = 30$	$\Sigma fX_c = 433$

The $\Sigma f = 30$ and is all the observations, or N

The $\Sigma fX_c = 433$, which is the sum of all the coded scores

$$\bar{X}_c = \frac{\Sigma fX_c}{N} = \frac{433}{30} = 14.4$$

$$\bar{X} = \bar{X}_c + k; \; \bar{X} = 14.4 + 10 = 24.4$$

coding procedure with the data given in example 4–1. Any constant can be used. The main idea is to simplify the arithmetic involved in the computation. With desk calculators it is usually faster to work with the original data.

Coding of scores can also be used with grouped frequency distributions. The constant to be added or subtracted is usually the midpoint of an interval. One additional step is added in this form of coding. The interval size is coded as one (1). Each interval is then coded as the number of interval steps from the reference interval (usually called the arbitrary origin). After the mean of this coded data has been determined, this mean is multiplied by the interval size, and the midpoint of the reference interval is subtracted or added (reverse of the original procedure) to determine the mean of the group data. This procedure is shown in example 4–4, using the data for example 4–2. In this example, an interval

---

Example 4–4 Illustration of the computation of the mean from grouped data using a coding procedure

The interval 15–19 has been used as the arbitrary origin. Each interval is shown as a one-step deviation from this reference point.

Score interval	f	x'	fx'
35–39	2	4	8
30–34	1	3	3
25–29	4	2	8
20–24	9	1	9
15–19	8	0	0
10–14	3	−1	−3
5–9	2	−2	−4
0–4	1	−3	−3
	$N = 30$		$\Sigma fx' = 18$

$$\bar{X}_c = \frac{\Sigma fx'}{N} = \frac{18}{30} = .6$$

$$\bar{X} = (\bar{X}_c)(i) + \text{Mid of interval}$$
$$\bar{X} = (.6)(5) + 17 = 3.0 + 17 = 20$$

---

in the middle of the distribution was used for the arbitrary origin. Often the lowest interval is selected for the arbitrary origin so that deviations with negative signs are not necessary. The advent of modern calculators has made coding procedures much less useful than in past years.

Combining means

Often, it is desirable to find the combined mean from two or more groups. One error that is frequently made is simply to determine the mean (average) of the different means. That is, add the obtained means together and find the mean of these means. It is only under the specific situation that *each mean is based on exactly the same number of cases* that this procedure of averaging means will yield the mean of the total. To obtain the total mean based on smaller samples, each mean of a sample should be multiplied by the number of cases in the sample ($N_1\bar{X}_1$, $N_2\bar{X}_2$, etc.). These should then be added together and divided by the total number of cases in all samples:

$$\bar{X}_t = \frac{N_1\bar{X}_1 + N_2\bar{X}_2 + N_3\bar{X}_3 + \cdots + N_k\bar{X}_k}{N_1 + N_2 + N_3 + \cdots + N_k}$$

The correct and incorrect procedures are both illustrated in example 4–5.

Example 4–5 Illustration of correct and incorrect methods of combining means from different groups of scores

Group 1	Group 2	Group 3
$\bar{X} = 17.3$	$\bar{X} = 17.9$	$\bar{X} = 15.2$
$N = 10$	$N = 10$	$N = 100$

The *incorrect* method of combining means:

$$\bar{X}_t \neq \frac{17.3 + 17.9 + 15.2}{3} = \frac{50.4}{3} = 16.8$$

This technique gives no more weight to the sample with 100 cases than to the sample with 10 cases; thus the mean is biased toward the smaller sample size. What is needed is to convert each of these computations of the mean back to the sum of scores on which the means are based.

The *correct* method of combining means: Since $\bar{X} = \dfrac{\Sigma X}{N}$, then $\Sigma X = N\bar{X}$; thus

the correct method becomes

$$\bar{X}_t = \frac{173.0 + 179.0 + 1{,}520.0}{10 + 10 + 100} = \frac{1{,}872}{120} = 15.6$$

Throughout the discussion of computing the mean of a distribution, we have been adding and subtracting numbers, multiplying, and dividing. When we apply these kinds of arithmetic to numbers, we are assuming that the numbers represent an interval or ratio scale (see chapter 3). If the data we have obtained are clearly ordinal in nature, then the determination of the mean as a measure of the location or central tendency may be misleading. In dealing with measurement which is clearly not interval or ratio, we shall have to use some type of statistics which are based on the *order* characteristics rather than *interval* characteristics.

The median

All measures of location are oriented toward describing the middle portion of a distribution. As was indicated above, the mean is that point in a distribution about which the sum of the deviations is zero. This indicates that if the sum of all deviations of scores above the mean were determined, it would be exactly equal (within rounding errors) to the sum of the deviations of all scores below the mean. In this sense the mean is the middle of the distribution in terms of deviations of the scores. The median is also the middle of the distribution, but only in the sense of order. That is, the median is that point in a distribution above which are located 50 percent of the scores and below which are located 50 percent of the scores. The median is the middle, not in terms of measured distance of scores, but rather in terms of the rank or order of scores.

Since the median is a specific percentile (50 percent), the computation procedure is identical to that for any percentile. Chapter 2 showed the computation of percentiles from grouped frequency data and from data presented graphically. It also included an example of the determination of the 50th percentile (the median).

For review purposes, the computation of the median is shown in example 4–6. The formula for the computation of any percentile (see chapter 2) has been made specific for the computation of the median and is given below:

$$\text{Median} = \frac{(N/2 - c.f.)}{f} \times i + l.r.l. \tag{4-4}$$

where:

$N/2$ is 50 percent of the cases
$c.f.$ is the cumulative frequency in the interval below
f is the frequency of cases in the interval
i is the width of the interval
$l.r.l.$ is the lower real limit of the interval

The mode

The mode as a measure of central tendency is somewhat different from the mean and the median, in that the mode does not necessarily indicate the middle portion of the distribution. The mode indicates the most fre-

Example 4–6

The data for this example are taken from table 2–8

Cumulative distribution of the test scores from 80 freshman students

Score interval	Frequency distribution	Cumulative frequency	Cumulative percentage
95–99	2	80	100
90–94	3	78	98
85–89	4	75	94
80–84	4	71	89
75–79	5	67	84
70–74	5	62	78
65–69	5	57	71
60–64	7	52	65
55–59	8	45	56
50–54	13	37	46
45–49	10	24	30
40–44	8	14	18
35–39	4	6	8
30–34	2	2	2

$$\text{Median} = \frac{(80/2 - 37)}{8} \times 5 + 54.5 = \frac{3}{8} \times 5 + 54.5 = 1.9 + 54.5 = 56.4$$

quently occurring score. If the data are ungrouped, the mode is the most frequent score. In the data presented in example 4–1, the score of 24 has the greatest frequency of occurrence (there are six scores of 24); hence the mode is 24.

If the data have been grouped, the mode is defined as the midpoint of the interval with the greatest number of cases. In example 4–6 the interval 50–54 has a frequency of 13; hence the mode would be 52—the midpoint of that interval.

The geometric mean and the harmonic mean

On rare occasions the student in the behavioral sciences may encounter the use of the geometric or harmonic mean. Both of these statistics require that the numbers being combined form a ratio scale.

The geometric mean is the *n*th root of the product of N scores. The general formula is:

$$GM = \sqrt[n]{\pi X_i} = \sqrt[n]{(X_1)(X_2)(X_3) \cdots (X_n)}$$

where:

π is an operator indicating multiplication

n is the number of scores; also indicates the root to be taken

The geometric mean of 4 and 9 is 6: $\sqrt{(4)(9)} = \sqrt{36} = 6$. The geometric mean of 2, 4, and 8 is 4: $\sqrt[3]{(2)(4)(8)} = \sqrt[3]{64} = 4$. The geometric mean is used in some areas of psychophysics.

The harmonic mean is the arithmetic mean of the reciprocals of the measures. The general formula for the harmonic mean is:

$$HM = \frac{1}{1/N[\Sigma(1/X_i)]} = \frac{1}{1/N(1/X_1 + 1/X_2 + 1/X_3 \cdots + 1/X_n)}$$

The harmonic mean of 4 and 9 is 5.56:

$$\frac{1}{1/2(1/4 + 1/9)} = \frac{1}{(.5)(.36)} = \frac{1}{.18} = 5.56$$

The harmonic mean is used in certain areas of engineering and also in economics, but like the geometric mean it has had only limited application in the behavioral sciences. Neither of these measures will be used further in this textbook.

Summary of the use of the mean, median, and mode

NOMINAL DATA

When the data can only be categorized but not ordered, it is not appropriate to speak of central tendency or location. The "central" concept implies some type of dimension or order underlying the data. One can always compute the various measures of central tendency with nominal data (numbers do not protest inadmisssible operations performed on them), but the resulting statistic has no relevant meaning and can be misleading. An apparent exception to this should be pointed out. Oftentimes, measures of central tendency are applied to the *frequencies* found in the categories of a nominal scale. One can determine the modal frequency of the different categories. When this is done, the operations do not describe the middle portion of the nominal scale, only the middle portion of the frequencies observed in the different categories of the nominal scale.

ORDINAL DATA

When the data meet the requirements of ordinal scaling, both the median and the mode may be used to indicate the location or central portion of the distribution. The median is to be preferred in most instances, since it supplies more information. The median locates that point above which is found 50 percent of the distribution and below which is found 50 percent of the distribution. The mode, on the other hand, locates the point at which the greatest number of scores fall, which may or may not be located near the middle part of the distribution.

INTERVAL OR RATIO DATA

When the data meet the requirements of interval scaling, all three measures of central tendency may be used to describe the location or central portion of the distribution. In most situations the mean is to be preferred. The mean utilizes all the information in the data, in the sense that the specific value of each score enters into the determination of the mean. The median uses only the ranking of the scores and is insensitive to distances above or below the middle. The mode locates the most frequent score or category. In addition to the utilization of information in the scores, the mean is to be preferred if we wish to make predictions from the set of scores to other sets of scores. If we selected a sample from some population of scores, and from this sample computed the mean, the median, and the mode, and then drew a second sample from the same population and again computed the mean, the median, and the mode, we would not expect the two means to be exactly the same, nor the two medians or modes, but we would expect them to be similar to each other. If these procedures were repeated for a large number of samples, we would find that the variation among the means would be less than the variation among the medians, which in turn would be less than the variation among the modes. The amount of variability among statistics computed from different samples of the same population is referred to as the *stability* of the statistic. Thus the mean is a more stable measure of central tendency than is the median, which in turn is more stable than the mode. If one were to draw a sample from the population and wished to make predictions concerning other samples to be drawn from the same population, the mean would be preferred, in that there should be less error in the prediction than would be the case using the median or the mode.

Central tendency in skewed distributions

The fact that the mean uses the data from each score makes it sensitive to the magnitude of each score in the distribution, which at times can be misleading in determining the location or central tendency of a set of scores. If the distribution is badly skewed (nonsymmetrical), then the few deviant scores tend to "pull" the mean in the direction of these scores. The mean will always reflect that point about which the sum of the deviations is equal to zero; but if the distribution is badly skewed, the mean could be very nontypical of the middle section of the distribution. In this situation the median is used as more representative or typical of the middle part of the distribution. A rough or crude test of the degree of skewness of a distribution is the difference between the mean and the median computed from the same set of data. The mean will always be pulled toward the "long" tail of the skewed distribution. Thus, if the data are positively skewed (long tail toward the high scores), the mean will be numerically larger than the median. If, on the other hand, the data are negatively skewed (long tail toward the lower scores), the mean will be numerically smaller than the median. In a symmetrical distribution the mean and median will be of the same numerical value. With a large number of scores, the mean, median, and mode for positively skewed, negatively skewed, and normal distributions, respectively, will be arranged as shown in figure 4–2.

To illustrate further the difference in these three measures of central

FIGURE 4–2
Illustration of the relationship among the mean, median, and mode in distributions of different types of skewness

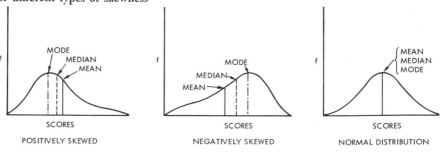

tendency when the data are at least interval in nature, we shall use a rather farfetched economic example. This is a small business which employs nine individuals. The president receives a salary of $200,000 per year, the vice president (brother-in-law) receives $50,000 per year, and the lowest paid employee receives $1,000 per year. Some of the employees request a raise, stating that "the average salary is $2,000 per year." The president replies that they are incorrect, that the average salary per year is in fact $30,000. Who is correct? The distribution of salaries is given in table 4–1.

TABLE 4–1
Distribution of salaries in factory
(annual salary of each employee)

$200,000
50,000
5,000
5,000
3,000
2,000
2,000
2,000
1,000

Both are correct. The term *average* is a broad term which can be used to cover various estimates of central tendency. To be more specific, the president of the company is correct when he states that the average salary is $30,000. It would be more precise to say that the mean salary is $30,000. The employees are correct when they say that the average salary is $2,000, but it would be more precise to state that the modal salary is $2,000. Notice that the mean is considerably higher than the median ($3,000), indicating that this distribution is positively skewed. The fact that the discrepancy is large would indicate that it is a markedly skewed distribution. In this situation the median is certainly more representative of the "typical" salary than is the mean. Perhaps this can be used to illustrate an old cliché that "figures don't lie, but liars can figure." Usually, the cliché is not true; it is simply a case of understanding what particular statistic is being used and then judging its appropriateness in the situation.

Statistical inference

The discussion of the various measures of central tendency could apply to either descriptive statistics (describing a population) or to inferential statistics (using a sample to make inferences about a population). When using samples to infer characteristics of populations some new considerations become necessary.

One minor problem concerns the symbols to be used to insure that there is no confusion as to which mean is being discussed—the mean of the sample, or the mean of the population from which the sample was drawn. The system that will be followed throughout this book was first adopted by the British and is sometimes referred to as the British notational system. This requires that letters from the Greek alphabet be used for population values and letters from the Latin alphabet be used to indicate sample values. Thus the mean of a population is symbolized by the greek letter μ (mu). This text will use a bar over the latin letter to indicate the mean of a sample drawn from a population of scores. Thus X will indicate a score from the population and $\bar{X}$ will indicate the mean of the sample of scores.

Sample values vary as to how well they estimate population parameters. Three of the characteristics of estimators deal with *bias, efficiency,* and *consistency.*

Bias

The mean of a sample ($\bar{X}$) may be larger than the population mean (μ) or it may be smaller than the population value. To the extent that a sample value is *systematically* larger (or smaller) than the population value, we say that the estimator is *biased.* If random sampling procedures are followed, then the sample mean ($\bar{X}$) is an unbiased estimator of the population mean (μ). This does not mean that a single sample mean will be an exact estimate of the population mean. Rather it indicates that the sample mean has the same chance (probability) of being too large as of being too small. In general, unbiased estimators are to be preferred to biased estimators.

Efficient

There is a rather old joke about how to count the number of sheep in a flock. Count the number of legs and divide by four. An efficient estimator is one that uses all the information that is included in the scores. In the sheep illustration, the estimator uses all the information, but much of it is redundant; hence, the estimator would be inefficient. If the scores in a sample meet the requirements for interval or ratio scaling, then the mean would be an *efficient* estimator of central tendency while the median would not be efficient. The median does not use all the information contained in the data. Incidentally, if some of the sheep had less than four legs, counting the legs and dividing by four would also be a *biased* estimator.

CONSISTENT

An estimator of a population parameter should also be consistent. That is, as the sample size increases, the sample statistic should be systematically closer in value to the population parameter. The sample mean (if the sample is drawn at random from the population) does converge on the population mean (μ) as the size of the sample increases.

Summary

One aspect of describing a set of scores is to specify the central portion or location of the distribution along some continuum. The three measures which are used most often to describe the location or central tendency of scores are the mean, the median, and the mode. All three are measures of central tendency, but each describes the location of the distribution in a different way.

The *mean* is that point in a distribution about which the sum of the deviation is equal to zero. The mean is based on all numbers in a set and hence uses all the information which may be contained in the numbers. If the underlying scale is nominal or ordinal, the mean uses information which *is not present* in the objects and may give erroneous information.

The *median* is a point in the distribution—that point above which are located 50 percent of the scores and below which are located 50 percent of the scores. The median is based on the rank or order of the observation and not on distance between objects.

The *mode* is the category or measurement which occurs most frequently in the set of scores.

If the objects being measured form a *nominal scale,* the concept of central tendency has no meaning. We cannot describe the middle of a distribution if there is no underlying continuum of categories.

When the objects being measured are on an *ordinal scale,* either the median or the mode may be used to describe the central portion of the distribution.

With *interval* or *ratio scaling,* the mean, median, and mode may all be used as measures of central tendency. If repeated samples were drawn from the same population and the mean, median, and mode of each sample were computed and compared, there would be less variation among the means than among the medians, and the modes would differ most from each other. The mean is the most *stable* measure of central tendency. In skewed distributions the median is the more preferred measure of central tendency. One rough indication of skewness is to compare the mean and the median for the same set of data. The mean, since it is influenced by the distance of each score, will be pulled in the direction of the long tail of the skewed distribution. The median—based on order of scores—will not be influenced. In a positively skewed distribution the mean will be numerically larger than the median. In a negatively skewed distribution the mean will be numerically smaller than the median.

When using a sample to estimate population values, the Greek alphabet is used to denote the population parameter and the Latin alphabet is used for the sample statistic. The mean of a population is symbolized by the Greek letter μ (mu). The sample mean will be symbolized by a bar over the Latin letter $(\bar{X})$ used to indicate the score.

A sample statistic used to estimate a population parameter should be *unbiased* (neither systematically too large or too small); *efficient* (utilize all of the information contained in the data); and *consistent* (as the sample size increases, the estimator becomes closer to the population value).

References

SKEWNESS AND KURTOSIS

QUINN MCNEMAR, *Psychological Statistics* (4th ed.; New York: John Wiley & Sons, Inc., 1969), pp. 25–30.

MEASURES OF CENTRAL TENDENCY

H. M. WALKER, *Mathematics Essential for Elementary Statistics* (rev. ed.; New York: Henry Holt & Co., Inc., 1951), chap. vii.

METHODS OF CODING DATA

QUINN MCNEMAR, *Psychological Statistics* (4th ed.; New York: John Wiley & Sons, Inc., 1969), pp. 16–25.

5

Measures of variability

FIGURE 5–1 shows two distributions which have a symmetrical "bell-shaped" appearance. Since the two distributions are symmetrical, the mean, median, and mode will be identical for each distribution. The two curves are drawn so that each distribution has the same central tendency, but

FIGURE 5–1
Illustration of two distributions with the same central tendency but different variability

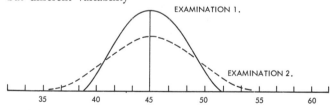

the two distributions are clearly different from each other. The one distribution has far more spread among the scores than does the other, or we could say that there is more variability in the one set of scores than in the other.

In a college course an examination is given, and the maximum score possible is 60 points. Student *A* receives a score of 51. Is this a good score? A poor score? The answer depends upon how well the rest of the class did on the examination. The mean of the class was 45 points on the examination. Now, is the score a good one? It is clearly above the average—in fact, the score is six points above the average (the student's score could be expressed as +6 to indicate that it deviates six points above the mean). A second examination is given, and once again the student scores 51 out of a possible 60 points, and the mean of the group on the second examination is also 45. Did our student *A* do better, poorer, or the same on the second examination? At first glance, it would appear that the student did the same on the second examination (again, his deviation score would be +6). Let us go back to figure 5–1 and interpret this as the distribution of scores on the two examinations. The solid line indicates the first examination; the dashed line, the second. Student *A* would rank very near the top on the first examination but not nearly so high on the second compared with the other students in the class. Notice that most of the students scored near the mean of 45 on the first examination (little variability in the distribution), whereas the scores were spread much more on the second examination. To describe these data more adequately, we need some measure or index of the degree of variation or spread of scores around the central portion of the distribution. This chapter will examine four measures of variability which may be used to indicate the amount of spread in a distribution of scores.

Determining variability for nominal data

By now the reader should be well aware that the types of operations which should be performed on the numbers obtained in a study are dependent upon the underlying characteristics being measured. If the data obtained are nominal in nature, this precludes using any statistic which utilizes order in numbers or distance between numbers.

The concept of variability implies a comparison to some reference point. Since nominal scaling does not involve the idea of some underlying continuum, it is not accurate to speak of variability in that sense. It is possible to compare the frequency of observations in different categories of the nominal scale. It would be possible to ask how much variation there is among the *frequencies* observed in the different *categories*. The measure which can be used to compare differences in frequencies among the categories of a nominal scale is the range.

The range tells us about the difference in frequencies of cases between the category with the fewest cases and the category with the largest number of cases. Unfortunately, the range when used with nominal data does not yield a great deal of information; but when we can only categorize observations on a nominal scale, we do not have a great deal of information about the phenomenon being measured.

Determining variability for ordinal data

If the data can be ordered, the range can be computed. The range is the difference between the highest ordered category and the lowest ordered category. This tells the spread or dispersion of the scores with reference to these two points—upper and lower. With ordering of the data, it is advantageous to compute other range-type statistics which are variations on the simple range.

THE SEMI-INTERQUARTILE RANGE

The most widely used measure of variability with ordinal data is the semi-interquartile range. Let us examine this series of words, *semi-interquartile range,* to determine the characteristics of this statistic. The term *range* should suggest that this measure is the difference between two points—a range. The word *quartile* refers to the division of a distribution into four equal parts. The median is often referred to as the 50th percentile, which would also be the second quartile. The expression *interquartile* refers to the distance or range between the 25th and the 75th percentiles (also referred to as the first and third quartiles). The method of computing the first quartile and the third quartile is the same as that used in determining any percentile. *Semi* means one half, so that the term *semi-interquartile range* means one half the distance between the first and third quartiles. The computation of this statistic is shown in example 5–1.

Notice that the semi-interquartile range is a range statistic; that is, it is one half the range (distance) between the first and third quartiles. It would be quite possible to use the interquartile range, which is the distance between the first and third quartiles, without the division by two; but in any case it is a range—the distance between two points. The reason for using the first and third quartiles rather than the highest and lowest scores is related to the concept of stability of a statistic which was mentioned in chapter 4. If repeated samples are drawn from the same population and the simple range is computed for each of these

---

Example 5-1   Computation of the semi-interquartile range

These data are from table 2-7.

Score	Frequency	Cumulative frequency	
95–99	2	80	
90–94	3	78	
85–89	4	75	
80–84	4	71	
75–79	5	67	
70–74	5	62	Interval of 75 percent
65–69	5	57	
60–64	7	52	
55–59	8	45	
50–54	13	37	
45–49	10	24	Interval of 25 percent
40–44	8	14	
35–39	4	6	
30–34	2	2	

$$\text{Percentile} = \left[\frac{(n\text{th case} - c.\ f.\ \text{in interval below})}{f \text{ in interval}} \right] \times i + l.r.l. \text{ of interval}$$

$$25\% = \left[\frac{(20-14)}{10} \right] \times 5 + 44.5 = \left(\frac{6}{10}\right) \times 5 + 44.5 = 3.0 + 44.5 = 47.5$$

$$75\% = \left[\frac{(60-57)}{5} \right] \times 5 + 69.5 = \left(\frac{3}{5}\right) \times 5 + 69.5 = 3.0 + 69.5 = 72.5$$

$$\text{Semi-interquartile range } (Q) = \frac{Q_3 - Q_1}{2} = \frac{75\% - 25\%}{2}$$

$$Q = \frac{72.5 - 47.5}{2} = \frac{25.0}{2} = 12.5$$

---

samples as well as the semi-interquartile range, the variation among the different ranges will be greater than the variation among the different semi-interquartile ranges. The semi-interquartile range is a more *stable* measure of spread or variation within a set of scores.

In 1929, L. L. Thurstone and E. J. Chave[1] developed one of the first successful approaches to the measurement of attitude. They selected attitudes toward the church as the area to be explored. Several hundred statements were prepared which might reflect differing attitudes ranging from extremely favorable to extremely unfavorable. These items were given to a group of several hundred judges who were asked to place each item in one of 11 piles ranging from pile 1, which was to represent

---

[1] L. L. Thurstone and E. J. Chave, *The Measurement of Attitude* (Chicago: University of Chicago Press, 1929).

the most unfavorable possible attitude, through neutral (position 6), to position 11, which would be the most favorable possible attitude toward the church. The judges were not asked whether or not they agreed with the statement, but rather their judgment as to the kind of attitude expressed by the item.

It would be somewhat presumptuous to assume that the ordered categories 1 to 11 represent an interval scale (the instructions used by Thurstone and Chave did ask the judges to divide the total distance between 1 and 11 into "equal-appearing" units). Even if this method of scaling did meet the assumptions of an equal-interval scale, certain other problems would be presented. The first step would be to determine, on the average, what degree of favorableness or unfavorableness is being expressed by each item. Some measure of central tendency would be used for this. In those cases where the item is judged to represent extreme attitudes, either pro or con, we run into an interesting characteristic of this type of scaling. Notice that if category 1 represents the most unfavorable attitude possible toward the church, the item cannot be placed in any category lower than 1. Yet some judges will place this item in higher categories. This would lead to a positively skewed distribution. At the opposite end of the scale, items will tend to have negatively skewed distributions. Since the items judged at the extremes of the continuum will tend to produce markedly skewed distributions, Thurstone and Chave used the median to determine the "average" scale position of each item.

Some of the items showed great agreement among the judges. That is, most of the judges placed the items in about the same positions along the scale. This would produce a "bunching" of the ratings. Other items might be placed almost equally at each of the 11 points on the continuum, indicating that the different judges interpreted these statements quite differently. Thus the degree of dispersion or "spread" of the judgments could be used as an indication of the degree of ambiguity of the attitude statements. The statements which were highly ambiguous would yield a large dispersion in the judgments of the raters. Thurstone and Chave used the semi-interquartile range as their measure of ambiguity. From the original pool of items, only those items were used which had medians ranging from close to 1 to close to 11. Each of these items had to have a small semi-interquartile range or low ambiguity index.

Modern attitude assessment has progressed a great deal from these early beginnings, but the article by Thurstone and Chave is well worth the effort to see the logic behind their research and the implementation of this logic in their methodology and data analysis.

Determining variability for interval and ratio data

Since data that are at least interval contain within the numbers information as to the distance between scores, measures of dispersion which
can utilize this information will be preferred to those measures which
can only utilize order-type information. Two of these measures utilizing
"distance" between numbers will be discussed in this section.

AVERAGE DEVIATION

Where the scores in a distribution are relatively close to the mean
of the distribution, we would expect that the deviations of these scores
from the mean should be relatively small. On the other hand, with scores
that spread very high and very low from the mean, we would expect
large deviations (some positive and some negative). Unfortunately, the
sum of the deviations about the mean will be zero for any distribution.
A distribution with very little spread about the mean and a distribution
with a great deal of spread about its mean will have exactly the same
sum of deviations—zero! We can utilize the size of the individual deviations if the direction of the deviation is ignored. If we treat each deviation
as if it were positive and sum all deviations together, the average of
this will reflect the average size of the deviations about the mean. Symbolically, this can be represented as

$$\text{Average deviations} = \frac{\Sigma|x|}{N}$$

where

Σ is the familiar summation operator
| | indicates the absolute rather than algebraic value
       (ignore the plus or minus sign, and treat all scores as positive)
x is the deviation of the score from the mean $(X - \bar{X})$
N is the total number of scores in the set

Although the average deviation utilizes information from each score
in the set (as compared to a range statistic which only utilizes two points
in the distribution), the statistic is quite limited, in that it has little use
in any further analysis of data. Descriptively, this could be a rather useful
statistic, but the standard deviation has as much descriptive value as the
average deviation and is not a terminal statistic. That is, the standard

deviation does enter into further analysis. The average deviation is presented here for two reasons: (1) to show the logic of using the deviation of the score about the mean in determining a measure of variability, and (2) because many older studies in the fields of psychology and education made rather extensive use of the average deviation as a measure of dispersion and the student should be familiar with it as a descriptive statistic.

STANDARD DEVIATION

The most widely used measure of dispersion for data that are interval or ratio in nature is the standard deviation. As the name implies, this measure is based on the deviation of each score about the mean. The formula for the standard deviation is

$$S = \sqrt{\frac{\Sigma x^2}{N}} \qquad (5\text{--}1)$$

where

S is the symbol for standard deviation[2]
x^2 is the squared deviation of the score about the mean $(X - \bar{X})^2$
Σ is the summation operator
N is the total number of scores
$\sqrt{}$ indicates the operation of obtaining the square root

A point of difficulty for some students is the confusion in mixed arithmetic operations (see chapter 3). Since we are required to square each deviation (x^2) and then add the squared deviations together, the formula is written as Σx^2. If we were to add the deviations first and then square the total sum, the expression would be written as $(\Sigma x)^2$.

Although this formula is excellent for understanding the steps involved in the computation of the standard deviation, it is a poor technique to use in actual computations. Example 5–2 shows the computation of the standard deviation using formula 5–1. Notice that there are many opportunities for error in this procedure. Since each deviation score (x) requires

---

[2] Modern practice in statistics uses the Latin letter to indicate sample values and reserves the Greek letters for population values. Thus the symbol S will be used to indicate the standard deviation from a sample, and the symbol σ will be reserved for the standard deviation of the population from which the sample was obtained. Older treatments often used the lowercase sigma (σ) to indicate sample values.

Example 5-2   Computation of the standard deviation by the deviation score method

The following eight scores are from the zero-second-delay group presented in Chapter 2: 42, 40, 43, 16, 42, 44, 28, 22.

$$x = X - \bar{X}$$

	x	x^2
$x_1 = 42 - 34.6 =$	7.4	54.76
$x_2 = 40 - 34.6 =$	5.4	29.16
$x_3 = 43 - 34.6 =$	8.4	70.56
$x_4 = 16 - 34.6 =$	−18.6	345.96
$x_5 = 42 - 34.6 =$	7.4	54.76
$x_6 = 44 - 34.6 =$	9.4	88.36
$x_7 = 28 - 34.6 =$	− 6.6	43.56
$x_8 = 22 - 34.6 =$	−12.6	158.76
$\Sigma X = 277$	$\Sigma x^* = 0.2$	$\Sigma x^2 = 845.88$

$$\bar{X} = \frac{\Sigma X}{N} = \frac{277}{8} = 34.625$$

$$S = \sqrt{\frac{\Sigma x^2}{N}} = \sqrt{\frac{845.88}{8}}$$

$$S = \sqrt{105.7350}$$

$$S = 10.28$$

* The Σx is not eactly zero because of rounding errors associated with the mean.

---

that the mean must be subtracted from the original score, there will be N different subtractions to be made, with the attendant possibility of errors. In practice, the standard deviation is seldom computed using the individual deviations.

WHOLE-SCORE METHOD FOR COMPUTING THE STANDARD DEVIATION. With the substitution of a few symbols for their equivalents and a little algebraic manipulation, it is possible to derive a whole-score method of computing the standard deviation which involves fewer operations and hence less opportunity for error. This whole-score method is especially well suited for use when a desk calculator is available. We shall begin the derivation of the whole-score computing formula by squaring both sides of the equation to remove the radical sign from the right-hand portion:[3]

$$S^2 = \frac{\Sigma x^2}{N}$$

For x, substitute the equivalent $X - \bar{X}$:

$$S^2 = \frac{\Sigma(X - \bar{X})^2}{N}$$

---

[3] The symbol S^2 (or σ^2 for the population value) is called the *variance*. The simplest way to conceptualize the variance is to think of it as the square of the standard deviation. Later, we shall have occasion to return to the concept of variance.

Squaring the term in parentheses:

$$S^2 = \frac{\Sigma(X^2 - 2X\bar{X} + \bar{X}^2)}{N}$$

Distributing the summation operator (see chapter 3):

$$S^2 = \frac{\Sigma X^2 - \Sigma 2X\bar{X} + \Sigma \bar{X}^2}{N}$$

Since summing a series involving the product of a variable and a constant is equal to the constant times the sum of the variable (see chapter 3), $-\Sigma 2X\bar{X} = -2\bar{X}\Sigma X$ (2 is a constant, and the mean of the distribution will be a constant for each score within the set). Also, $\Sigma \bar{X}^2$ is equivalent to $N\bar{X}^2$. Substituting:

$$S^2 = \frac{\Sigma X^2 - 2\bar{X}\Sigma X + N\bar{X}^2}{N}$$

Breaking the equation into three fractions:

$$S^2 = \frac{\Sigma X^2}{N} - \frac{2\bar{X}\Sigma X}{N} + \frac{N\bar{X}^2}{N}$$

Note that the N's in the numerator and denominator of the last term will cancel; also, in the second term, that $\dfrac{\Sigma X}{N} = \bar{X}$:

$$S^2 = \frac{\Sigma X^2}{N} - 2\bar{X}^2 + \bar{X}^2$$

Combining the last two terms:

$$S^2 = \frac{\Sigma X^2}{N} - \bar{X}^2$$

Extracting the square root of each side:

$$S = \sqrt{\frac{\Sigma X^2}{N} - \bar{X}^2} \qquad (5\text{--}2)$$

The computation of the standard deviation using the whole-score method is simple and straightforward. The equation directs us to square each whole score (X^2), sum all these squared scores (ΣX^2), divide this sum by the number of scores

$$\frac{\Sigma X^2}{N},$$

from this subtract the squared mean ($\bar{X}^2$), and finally extract the square root of the result of the last step. An illustration of this procedure is given in example 5–3. Squaring and summing the whole scores is a simple operation if a desk calculator is available; if not, the method is still simple if a table of squares is used to obtain the square of each whole score. See table A in the Appendix.

With a little more algebra, formula 5–2 can be modified to a form which is particularly useful with a desk calculator:

$$S = 1/N \ \sqrt{N\Sigma X^2 - (\Sigma X)^2} \qquad (5\text{–}3)$$

The use of this formula is also shown in example 5–3.

Example 5–3 Computation of the standard deviation by the whole-score method

The eight scores used in this example are the same as those used in example 5–2.

X	X²
42	1,764
40	1,600
43	1,849
16	256
42	1,764
44	1,936
28	784
22	484
$\Sigma X = 277$	$\Sigma X^2 = 10,437$

$\bar{X} = \dfrac{\Sigma X}{N} = 34.625$

$\bar{X} = 34.6$

$S = \sqrt{\dfrac{\Sigma X^2}{N} - (\bar{X})^2}$

$S = \sqrt{\dfrac{10,437}{8} - (34.625)^2}$

$S = \sqrt{1,304.625 - 1,198.8906}$

$S = \sqrt{105.7344}$

$S = 10.28$

Using formula 5–3 with the data given in example 5–3

$$S = 1/N \ \sqrt{N\Sigma X^2 - (\Sigma X)^2} = 1/8 \ \sqrt{(8)(10,437) - (277)^2}$$
$$= 1/8 \ \sqrt{83,496 - 76,729} = 1/8 \ \sqrt{6,767} = (1/8)(82.26)$$
$$= 10.28$$

GROUPED-DATA METHOD FOR COMPUTING THE STANDARD DEVIATION. If the data have been placed in a grouped frequency distribution, a slight change in formula 5–2 will result in a suitable computing formula for grouped data:

$$S = \sqrt{\dfrac{\Sigma f(\text{Midpoint})^2}{N} - \left[\dfrac{\Sigma f(\text{Midpoint})}{N}\right]^2} \qquad (5\text{–}4)$$

In using grouped data, we are once again faced with the problem of the loss of the identity of individual scores, so we do the usual thing of using the midpoint of the interval as the best estimate of each score. The computation of the standard deviation from grouped data is shown in example 5–4. Formula 5–4 can also be modified for more convenient use with a desk calculator. This modified formula is presented below and its use is included in example 5–4.

$$S = 1/N \sqrt{N\Sigma f(\text{Midpoint})^2 - [\Sigma f(\text{Midpoint})]^2} \qquad (5\text{-}5)$$

Example 5-4   Computation of the standard deviation from grouped data

The data are those used in example 4–2.

Score	f	Mid	f Mid	f (Mid)²
35–39	2	37	74	2,738
30–34	1	32	32	1,024
25–29	4	27	108	2,916
20–24	9	22	198	4,356
15–19	8	17	136	2,312
10–14	3	12	36	432
5–9	2	7	14	98
0–4	1	2	2	4
Sums	30		600	13,880

$$\bar{X} = \frac{\Sigma f \text{Mid}}{N}$$

$$\bar{X} = \frac{600}{30} = 20$$

$$\bar{X}^2 = \left(\frac{\Sigma f \text{Mid}}{N}\right)^2 = (20)^2 = 400$$

$$S = \sqrt{\frac{\Sigma f(\text{Mid})^2}{N} - \left(\frac{\Sigma f \text{Mid}}{N}\right)^2}$$

$$S = \sqrt{\frac{13{,}880}{30} - \left(\frac{600}{30}\right)^2} = \sqrt{462.6667 - 400.0} = \sqrt{62.6667}$$

$S = 7.916$, rounding to
$S = 7.92$

Using formula 5–5 with the data given in example 5–4

$$S = 1/N \sqrt{N\Sigma f(\text{Midpoint})^2 - [\Sigma f(\text{Midpoint})]^2}$$
$$S = 1/30 \sqrt{(30)(13{,}800) - (600)^2} = 1/30 \sqrt{416{,}400 - 360{,}000}$$
$$S = 1/30 \sqrt{56{,}400} = (1/30)(237.49)$$
$$S = 7.92$$

CODING SCORES FOR COMPUTATION.   As was mentioned in chapter 4, if desk calculators are not available coding data can markedly reduce the computational labor involved in determining the mean of a distribution of scores. This same coding will also reduce the labor in computing the standard deviation.

Adding or subtracting a constant value in a distribution has the effect of shifting the mean by the amount of the constant. Adding or subtracting a constant from all scores in a distribution will have *no* effect on the standard deviation. If a constant is subtracted from every score, then the coded mean will be smaller by the amount of that constant. The deviation score will be *exactly the same*. Thus, the standard deviation will not be effected by the addition or subtraction of a constant to every score.

When the data have been grouped, the coding procedure described in chapter 4 can be extended to include the standard deviation. The interval width was coded as unity (one) which has the effect of dividing the scores in the distribution by the interval width (i), and division (or multiplication) will effect the size of the standard deviation. Formula 5–6 is for computing the standard deviation from grouped data which has been coded. Notice that the interval width is included in this formula and serves to "correct" for the use of the coded interval.

$$S = 1/N \sqrt{i^2[N\Sigma fx'^2 - (\Sigma fx')^2]} \qquad (5\text{–}6)$$

The use of this coding system and formula 5–6 is shown in example 5–5. Also included in the example is the use of Charlier's check on the accuracy of this particular layout for coding scores. The last column of example 5–5 includes the entry $f(x' + 1)^2$. This directs one to add 1 to the coded value of the interval, square this value and then multiply by the frequency in the interval. The sum of this value is equal to the sum of the three columns: $\Sigma fx'^2$, $2\Sigma fx'$, and Σf.

To this point, we have presented different procedures in computing the standard deviation without any discussion of the meaning of the statistic. The standard deviation is a measure of the spread or variability of the numbers and utilizes information from all scores in a distribution. The standard deviation is always in the units of the original measurement; thus, if we determined the height of a group of individuals in inches, the standard deviation would express variation in inches. If the mean height of the group were 68 inches and the standard deviation was determined to be seven, this would be 7 inches. The standard deviation is always a measure of dispersion about the mean expressed in the measurement unit being used. One of the great advantages of the standard deviation is that it not only expresses the amount of spread or dispersion

Example 5-5   Computation of the standard deviation from grouped data using a coding procedure

The data used in examples 2 and 4 of chapter 4 and example 4 of chapter 5 are used for this example.

Score interval	f	x'	fx'	fx'^2	$f(x' + 1)^2$
35–39	2	4	8	32	50
30–34	1	3	3	9	16
25–29	4	2	8	16	36
20–24	9	1	9	9	36
15–19	8	0	0	0	8
10–14	3	−1	−3	3	0
5–9	2	−2	−4	8	2
0–4	1	−3	−3	9	4
	$\Sigma f = N = 30$		$\Sigma fx' = 18$	$\Sigma fx'^2 = 86$	$\Sigma f(x' + 1)^2 = 152$

$$\bar{X} = (\bar{X}_c)(i) + \text{Mid of Int.} = \frac{\Sigma fx'}{N} i + M.I. = \left(\frac{18}{30}\right)(5) + 17 = (.6)(5) + 17 = 20$$

$$S = 1/N \sqrt{i^2[N\Sigma fx'^2 - (\Sigma fx')^2]} = 1/30 \sqrt{(5^2)[(30)(86) - (18)^2]}$$

$$S = 1/30 \sqrt{(25)(2580 - 324)} = 1/30 \sqrt{56400} = (1/30)(237.49)$$

$$S = 7.92$$

Charlier's check on the accuracy of columns fx' and fx'^2

$$\Sigma f(x' + 1)^2 = \Sigma fx'^2 + 2\Sigma fx' + \Sigma f$$
$$152 = 86 + (2)(18) + 30$$
$$152 = 86 + 36 + 30 = 152$$

in a set of scores, but we can obtain more or less precise information as to the *number of scores* within varying distances from the mean.

TCHEBYSHEFF'S[4] INEQUALITY.   One of the famous theorems of mathematical statistics is referred to as *Tchebysheff's inequality*. This is a proof of the proportion of cases which will be found within varying standard deviation units of the mean, *regardless of the shape of the distribution*.

If we know the mean, the standard deviation, and the total number of scores in a set ($\bar{X}$, S, N), we can state that *at least* a certain proportion of the scores will be within any number of standard deviation units from the mean. Or the reverse statement will also be true. We can state that *no more than* a certain proportion of the scores will deviate further than any specific number of standard deviation units from the mean.

Tchebysheff's inequality may be stated as:

$$P\{|X - \bar{X}| > k\} < \frac{1}{k^2}$$

[4] Alternate spelling, Chebyshev.

where:

P	is the probability of the occurrence of the event in the brackets
$\lvert X - \bar{X} \rvert$	is the deviation of a score about the mean without regard to the direction of the deviation
k	is the number of standard deviation units (must be greater than one).

Thus the statement says that the probability that a score will deviate more than so many standard deviations (k) from the mean is less than the reciprocal of the square of that many standard deviations.

If the mean height of a group of 100 individuals is 68 inches and the standard deviation is 7 inches, what is the minimum number of people in the distribution who are between 54 inches and 82 inches tall (plus and minus two standard deviations from the mean of the group)? From Tchebysheff's inequality, it is possible to determine that *at least* 75 members of the group must be between $\pm 2S$ of the mean in height. Between $\pm 3S$, there must be *at least* 89 of the 100 individuals (the interval 47 inches to 89 inches) *regardless of the shape of the distribution.*

$$P\{\lvert X - \bar{X} \rvert > 2\} < \frac{1}{4}; \quad P\{\lvert X - \bar{X} \rvert > 3\} < \frac{1}{9}$$
$$P < .25; \quad P < .11$$

It is Tchebysheff's inequality which makes the standard deviation a more valuable measure of the spread of a distribution of scores than the average deviation. Both use the same amount of information, both indicate a wider or narrower spread of scores; but the standard deviation allows us to go further and state at least crude limits within which we expect to find a given percentage of the scores in a distribution, regardless of the shape of that distribution. If we know more about the shape of a distribution, we can make more exact statements concerning the frequency of cases which lie within given standard deviations about the mean.

THE NORMAL CURVE. If the distribution is of the Gaussian or normal-curve type, then we can state that 68.2 percent of the scores will lie in the interval $\pm 1S$ from the mean, 95.4 percent of the cases in the interval $\pm 2S$ around the mean, and 99.7 percent of the cases within $\pm 3S$. Figure 5–2 shows the relationship between standard deviation distance on the abscissa of the normal curve and areas under the curve.

If 99.7 percent of all observations in a normal curve will be in the interval from three standard deviations below the mean to three standard deviations above the mean, then we can expect that the range of scores (highest minus

FIGURE 5-2
The normal curve and standard deviation units

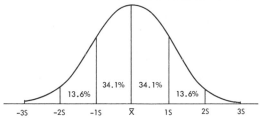

lowest plus one) will be approximately equal to six standard deviations, in the normal curve. Using the distribution of height referred to above ($\bar{X} = 68$ inches, $S = 7$ inches, $N = 100$) and knowing that the distribution is normal, how many individuals will be at least 61 inches tall but not taller than 75 inches? The answer is 68.2 percent of 100, or 68 persons. What would be the approximate range from the shortest person to the tallest in the group? Since the distribution is normal, we would expect the range to be equal to approximately six times the standard deviation, or 42 inches. We would expect the shortest individual to be about 47 inches ($-3S$ or -21 inches) and the tallest to be about 89 inches ($+3S$ or $+21$ inches).

The relationship of six standard deviations in the range of a normal distribution will hold only for rather large sample sizes. With smaller samples we are not as apt to select extremely deviant individuals, and hence both the range and the standard deviation will tend to be smaller with small samples. Since the standard deviation is based on the square of the deviation, the lack of extremely deviant scores in a distribution tends to reduce the size of the standard deviation quite markedly. Tippet[5] has determined the relationship between the number of standard deviations in the range for different size samples from a normal population. The relationship is as follows:

N	*Number of standard deviations in the range*
10	3.1
20	3.7
30	4.1
40	4.4
50	4.5
100	5.0
500	6.1
1000	6.5

[5] L. H. C. Tippet, "On the Extreme Individuals and the Range of Samples Taken from a Normal Population," *Biometrika* 17 (1925):364–87.

Summary of use of range, semi-interquartile range, average deviation, and standard deviation

NOMINAL DATA

Since nominal-type data do not include information as to distance between measurements or even order of magnitude, only the range can be used to assess the variability of frequencies in the different categories.

ORDINAL DATA

When the numbers assigned to the observation contain the information on order, various types of range statistics which utilize order may be used. The most common statistic used in this situation is the semi-interquartile range.

INTERVAL OR RATIO DATA

When the numbers convey the information as to distance between scores, all four measures of variability discussed could be used. In general, it is preferable to use those measures which utilize all the information contained in the numbers. With interval or ratio data, these statistics would be the average deviation or the standard deviation.

The average deviation utilizes all the information in the numbers, and the magnitude of the average deviation indicates the amount of variability (large average deviation, greater variability; small average deviation, less variability); however, the average deviation is a terminal statistic. It is not suitable for further analysis of the data.

The preferred measure of variability with data which approach the interval or ratio type is the standard deviation. Like the average deviation, it utilizes all the information in the numbers and is a measure of distance or spread of the scores, expressed in the original measuring units. The standard deviation has marked advantages over the average deviation, in that the standard deviation can be used for further data analysis. As was noted earlier, Tchebysheff's inequality can be used with the standard deviation to set limits on the proportion of cases to be found within varying distances from the mean of the distribution. Although Tchebysheff's inequality holds for *any distribution,* the standard deviation becomes maximally useful when the shape of the distribution can be specified.

In those situations when the obtained data are markedly skewed, the semi-interquartile range is to be preferred to the standard deviation, even though the data are interval in nature. Since the standard deviation is based on each observation, it is also influenced by each observation. If the situation were such that the data were markedly skewed, the standard deviation would become extremely large. (Remember that each deviation is squared in the computation of the standard deviation; thus, extreme deviations—either positive or negative—exert a very large effect on the magnitude of the standard deviation.) In general, in those situations in which the median is the preferred measure of central tendency, the semi-interquartile range becomes the preferred measure of variability.

In comparing the range, semi-interquartile range, and standard deviation for use with interval data, the notion of *stability* becomes an important consideration. In chapter 4, this concept was introduced in connection with measures of central tendency. The same situation holds with respect to variability. If repeated samples were drawn from the same population, and for each sample the range, semi-interquartile range, and standard deviation were computed, and all ranges from all samples were compared, all semi-interquartile ranges were compared, and all standard deviations were compared, there would be less difference among the different standard deviations than among the semi-interquartile ranges, and the greatest differences would be found among the ranges. Thus the standard deviation is the most stable measure of variability, semi-interquartile range is the next most stable, and the simple range is the least stable measure of variability.

Statistical inference

Most of the discussion of variability has been directed at descriptive statistics. When statistics such as the standard deviation are used to make inferences about populations from which the sample has been drawn, we again run into the problems of symbols and characteristics of estimators.

As was mentioned earlier, the Greek alphabet will be used to indicate population parameters, and Latin letters will be used for sample statistics. The formula for the standard deviation should be as follows:

$$\sigma = \sqrt{\frac{\Sigma(X - \mu)^2}{N}} \tag{5-7}$$

Notice that the Greek letters sigma (σ) and mu (μ) have been used in formula 5–7. This indicates that this is the standard deviation of a population. But notice also that the deviation of the score (X) is taken about the population mean (μ). When a sample is drawn, the population mean is unknown and hence cannot be used to compute the standard deviation. Instead we have to use an estimate of μ, namely $\bar{X}$, in the computation of the standard deviation. With smaller samples this tends to interfere with the estimate of $\Sigma(X - \mu)^2$ since the scores (X) must be used to estimate μ. The concept of freedom of sample values to vary is referred to as "degrees of freedom." A more complete discussion of degrees of freedom must be postponed until chapter 10.

BIAS

The standard deviation computed with N in the denominator is a *biased* estimator of σ. The degree of bias is related to the number of degrees of freedom that are used to estimate σ. With small samples, the chance of selecting extremely deviant scores is much reduced; thus $\Sigma(X - \bar{X})^2$ will tend to be systematically smaller than it should be. As the size of the sample increases, the degree of bias will become smaller. By using the sample size less one ($N - 1$) which is also the degrees of freedom, we can achieve an estimator of σ which is unbiased. This unbiased estimator is given in formula 5–8 below:

$$s = \sqrt{\frac{\Sigma(X - \bar{X})^2}{N - 1}} \qquad (5\text{–}8)$$

EFFICIENT

The computation of s using formula 5–8 (or the other formulas in this chapter) will yield an efficient estimator of the variability of a set of data. Obviously the semi-interquartile range is not an efficient estimator of variability with interval or ratio data, since the semi-interquartile range does not make use of all of the information contained in the scores.

CONSISTENT

The standard deviation, either the biased or 'the unbiased form, is consistent. As the sample size increases, the value of the estimator becomes closer and closer to the population value.

Throughout this text, three different symbols will be used to indicate the standard deviation. The symbol σ (lowercase Greek letter sigma) will indicate a population value, either known (descriptive statistics) or to be estimated (inferential statistics). The symbol S (uppercase Latin letter) will be used to indicate a standard deviation computed from a sample in which the total number of observations (N) is used in the divisor. The symbol s (lowercase Latin letter) will be used to indicate a standard deviation for a sample in which the degrees of freedom ($N-1$) is used in the divisor. Thus s will indicate an unbiased estimate of σ.

The Z score

In discussing the standard deviation, we have stated that a given score is one standard deviation above the mean of the group or is two standard deviations below the mean. In the introduction to this chapter, distributions on two tests were presented. It was mentioned that a given student received a score of 51 on each of the two tests and the mean of each test was 45. Thus the student scored six points above the mean on each test. But the two distributions were markedly different in variability. On test 1, a score of 51 was one of the highest scores in the class; on test 2, the score of 51 was not one of the highest. By using the standard deviation, we could express each of these scores as deviations about the mean in terms of the number of standard deviations. The statement of the amount of deviation about the mean in terms of standard deviations is called a Z score. The formula for computing a Z score is

$$Z = \frac{X - \bar{X}}{S}$$

For the two examinations discussed above, the deviation ($X - \bar{X}$) in each case is $+6$. If the standard deviation on test 1 was $S_1 = 2$ and the standard deviation on test 2 was $S_2 = 5$, the Z score for our student on the first examination would be:

$$Z = \frac{51 - 45}{2} = \frac{+6}{2} = 3$$

We can state that the student scored three standard deviations above the mean of his group on the first test (an extremely high score). The same score of 51 on the second test would yield a Z score of

$$Z = \frac{51 - 45}{5} = \frac{+6}{5} = 1.2$$

On the second test the student scored only 1.2 standard deviations above the mean of the group (a good score but certainly not as extreme as his score on test 1).

COMPARING MEASUREMENTS FROM DIFFERENT DISTRIBUTIONS

If we wish to compare different scores within the same group, the deviation above or below the mean of the group will be sufficient. But if we wish to compare scores on different examinations, we need some method to equate not only for difference in mean between the two tests, but also for difference in variability. The Z score does this by changing all scores into deviations about their respective means and then by expressing this deviation in terms of standard deviations of that specific group.

Is John taller than he is heavy? At first, this would appear to be an unanswerable question. How can we compare height and weight? This type of comparison can be made if the measurements are converted into Z scores. Let us assume that John weighs 185 pounds and stands 6 feet tall. Let us further assume that for the group with which we wish to make comparisons the mean weight is 160 and the mean height is 5 feet 8 inches. We can say that John is 25 pounds heavier than average and 4 inches taller than average. But is he taller than he is heavy? This question can be answered if we convert the weight measurement into a Z score and compare it to the Z score in height. In order to do this we need to know the standard deviation of weight and the standard deviation in height of this group. Let us assume that the S of weight is 15 pounds and the S of height is 4 inches. John's Z score for weight is

$$Z = \frac{185 - 160}{15} = \frac{25}{15} = 1.67$$

John's Z score for height is

$$Z = \frac{6' - 5'8''}{4''} = \frac{4''}{4''} = 1.0$$

We can state that John is 1.67 standard deviations above the mean weight for his group but only 1.0 standard deviation above the mean height for his group, or that he is heavier than he is tall when compared to the rest of the group.

Notice that the Z score transformation of the original scores does two things: (1) It expresses all scores in terms of deviations from their respective means, which makes the mean of each group of deviation scores equal to zero. Thus, we can say that the distributions have been equated for central tendency—all converted to deviation scores, so that the mean of each set of deviation scores is equal to zero. (2) It converts the original measurements from the original units (inches, pounds, test items correct, etc.) into standard deviation units. Notice that when it was stated that John's Z score in weight was 1.67, this was 1.67 standard deviations, not 1.67 pounds.

After a set of scores have been converted to Z scores, what effect does this have on the mean of these transformed scores? The standard deviation? A little algebraic manipulation will show one of the important characteristics of any set of Z scores—that the *mean of the Z scores is always equal to zero* and that the *standard deviation of the Z scores is always equal to one.*

MEAN OF Z SCORES

The mean of any distribution can be found by summing all values and dividing by the number of observations. Thus the mean of a set of Z scores can be expressed as

$$\bar{Z} = \frac{\Sigma Z}{N}$$

But

$$Z = \frac{x}{S}$$

Substituting:

$$\bar{Z} = \frac{\Sigma(x/S)}{N}$$

But S is a constant for this distribution of scores, and a variable divided by a constant is equal to the sum of the variable and then dividing the sum by the constant (see chap. 3). Thus

$$\Sigma \frac{x}{S} = \frac{\Sigma x}{S}$$

But

$$\Sigma x = 0$$

Substituting:

$$\bar{Z} = \frac{\dfrac{0}{S}}{N} = 0$$

Thus the mean of any set of Z scores is *always* equal to zero.

STANDARD DEVIATION OF Z SCORES

The standard deviation of any distribution can be found by determining each deviation from the mean of the distribution, squaring each deviation, summing all squared deviations, dividing by the number of observations, and extracting the square root. For Z scores, this can be expressed as

$$S_Z = \sqrt{\frac{\Sigma z^2}{N}}$$

Squaring both sides of the equation and substituting $z = Z - \bar{Z}$:

$$S_Z{}^2 = \frac{\Sigma(Z - \bar{Z})^2}{N}$$

But

$$\bar{Z} = 0$$

$$S_Z{}^2 = \frac{\Sigma Z^2}{N}$$

Substituting:

$$Z^2 = \left(\frac{x}{S}\right)^2 = \frac{x^2}{S^2}:$$

$$S_Z{}^2 = \frac{\dfrac{\Sigma x^2}{S^2}}{N}$$

Multiplying top and bottom of the fraction by $1/N$ to simplify the expression:

$$S_Z{}^2 = \frac{\dfrac{\Sigma x^2}{S^2} \cdot \dfrac{1}{N}}{N \cdot \dfrac{1}{N}} = \frac{\Sigma x^2}{S^2 N}$$

Note that

$$\frac{\Sigma x^2}{N} = S^2$$

$$S_Z{}^2 = \frac{S^2}{S^2} = 1$$

Extracting the square root of both sides of the equation:

$$S_Z = 1$$

Thus the standard deviation (also the variance) of any set of Z scores is *always* equal to one.

The Z score transformation of data permits direct comparison of measurements which have been originally expressed in different scales. It is possible to compare height and weight, clerical aptitude and numerical aptitude, etc. The Z score transformation is considered a *linear* transformation, since it does not change the interrelationships among the scores within each distribution. The transformation has two parts and two different effects: (1) Converting each observation into a deviation score about the mean ($x = X - \bar{X}$) shifts the mean of the distribution to zero. This is also referred to as *shifting the location of the distribution*. Adding or subtracting any constant from all scores in a distribution will have the effect of shifting the location, but it will not have any other effect. Subtracting the mean from each observation has the effect of shifting the location of the distribution to zero. (2) Dividing each deviation score by the standard deviation (x/S) has the effect of changing the standard deviation of the original scores from its value to the value of one. This is referred to as *changing the scale of measurement*. The Z score transformation does not change the shape of the distribution of scores. Thus the Z score transformation permits comparisons of scores from two or more distributions by transforming each distribution to the same location (zero) and to the same scale (one) without changing the shape of the original distributions. The Z score is a powerful tool for comparing measurements from different scales.

Summary

Four different measures of variability were presented in this chapter. In all cases, these measures tell us something about the degree of bunching or dispersion of the scores.

The *range* is the spread or distance between the minimum and maximum scores in the set and is based on only two scores.

The *semi-interquartile range,* like the simple range, is also based on two scores in the distribution, but they are the 25th percentile and the 75th percentile. This interquartile range (75th percentile minus the 25th percentile) is then divided by two to obtain the semi-interquartile range.

The *average deviation* is based on all scores in the distribution and is the mean of the absolute values of the deviation scores.

The *standard deviation,* like the average deviation, is based on all scores in the distribution. This statistic utilizes the square of each deviation score.

Since nominal scaling does not imply any continuum underlying the categories, the concept of variability along a continuum has no relevance. When objects form a nominal scale, the computation of a measure of variability can be misleading.

Either the range or the semi-interquartile range are appropriate measures of variability when the measurement is ordinal in nature. The semi-interquartile range is to be preferred, since it is a more stable measure.

With interval or ratio scaling, all measures of variability may be used. The standard deviation is generally the preferred measure, since it is based on all scores (as is the average deviation), and, in addition, may be used in further description and analysis of the data. The determination of the standard deviation permits the use of *Tchebysheff's inequality* to make statements of the *minimum* number of scores which will be found within specific standard deviation units of the mean in any distribution, regardless of the shape of that distribution. If the shape of the distribution can be specified, the standard deviation may be used to determine the *specific* number of cases which will be found within specified standard deviation units of the mean.

The standard deviation is not a point. It is a measure of distance or spread and is *always* expressed in the units of the original measurement. Because the mean and standard deviation are in units of the original measurements, it is not possible to compare directly scores from different measurements. By equating the means of two different distributions and the standard deviations of these distributions, it is possible to make comparisons between different measuring instruments. The transformation of all scores into Z *scores* permits this type of comparison.

The Z score is an expression of the deviation score in terms of standard deviation units $[x/S = (X - \bar{X})/S]$. This transformation results in a mean of the Z score distribution which is always equal to zero, and a standard deviation of the Z score distribution which is always equal to one. With the Z score transformation, it is possible to state how far each individual score deviates from the mean in terms of standard deviation units and to compare this deviation with one obtained from a different distribution, also expressed in Z score form.

References

VARIABILITY

QUINN MCNEMAR, *Psychological Statistics* (4th ed.; New York: John Wiley & Sons, Inc., 1969), pp. 19–25.

Z SCORES

PAUL BLOMMERS and E. F. LINDQUIST, *Elementary Statistical Methods in Psychology and Education* (2d ed.; Boston: Houghton Mifflin Co., 1960), chap. vii.

TCHEBYSHEFF'S INEQUALITY

JOHN E. FREUND, *Mathematical Statistics* (Englewood Cliffs, N.J.: Prentice–Hall, Inc., 1962), pp. 96–98.

6

The normal curve and probability

The normal curve

IN CHAPTER 5, we examined the assessment of variability within a set of scores. One way to describe the relation of a score to the distribution was to use the deviation of the individual score from the mean of the group in terms of standard deviations. This Z score allowed us to make comparisons among different tests and differing measurement units. The use of the standard deviation as a unit of deviation is also valuable when one considers Tchebysheff's inequality. By determining the standard deviation and Tchebysheff's inequality, we can establish the *minimum* number of scores which will fall within a given interval about the mean. Tchebycheff's inequality holds for *any* distribution, but if we can specify the specific distribution, we can improve on the determination of the *minimum* proportion of scores within an interval to the determination of the exact proportion within that interval.

THE NORMAL CURVE AND HUMAN CHARACTERISTICS

There are many different types of curves which are of interest to the statistician. One of these, the *normal curve,* is of much more importance than the others. Earlier, it was pointed out that the "normal curve" should not be considered usual or typical nor should any other type of distribution be regarded as "abnormal." There are many characteristics of living matter which are distributed in a nonnormal manner, but there are also many characteristics which do follow the Gaussian distribution.

In the 1800s a Belgian by the name of Lambert Quételet (1796–1874) measured the height of soldiers. In examining his frequency distribution, he discovered that a few soldiers were quite short, and a few were quite tall; but the great majority of the soldiers were in the middle of the height distribution. His distribution when graphed was bell shaped. He also measured the chest expansion of soldiers and once again obtained a bell-shaped distribution. He found that his empirical distributions closely approximated the mathematical distribution known as the Gaussian distribution, which has the following formula:

$$y = \frac{1}{\sigma \sqrt{2\pi}} e^{-\frac{x^2}{2\sigma^2}}$$

where

y is the height or ordinate of the curve at any point
σ is the standard deviation of the population
π is the mathematical constant 3.1416
e is the mathematical constant 2.7183
x is the deviation of any X value about the population mean (μ)

Specifying the mathematical formula for a distribution of scores permits us to determine directly the height of the ordinate (y) at any point x by solution of the equation. But perhaps more important, by using integral calculus, it is possible to determine the number of cases that fall within any distance from the mean of the distribution. In place of Tchebysheff's inequality to establish the *minimum* number of scores to be found within any Z score distance of the mean, it is possible to determine the *exact* frequency within these Z score distances.

Sir Francis Galton, working in England, extended Quételet's work and measured almost every possible physical characteristic of people; he discovered that, in general, the scores tended to fall into a close approximation of the Gaussian distribution. Since so many human (and animal) characteristics seem to fall into this bell-shaped curve, some individuals proposed that the average (mean) of the distribution was nature's ideal and that deviations from the mean represented "errors" on the part of nature. We shall have occasion later to work with certain statistics which refer to expected variations about the mean which are called "error" terms—a holdover from the concept of variation about the mean implying mistakes or errors of nature.

AREA OF THE NORMAL CURVE

The total area lying under the normal curve can be considered to represent 100 percent of the cases that make up the frequency distribution. We can then use the Z score to determine what proportion of the total area lies within specified standard deviation limits. If we have no information concerning the shape of the frequency distribution, we still know that at least 75 percent of the area under the curve lies within plus and minus two standard deviations about the mean (Tchebysheff's inequality). But if our empirically obtained frequency distribution is a reasonably close approximation to the mathematical model of the normal curve, we can state that 95 percent of the area under the curve will be within plus or minus two standard deviations of the mean; hence, 95 percent of the cases will fall in this interval.

By integrating the equation for the normal curve between any specified limits, it is possible to determine the proportion of the total area lying within these limits. Since the normal curve has a specific equation, it is not necessary for each person to integrate the equation every time he wishes to determine an area under the curve. This integration has been done and is presented in table B of the Appendix. A portion of table B is given in table 6–1; the dotted lines indicate that a portion of the table has been deleted.

TABLE 6–1
Proportion of the total area under the normal curve between the mean and any standard deviation distance from the mean

$\dfrac{x}{S}$	.00	.01	.02	.03	. . .*	.09
0.0	00.00	00.40	00.80	01.20	. . . . .	03.59
0.1	03.98	04.38	04.78	05.17	. . . . .	07.53
0.2	07.93	08.32	08.71	09.10	. . . . .	11.41
. . .	. . . . .	. . . . .	. . . . .	. . . . .	. . . . .	. . . . .
1.0	34.13	34.38	34.61	34.85	. . . . .	36.21
1.1	36.43	36.65	36.86	37.08	. . . . .	38.30
. . .	. . . . .	. . . . .	. . . . .	. . . . .	. . . . .	. . . . .
2.0	47.72	47.78	47.83	47.88	. . . . .	48.17
. . .	. . . . .	. . . . .	. . . . .	. . . . .	. . . . .	. . . . .
3.0	49.87					

* Columns .04, .05, .06, .07, and .08 omitted.

Notice that the entry in the extreme upper left-hand corner of the table is a Z score (x/S). In order to use the table of areas under the normal curve, any score must first be converted into a Z score. If we wished to find the area under the normal curve between the mean and a Z score of 1.03, we would look in the first column on the left side of the table for the Z score of 1.0 and then move across the table to the column headed .03. The entry in this cell (34.85) gives the proportion of the normal curve between the mean and the Z score of 1.03. The first column to the left gives the Z score to tenths, and the first row shows the Z score to hundredths. The area between the mean and one Z score is 34.13, or we can say that we would expect 34.13 percent of the scores in a normal distribution to be within the interval from the mean to $+1.00$ Z scores.

What percentage of the area under the normal curve lies within the interval for the mean to -2.00 Z scores? Notice that the table of the normal curve does not show any entries for negative Z scores. Since the normal curve is symmetrical about the mean, a table for negative Z scores would be identical to the table shown; thus, only one half the table of the normal curve is presented. The area between the mean and -2.00 Z scores is exactly the same as the area between the mean and $+2.00$ Z scores (47.72 percent).

USES OF THE TABLE OF THE NORMAL CURVE

If a distribution of scores is a close approximation to the normal distribution, then the determination of the mean and the standard deviation along with the use of the table of the normal curve is all that is needed to solve a great variety of problems. Suppose that we have a distribution of test scores with $\bar{X} = 50$, $S = 10$, and $N = 150$, and the distribution is a close approximation to the normal curve. How many students in the group scored at or above 65? To solve this problem, we must convert the score of 65 into a Z score before we can use the table of the normal curve:

$$Z = \frac{X - \bar{X}}{S} = \frac{65 - 50}{10} = \frac{+15}{10} = +1.50$$

From table B in the Appendix, we can determine that 43.32 percent of the area under the curve is between the mean and a Z score of 1.50. Since we are interested in the number of students who scored above a Z score of 1.50 ($X = 65$), we would have to subtract 43.32 percent from 50 percent (remember that 50 percent of the group will score above

FIGURE 6–1
Graphic representation of percentage of people
scoring at or above a Z score of 1.50

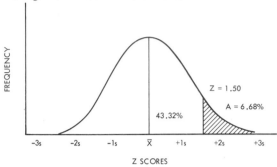

the mean). Thus, 50.00 percent — 43.32 percent = 6.68 percent of the group that scored at or above 65. But the original question was in terms of the number of people who scored above 65. We can multiply 150 (*N*) by the percentage scoring above 65 (6.68 percent); thus, 150 × .0668 = 10.02, or 10 people who scored at or above 65 on the examination. Figure 6–1 shows the portion of the normal curve involved in this particular problem. It is a wise practice to sketch a picture of the normal curve for use in the solution of these types of problems.

A similar type of question might be: How many students scored between 55 and 65 on the examination? Notice that both scores are above the mean. The score of 55 is equivalent to a Z score of +.5, and the score of 65 is the equivalent of a Z score of +1.5. The graphic representation of this problem is shown in figure 6–2. Notice that table B in the Appendix does not directly give the proportion of the area in the interval

FIGURE 6–2
Graphic representation of percentage of people
scoring between Z = +.5 and Z = +1.5

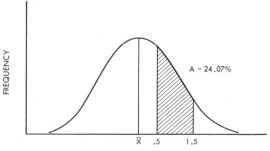

between +.5 and +1.5 Z scores. All areas are given in Z score distances from the mean. In this case it would be necessary to determine the area between the mean and Z = +1.5, which is 43.32 percent of the total area. Next, determine the area between the mean and Z = +.5, which is 19.15 percent of the total area. The area in which we are interested is the difference between these two (43.32 − 19.15 = 24.17 percent). We have determined that 24.25 percent of the group scored between 55 and 65 on the examination. To determine the number of people in this group, we need only to multiply the total N (150) by the percentage in this range (24.17 percent) for 36.2550, or 36 people between 55 and 65.

Some instructors use the normal curve in assigning grades on examinations. This assumes that the scores of the students are distributed in a close approximation to the normal distribution. Where this technique is used, the interval from −.5 to +.5 Z scores is generally the C range. Using the data presented earlier ($\bar{X}$ = 50, S = 10, N = 150), how many students would receive C's on this examination? This problem is presented graphically in figure 6–3. From table B in the Appendix, we can determine that 19.15 percent of the area of the normal curve is between the mean and a Z score of −.5. Similarly, 19.15 percent of the area is between the mean and a Z score of +.5; thus, 38.30 percent of the area lies in the interval from −.5Z to +.5Z. Multiplying 38.30 percent by 150 yields 57.4500, or 57 people scoring between 45 and 55 on the examination.

Notice in figure 6–2 and in figure 6–3 that the distance involved in both cases is one standard deviation. In figure 6–2, we found the difference in areas associated with a Z score of +1.5 and +.5 to be 24.07 percent of the total area. In figure 6–3, we found that the sum of the areas associated with a Z score of −.5 and a Z score of +.5 was 38.30 percent

FIGURE 6–3
Graphic representation of percentage of people scoring between −.5 and +.5 Z scores

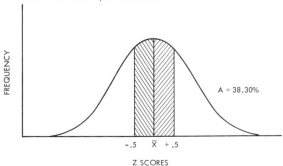

of the total area of the curve. In both cases the distance along the abscissa is one standard deviation; yet if we look in table B of the Appendix for the area between the mean and one standard deviation from the mean, we find that it is 34.13 percent of the total area under the curve. In solving normal-curve problems of this type, we must determine that area under the curve corresponding to the given Z score and then add or subtract *areas*. The area between the mean and $+1Z$ is 34.13 percent of the total area. The area between $+1Z$ and $+2Z$ is 47.72 percent -34.13 percent, or 13.59 percent.

To this point, we have been concerned with determining what percentage or number of individuals scored above a certain score, or scored between given scores. We can also use the normal curve to answer questions such as what the score is above which only 5 percent of the class scored. In this situation we start with an area of the normal curve and must work backward to the Z score and then convert the Z score into a whole score. Using the data $\bar{X} = 50$, $S = 10$, and $N = 150$ we shall illustrate this use of the normal curve to determine the score above which only 5 percent of the group score. This is presented graphically in figure 6–4. Since 5 percent score above this

FIGURE 6–4
Graphic representation of determination of whole score above which 5 percent of the group score

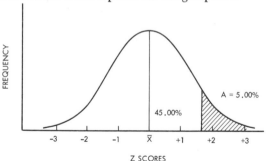

point, we need to look in the body of table B in the Appendix to find the entry of 45.00. In table B, in the row for a Z score of 1.6 and the column headed .05, is the entry 45.05, which is a little more area than we wish. In row 1.6 and column .04 is the entry 44.95, which is slightly less than the entry of 45.00. The Z score we are attempting to find must be between 1.64 and 1.65. Since one tabled entry is .05 percent too large and the other is .05

percent too small, we shall take one half the distance between the two, or a Z score of 1.645. The formula for finding a Z score is

$$Z = \frac{X - \bar{X}}{S}$$

By rearranging terms, we can solve for the whole score as

$$X = (Z)(S) + \bar{X}$$

Substituting in this equation:

$$X = (1.645)(10) + 50 = 16.45 + 50 = 66.45$$

Thus, we know that less than 5 percent of the students scored 67 or above on this examination.

Equipment designers and manufacturers are often interested in a slightly different use of the normal curve. Most equipment (cars, lawn mowers, etc.) must be built so that people of "average" height can utilize them. The modern automobile is built with adjustable seats so that individuals varying greatly in height can drive the same automobile. The designer might be interested in the question: What are the limits within which the "middle" 95 percent of the adults in the United States fall? The designer of military equipment might be interested in the question: What are the limits within which the "middle" 95 percent of the adult male population of the United States fall with respect to height? Using $\bar{X} = 68$ inches and $S = 2.5$ inches (which are fairly close to the results of several studies), what are the limits within which 95 percent of American men will fall with respect to height? An alternative question might be: What heights are so extreme that only

FIGURE 6–5
Graphic representation of limits within which the "middle" 95 percent of a group will be found

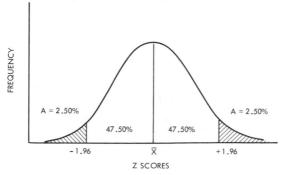

5 percent of the men are beyond these limits? This type of problem is presented graphically in figure 6–5. Since the normal curve is symmetrical, we are interested in that 5.0 percent of the population who are in the "tails" of the distribution; the tallest 2.5 percent of the population and the shortest 2.5 percent of the population. Looking in the body of table B in the Appendix, we find that the entry 47.50 is in the row of the Z score of 1.9 and in the column headed .06. The Z score is 1.96. Since we are interested in both sides of the distribution, we must be concerned with $Z = +1.96$ and $Z = -1.96$. Using the equation $X = (Z)(S) + \bar{X}$ and substituting the values given above:

$$X = (+1.96)(2.5) + 68 = +4.9 + 68 = 72.9 \text{ inches}$$
$$X = (-1.96)(2.5) + 68 = -4.9 + 68 = 63.1 \text{ inches}$$

We would expect that 95 percent of the American adult males would be at least 63.1 inches tall, but no taller than 72.9 inches. We would expect 95 out of every 100 American adult males to be within the limits of 63.1–72.9 inches.

Certain Z scores and the area under the normal curve associated with these Z scores should be committed to memory. The area of the normal curve between the mean and one Z score from the mean will be 34.13 percent of the total area under the curve; thus, between $\pm 1Z$ about the mean will be found 68.26 percent of the area. The Z score of ± 1.96 indicates the middle 95 percent of the area of the normal curve—2.5 percent of the area is in each tail beyond the Z score of 1.96. The middle 99 percent of the area under the normal curve will be found within ± 2.58 Z scores about the mean. Only 1 percent of the scores will be so extreme as to be more than ± 2.58 Z scores from the mean.

The normal curve permits us to do more than just establish an interval about an estimated mean. We can also use the normal curve to assign a probability statement to our chances of being right or wrong in our estimates! What are the chances that the next adult American male we meet will be between 63.1 and 72.9 inches in height? Ninety-five out of 100.

If we were to measure the height of a large number of American adult males, and if before measuring each individual we stated that this person would be at least 63.1 inches tall but no taller than 72.9 inches, we would be correct 95 percent of the time, and we would be wrong 5 percent of the time. If our sample is representative of the entire population, we can use our information from the sample to make probability statements about the population of which it is a sample.

Probability

A coin is often used when a two-choice decision is to be made. The decision as to who buys coffee is often settled by tossing a coin. What is involved in using a coin in this way? It appears to be a "fair" way to decide, but what is involved in this "fairness"? Since the coin has two sides, it is assumed that one side is as apt to be "up" on the toss as the other side. It could be stated that each of the two people involved in the coin tossing has the same chance of winning or losing—the odds are even. One way of stating this is that a person has a 50–50 chance of winning (or losing). What is implicit in this ritual of coin tossing is that the probability of getting a head is equal to the probability of getting a tail.

Insurance is not very democratic. A young man (under 25 years of age) who is unmarried must pay a higher premium for his automobile insurance than a 50-year-old woman who is married. Why this difference? Life insurance for certain occupations is much higher than for others. Why? In the case of insurance premiums, the young, unmarried male driver has a higher probability of being involved in an accident than the older, married woman. The chance (probability) of death or accident is higher in certain "dangerous" occupations.

At a horse race, the odds of three to one are given for one horse. Does this mean that the horse has one chance in three of winning? How is this probability determined? What is meant by the term probability? The terms "chance," "probability," and "odds" are used in a variety of different ways.

To the individual who is just beginning a serious study of statistics and probability, the various ways the terms chance, odds, and probability are used can be very confusing. We shall differentiate between three similar, but quite different, uses of the term probability. These are *actuarial, subjective,* and *theoretical* probability.

ACTUARIAL PROBABILITY.   In the case of automobile insurance, we are dealing with actuarial probability. We look at the actual number of accidents among different age groups of drivers. We look at the death rate in different occupations. Actuarial probability is a factual history of the occurrence of the kinds of events in which we are interested. Actuarial probability is extremely important to our society, not just for insurance purposes but for a great variety of long-range-planning efforts.

SUBJECTIVE PROBABILITY.   In the case of horse racing (or football games) the odds given reflect the *beliefs* of people. In paramutual betting, the "odds" reflect the amount of money people are betting. If the "odds" are three to one, this indicates that for every one dollar that is bet on horse A to win, three dollars are being bet on other horses to win. Subjective probability has, in recent years, attracted the attention of a variety of scholars. Some scholars study children to determine how the concept of probability develops as the child matures. Others are interested in the discrepancies between people's beliefs about chance operations and the way in which chance actually operates in a given situation.

THEORETICAL PROBABILITY. Theoretical probability is based on mathematics but had its origins in both actuarial and subjective probability. The first major work in theoretical probability was published in 1713 by the Swiss mathematician Jacob Bernoulli and was entitled *The Art of Conjecture!* Theoretical probability is the basis for all inferential statistics. While actuarial and subjective probability are interesting and important in their own right, further discussion of probability will be limited to an introduction to theoretical probability.

DEFINITION OF PROBABILITY

Probability is simply a ratio of the number of ways a specified event can happen divided by the total number of ways all events under consideration can occur. When a coin is tossed once, there is only one way a head can occur. But in tossing the coin, a head could occur, and a tail could also occur (ignoring the extremely remote possibility of the coin standing on edge). Thus, there are two different events that could occur. The probability of obtaining a head on one toss of a coin is one divided by two (1/2 or .50). The probability of any event can be determined if we can count the number of ways the specified event can occur (r) and divide this by the total of all events that can occur (N), or $p = r/N$.

What is the probability of obtaining a four on one throw of a die (one half of a pair of dice)? Since a die has six sides, each marked with a unique set of spots from one spot to six, there are six different events that can occur, but only one side has four spots on it. The probability of obtaining a four on one throw of the die is one divided by six ($p = r/N = 1/6$).

The problem of determining probability remains, in principle, the same as we add more items, but the determination of the possible events becomes somewhat more involved. Let us illustrate by tossing one coin twice (equivalent to tossing two coins once each). What is the probability

of obtaining two heads when a coin is tossed twice? We need to determine
(1) the number of ways that two heads in succession can be obtained
and (2) the total number of different events that can occur. Figure 6–6

FIGURE 6–6
**Different combinations of heads and tails
in tossing one coin twice**

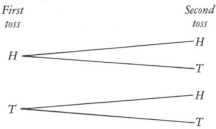

shows the different ways in which the heads and tails could occur. On
the first toss the coin may come up heads, or it may come up tails. On
the second toss the coin may come up heads, or it may come up tails;
thus the coin could have the following arrangements: (1) a head followed
on the second toss by a head, (2) a head followed by a tail, (3) a
tail on the first toss followed by a head, and (4) a tail followed by
a tail. There are four different ways in which the coin could be arranged
on two tosses of the coin. What is the probability of obtaining two heads
in two tosses of the coin? Since the sequence of head followed by a
head can occur in only one way and there are four different arrangements
that are possible, the probability of two heads on two tosses of the coin
is 1/4. What is the probability of obtaining a head on the first toss
followed by a tail on the second toss? The sequence H, T (see figure
6–6) can occur in only one way ($r = 1$), and all events can occur in
four ways ($N = 4$); thus, $p = 1/4$. What is the probability of obtaining
a head and a tail in the two tosses of the coin? Notice that this question
does not require any particular order in the head or tail. We could obtain
the sequence H followed by T or the sequence T followed by H. There
are two different ways in which we could obtain a head and a tail ($r = 2$),
and there are four different sequences which are possible ($N = 4$); thus
the probability of obtaining a head and a tail on tossing a coin twice
is 2/4 = .50.

THE ADDITIVE THEOREM OF PROBABILITY

In the last example we saw that the probability of obtaining one head and one tail in two tosses of the coin was 2/4. We obtained this by counting the different orders in which a head and a tail occurred and dividing this by the total number of sequences. This same result could have been obtained by determining the probability of obtaining the sequence H followed by T, which is 1/4, and to this *adding* the probability of obtaining the sequence T followed by H, which is also 1/4. The probability of obtaining either H followed by T, or T followed by H, is the sum of their separate probabilities, $1/4 + 1/4 = 2/4 = .50$.

The additive theorem of probability may be stated as: *The probability that any one of several mutually exclusive events will occur is the sum of their separate probabilities.* Notice that one of the requirements to add probabilities is that the events must be mutually exclusive. This means that if one of the events *does* occur, the others *cannot* occur. In our example, if we did obtain the sequence H followed by T, the other events (H followed by H, T followed by H, or T followed by T) cannot occur. The various orders in which the coins can arrange themselves are mutually exclusive.

What is the probability of obtaining either a one or a two or a three in one throw of a die? The probability of a one is 1/6, the probability of a two is 1/6, the probability of a three is 1/6; thus the probability of a one *or* a two *or* a three is $1/6 + 1/6 + 1/6 = 3/6 = .50$.

Figure 6–7 is a diagram of the possible combinations in tossing four coins (or one coin four times). On each toss of the coin, either a head or a tail may occur. Each column of figure 6–7 shows the possible arrangements of the heads and tails. On the first toss, either a head or a tail may occur. On the second, if a head occurred on the first, we can obtain either a head or a tail; and if a tail occurred on the first toss, we can obtain either a head or a tail on the second toss. The same possibilities occur for the third toss and the fourth, resulting in 16 different orders which might occur. What is the probability of obtaining four heads in a row? Since $r = 1$ and $N = 16$, the $p = 1/16$. What is the probability of obtaining three heads and one tail? Notice that the order of occurrence of the heads and the tail is not specified. Orders number 2, 3, 5, and 9 all contain three heads and one tail. We could obtain this probability either by counting the number of different ways the event $3H$ and $1T$ (r) can occur and dividing by the total number of different orders (N), or we can determine the probability of each of the sequences 2, 3, 5,

FIGURE 6–7
Different combinations of heads and tails in tossing
one coin four times

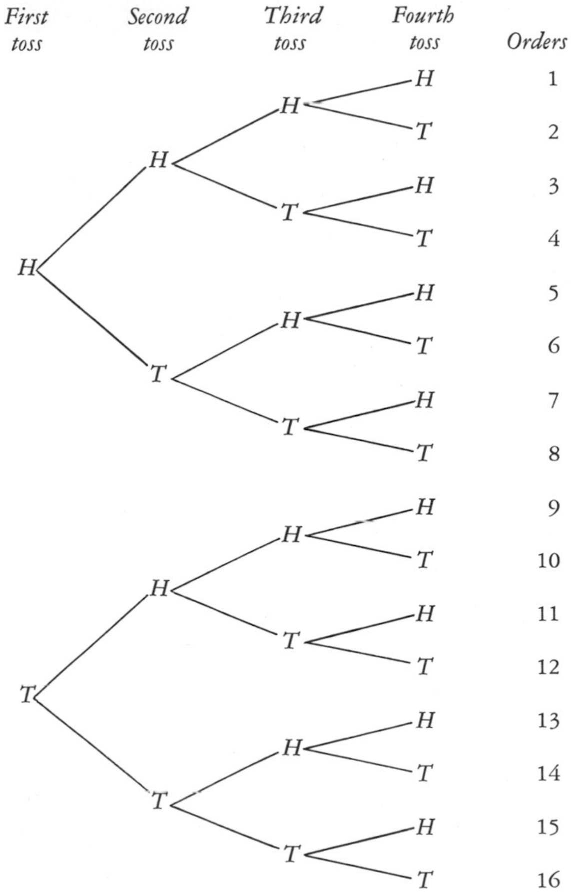

First toss	Second toss	Third toss	Fourth toss	Orders

and 9 and add their probabilities together. In the one case we obtain

$$p = r/N = 4/16 = .25.$$

In the other case we obtain

$$p(t) = p(2) + p(3) + p(5) + p(9)$$
$$= 1/16 + 1/16 + 1/16 + 1/16 = 4/16 = .25.$$

The two techniques are equivalent.

In tossing four coins, which is most probable—to get an even split of two heads and two tails, or to get a 3–1 split? We can use figure

6–7 to solve this problem. Notice that either combination of three heads and one tail or three tails and one head will meet the conditions above:

$$p(H3,1) = 4/16 \text{ (orders number 2, 3, 5, and 9)}$$
$$p(T3,1) = 4/16 \text{ (orders number 8, 12, 14, 15)}$$

The probability that the four coins will split 3 to 1 will be

$$p(H3,1) + p(T3,1) = 4/16 + 4/16 = 8/16 = .50$$

There are six ways that the four coins can be arranged to yield two heads and two tails (orders number 4, 6, 7, 10, 11, and 13); thus,

$$p(2,2) = 6/16 = .375$$

A 3–1 split is more probable than a 2–2 split. To follow this one step further, what is the probability of any uneven split occurring among the four coins? Since the probability of an even split is .375, the probability of an uneven split must be .625, or

$$p = p(0,4) + p(4,0) + p(1,3) + p(3,1)$$
$$= 1/16 + 1/16 + 4/16 + 4/16 = 10/16 = .625$$

THE MULTIPLICATIVE THEOREM OF PROBABILITY

In the previous examples we have counted the number of ways in which certain sequences of events could occur and then determined the probability of those events. The probability of obtaining four heads in four tosses of a coin is $1/16$ (the number of ways in which the event of four heads in succession can occur, divided by the total of all possible orders that could occur). This result could be obtained in a somewhat different way by utilizing the multiplicative theorem of probability. What is the probability of obtaining a head on the first toss, followed by a head on the second toss? From figure 6–6, we can easily ascertain that the correct answer is $1/4$. But we can also obtain this by multiplying the separate probabilities of these two events:

$$p(2H) = p(II) \times p(H) = 1/2 \times 1/2 = 1/4$$

The probability of obtaining four heads in four tosses of the coin would be

$$1/2 \times 1/2 \times 1/2 \times 1/2 = 1/16$$

The multiplicative theorem of probability may be stated as: *The probability that a combination of independent events will occur is the product of their separate probabilities.* What is the probability of obtaining a four on the throw of one die and then obtaining a three on the throw of the second die?

$$p(4,3) = 1/6 \times 1/6 = 1/36$$

Note that the probability of a four on the first die is $1/6$ and the probability of a three on the second die is also $1/6$; thus the probability of obtaining a four on the first die followed by a three on the second die is the product of the two probabilities, or $1/36$. Suppose we rephrase the question and ask for the probability of obtaining a seven on one throw of the dice. What are the different combinations which add to seven? One and six, two and five, three and four, four and three, five and two, and six and one are all acceptable combinations. Since we are willing to accept any combination of the dice which adds to seven, we shall need to use both the additive theorem and the multiplicative theorem of probability:

$$p(7) = p(1,6) + p(2,5) + p(3,4) + p(4,3) + p(5,2) + p(6,1)$$

The probability of obtaining a one followed by a six is the product of the separate probabilities $1/6 \times 1/6 = 1/36$. The same procedure would be followed for determining the probabilities of each of the acceptable combinations:

$$p(7) = 1/36 + 1/36 + 1/36 + 1/36 + 1/36 + 1/36 = 6/36 = 1/6$$

One more example will be presented to demonstrate the use of both the additive and the multiplicative theorems of probability. What is the probability of obtaining either a two or an eleven with one roll of the dice? Note that two different sums are acceptable: (1) any combination which adds to two and (2) any combination which adds to eleven. Only one combination of the dice will yield two—namely, a "one spot" on each die. The probability of obtaining a two on one roll of the dice is the product of the separate probabilities of obtaining a one on each die, or

$$1/6 \times 1/6 = 1/36$$

The sum of eleven can be obtained by having the first die show five and the second six, or by having a six on the first and a five on the second:

$$p(11) = p(5,6) + p(6,5) = 1/6 \times 1/6 + 1/6 \times 1/6 = 1/36 + 1/36 = 2/36$$

Adding together the probability of obtaining a two and the probability of obtaining an eleven:

$$p(2 \text{ or } 11) = p(2) + p(11) = 1/36 + 2/36 = 3/36 = .083$$

CONTINGENT PROBABILITY

To this point we have discussed situations in which the events under consideration are mutually exclusive—if a head occurs, a tail cannot occur. There are many situations in which the events are not mutually exclusive. Suppose there were a meeting of 100 people. Forty of these were women, and 30 were college graduates. What is the probability that an individual selected at random from this group of 100 will be either a woman or a college graduate? If the event *woman* and the event *college graduate* were mutually exclusive, this would be a simple additive situation:

$$p(W \text{ or } G) = p(W) + p(G) = .40 + .30 = .70$$

However these events are not mutually exclusive since some of the women might have been college graduates. These individuals would have been counted twice—once in the category *women* and once in the category *college graduates*. This can be corrected easily by subtracting the number of individuals who fit in both categories, thus:

$$p(W \text{ or } G) = p(W) + p(G) - p(W \text{ and } G)$$

If we determine that eight of the women are college graduates, then the probability that an individual selected at random will be either a woman or a college graduate becomes:

$$p(W \text{ or } G) = .40 + .30 - .08 = .62$$

This situation is diagrammed in figure 6–8.

The language of set theory is particularly useful in this context. The situation described above could be symbolized as:

$$p(A \cup B) = p(A) + p(B) - p(A \cap B)$$

and reads: The probability of the occurrence of the union of sets A and B is equal to the probability of the occurrence of set A plus the probability of the occurrence of set B, less the probability of the occurrence of the intersection of sets A and B. If the intersection of sets A and B is the null set, $\emptyset$, then events A and B are mutually exclusive.

We could ask the question: What is the probability that an individual selected at random will be both a woman and a college graduate? Since

FIGURE 6-8
Venn diagram showing a sample space of
100 individuals—40 women and 30 college
graduates (eight individuals are women and
college graduates)

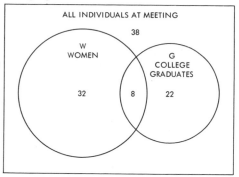

there are 100 individuals in the group and we know that 8 women are
college graduates, we know that the answer should be .08. But let us
attempt to use the multiplicative rule of probabilities

$$p(W \text{ and } G) = p(W)p(G) = (.4)(.3) = .12$$

which is not .08. Something is wrong and that *something* is the independence of the categories *women* and *college graduate*. They are related;
they are not independent categories. The probability of obtaining both
event A and event B when A and B are not independent is the probability
of the one event multiplied by the conditional probability of the one
event, given that the other event has occurred. Symbolically this can be
stated as:

$$p(A \text{ and } B) = p(A)p(A/B) = p(B)p(B/A)$$

where:

$p(A \text{ and } B)$ is the probability of both events A and B occurring jointly
$p(A)$ is the probability of the occurrence of event A
$p(B)$ is the probability of the occurrence of event B
$p(A/B)$ is the probability of the occurrence of event A given that event
      B has occurred
$p(B/A)$ is the probability of the occurrence of event B, given that event
      A has occurred

Returning now to the question of the probability that an individual selected at random will be both a woman and a college grad-

uate. We know that $p(W) = .40$ and since 8 women in the group are college graduates, $p(G/W) = 8/40 = .20$, and $p(W/G) = 8/30 = .267$, thus $p(W$ and $G) = p(W)p(G/W) = (.40)(.20) = .08$, or $p(G)p(W/G) = (.30)(.267) = .08$. Note that if the events A and B were independent then $p(A/B) = p(A)$ and $p(B/A) = p(B)$.

Again set language is particularly useful in working with contingent probabilities. The probability of the joint occurrence of two events is the probability of the intersection of the two sets. Symbolically

$$p(A \text{ and } B) = p(A \cap B)$$

As was stated earlier, theoretical probability is the basis for modern inferential statistics. Probability is an important subject in its own right, but we can do no more than introduce the topic and show the relationship between probability theory and modern statistics.

THE BINOMIAL EXPANSION AND PROBABILITY

In dealing with only two coins, we found that there were four different ways the coins could arrange themselves: HH, HT, TH, or TT. If we disregard the order of heads and tails, we could say that we can obtain one HH combination, two HT combinations, and one TT combination. In working with four coins (see figure 6–7), we can obtain one $HHHH$ combination, four $HHHT$ combinations, six $HHTT$ combinations, four $HTTT$ combinations, and one $TTTT$ combination.

This similarity should be apparent to anyone who has worked with the binomial expansion. The binomial expansion consists of two terms (*bi-nomial*), and the expansion refers to raising this expression to some power. Using a and b as the two terms in the binomial and raising the binomial to the second power:

$$(a + b)^2 = a^2 + 2ab + b^2$$

Similarly,

$$(a + b)^4 = a^4 + 4a^3b + 6a^2b^2 + 4ab^3 + b^4$$

Note that the coefficients of the binomial expansion are the same as the number of combinations of heads and tails in tossing coins. When two coins are tossed, we obtain one HH, two HT, and one TT; with four coins, one H^4, four H^3T, six H^2T^2, four HT^3, one T^4. The binomial expansion is clearly related to the determination of probability in a two-choice situation, such as tossing coins. To determine the probability of any event, all that is needed is to specify the number of ways the event can occur and divide this by the total of all possible events. The coefficients of the binomial expansion can be used to determine the number of times the particular combinations will occur, and the sum of the coefficients of the binomial expansion give the total number of all events that can occur. Thus, from the binomial expansion it is possible to compute directly the probability of the occurrence of any combination of events. Note further that the exponents of the terms indicate the particular kind of event. The term a^4 indicates occurrence of the event a four times without event b occurring; in a four-coin situation, this can occur only one way.

The term $4ab^3$ may be interpreted to mean that the combination of one event of the a type and three events of the b type (exponents of the term) can occur together in combination four ways (the coefficient of the term). The probability of obtaining one head (event a) and three tails (event b) in tossing four coins is $4/16 = .25$.

The coefficient of any term in the binomial expansion may be determined from the following expression:

$$\binom{n}{x} = \frac{n!}{x!(n-x)!} \qquad (6\text{-}1)$$

where

$\binom{n}{x}$ is the combination of n objects taken x at a time

$n!$ (n factorial) is the product of n times all integers smaller than n[1]

$x!$ (x factorial) is the product of x times all integers smaller than x

If ten coins are tossed, how many ways is it possible to obtain six heads and four tails? From formula 6–1, with $n = 10$ and $x = 6$:

$$\binom{10}{6} = \frac{10!}{6!(10-6)!} = \frac{(10)(9)(8)(7)(6)(5)(4)(3)(2)(1)}{(6)(5)(4)(3)(2)(1)\ (4)(3)(2)(1)} = 210$$

There are 210 ways in which 10 coins can be arranged to yield 6 heads and 4 tails. Since there are 1,024 different ways in which 10 coins can be arranged—$(1/2)^{10}$, the probability of obtaining 6 heads and 4 tails is $210/1024 = .205$.

Although formula 6–1 may be used to determine the coefficient of any term in the binomial expansion, these values may be generated rather easily by use of Pascal's triangle which is shown in table 6–2. Notice that the sum of two adjacent values in one row is the value to be entered

---

[1] The term 0! (zero factorial) is by definition equal to one ($0! = 1$).

Introduction to statistics for the behavioral sciences

TABLE 6–2
Pascal's triangle

n	Binomial coefficients							Sum of coefficients
1				1	1			2
2			1	2	1			4
3		1	3	3	1			8
4	1	4	6	4	1			16
5	1	5	10	10	5	1		32
6	1	6	15	20	15	6	1	64

in the row below and between the two added values. Referring to table 6–2, two triangles have been drawn. The first indicates that the entries 1 and 2 in row 2 when added yield 3, the entry in row 3. The second triangle indicates that if the entries 10 and 5 of row 5 are added they equal the entry 15 in row 6. This procedure can be used to generate the coefficients of the binomial expansion up to any value of n. For convenience, the values of the coefficients up to $n = 20$ are provided in table C of the Appendix.

The binomial expansion can be used to determine the probability of the occurrence of any discrete event, provided we can cast the problem into two mutually exclusive categories. The two categories *do not* have to have equal probabilities $(p \neq q)$, but the sum of the probabilities of the two categories must equal one $(p + q = 1)$.

The probability of the occurrence of a three on one die is $1/6$ $(p = 1/6)$. The probability of the occurrence of all events other than a three in one throw of a die is:

$$q = 1 - p = 1 - 1/6 = 5/6$$

What is the probability of obtaining *no* threes in one throw of a pair of dice? This problem would be of the form:

$$(p + q)^2 = p^2 + 2pq + q^2$$

The term p^2 would indicate the occurrence of two threes, pq the occurrence of one three and one event other than a three, and q^2 the occurrence of no threes. By substituting $q = 5/6$ into the expression q^2, we can obtain the probability of obtaining no threes in one roll of the pair of dice. This probability is .69:

$$q^2 = (5/6)^2 = 25/36 = .69$$

What is the probability of obtaining one three and any other number (except a three) in one roll of a pair of dice? The event of interest for this problem would be the combination p^1q^1. From equation 6–1, we can determine that the coefficient of this term should be:

$$\binom{2}{1} = \frac{2!}{1!(2-1)!} = \frac{(2)(1)}{(1)(1)} = \frac{2}{1} = 2$$

The term is $2pq$, and substituting the probabilities into the expression yields:

$$(2)(1/6)(5/6) = (2)(5/36) = 10/36 = .28$$

The probability of obtaining a three and a number other than three on one roll of a pair of dice is .28.

THE BINOMIAL EXPANSION AND THE NORMAL CURVE

As the power to which the binomial is raised increases, the number of possible events increases at an extremely rapid rate. Table 6–3 shows

TABLE 6–3
The different ways in which heads and tails can occur in tossing 16 coins

Number of heads	Number of tails	Number of ways for event to occur
16	0	1
15	1	16
14	2	120
13	3	560
12	4	1,820
11	5	4,368
10	6	8,008
9	7	11,440
8	8	12,870
7	9	11,440
6	10	8,008
5	11	4,368
4	12	1,820
3	13	560
2	14	120
1	15	16
0	16	1
		$N = 65,536$

the number of ways in which 16 coins can be arranged and the number of ways each specified event can occur. These numbers can be derived by expanding the binomial. The numbers in the column headed *number of ways for event to occur* are the coefficients of the terms in the expansion. The numbers in the columns headed *number of heads* and *number of tails* are the exponents of the various terms in the expansion. If p is used for the probability of a head occurring and $q = 1 - p$, the fifth term in the expansion would be $1{,}820p^{12}q^4$ and would indicate that the combination of 12 heads and 4 tails could occur in 1,820 different ways in tossing 16 coins. A histogram of the distribution of heads in tossing 16 coins is shown in figure 6–9. Notice that the histogram is constructed

FIGURE 6–9

Normal curve superimposed on a histogram of the distribution of the coefficients of the binomial expansion when $n = 16$

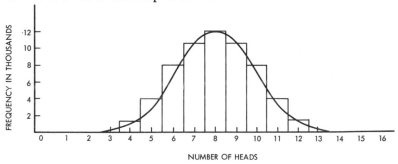

NUMBER OF HEADS

as if the distribution of heads were a continuous variable, which of course it is not, since it would be impossible to obtain 7.3 heads on any toss of the coins. The distribution is plotted as a continuous variable to allow a polygon to be superimposed on the histogram. Notice that the polygon has the familiar bell shape associated with the normal distribution. In fact, as n of the binomial expansion approaches infinity, the distribution of the coefficients of the binomial expansion will form the normal curve.

If we can think of the normal curve as being the frequency distribution of the coefficients of the binomial expansion, then all of the area under the normal curve would be equivalent to the sum of all of the coefficients of the binomial. In determining probability from the binomial, all that is needed is two items of information: (1) the number of ways the specified event can occur (r), and (2) the number of ways that all events can occur (N). Both of these may be obtained from the table of the normal curve. What is the probability of obtaining 12 or more heads on one toss of 16 coins? We shall solve this problem in two ways—(1) by using the coefficients of the binomial and (2) by using the normal curve.

The probability of obtaining 12 or more heads on the toss of 16 coins would be the sum of the probabilities of obtaining 12 heads, 13 heads, 14 heads, 15 heads, or 16 heads, or

$$p(12 \text{ or more } H) = p(12H) + p(13H) + p(14H) + p(15H) + p(16H)$$

Referring to Table 6–3, we can find the number of ways each of the above events can occur; and dividing each by 65,536, we can obtain the probability of each event:

$p(12 \text{ or more } H)$
$$= 1{,}820/65{,}536 + 560/65{,}536 + 120/65{,}536 + 16/65{,}536 + 1/65{,}536$$
$$= 2{,}517/65{,}536 = .0384$$

How can we use the normal curve to solve this same problem? The total area under the normal curve is equivalent to the 65,536 possible arrangements of the 16 coins. We need to find that part of the total area that is the equivalent of 12 or more heads. To enter the table of the normal curve, we must convert our observations to a Z score. This means that we shall have to determine the mean number of heads and the standard deviation of the number of heads that can be obtained in tossing 16 coins. On the average, we would expect to get 8 heads and 8 tails in tossing the 16 coins. We wouldn't expect that each toss would yield eight heads and eight tails; but over many different tosses, on the average we would expect eight heads. This most individuals "know" on an "intuitive" basis. It can be shown that the mean event in the binomial expansion is

$$\mu = np \qquad\qquad (6\text{-}2)$$

where

 n is the power to which the binomial has been raised
 p is the probability of occurrence of event a

In the present case the probability of a head occurring is .50 ($p = 1/2$), and the binomial has been raised to the 16th power

$$(n = 16); \text{ thus, } \mu = (16)(.50) = 8 \text{ heads}$$

The standard deviation of the events in the binomial expansion is

$$\sigma = \sqrt{npq} \qquad\qquad (6\text{-}3)$$

where

   n is the power to which the binomial has been raised
   p is the probability of occurrence of event a
   q is $1 - p$

In this case, $\sigma = \sqrt{(16)(.50)(.50)} = \sqrt{4} = 2$. Since we are interested in the probability of the occurrence of 12 or more heads, perhaps we should use 12 as the score in our Z score formula. Remember that the normal curve is a continuous function, whereas the distribution of the binomial is discrete. When using the normal curve to obtain the probability of a discrete event, we must treat the discrete event as if it were continuous. For normal-curve purposes the 12 heads represent the midpoint of the interval 11.5–12.5 heads; so in the Z score formula we shall use 11.5, the lower limit of the interval containing 12:

$$Z = \frac{X - \mu}{\sigma} = \frac{11.5 - 8}{2} = \frac{3.5}{2} = 1.75$$

Turning to table B in the Appendix, we find that the area from the mean to a Z score of $+1.75$ is 45.99 percent of the total area. Since we are interested in the area beyond a Z score of $+1.75$, we need to subtract 45.99 percent from 50 percent, with the result of 4.01 percent of the total area. Since the total area is equivalent to the total number of ways the coins can be arranged, and total probability is always one, the percentage figure can be changed directly into a probability statement. The probability of obtaining 12 or more heads in one toss of 16 coins is .0401. Referring back to the direct solution of this problem using the binomial expansion, we found that the exact probability of 12 or more heads was .0384 as compared to the normal-curve solution of .0401, or a difference of .0017 in the two statements. This very small "error" in using the normal curve to solve binomial probability problems is due to the fact that the distribution of the coefficients of the binomial becomes the normal distribution only as n approaches infinity. But even with $n = 16$, the amount of error is negligible. If we were to approach a similar problem with $n = 25$, there would be practically no difference in the normal-curve approximation to the exact probability determined from the binomial expansion.

   The above illustration indicates that we can use the normal curve to solve probability-type problems, but we can turn this statement around and say that the normal curve itself shows probability. In the problems presented in connection with figure 6–1, we stated that 6.28 percent of the group of students scored at or above 65 on the examination. We could also state that the probability of scoring 65 or above on the examination is .0628, or that there are approximately 6 chances in 100 that a particular examination paper will have a score of 65 or more.

Summary

Although many characteristics of human beings are distributed in a non-Gaussian manner, a large number of characteristics do tend to follow the Gaussian or normal curve. The distribution of the normal curve can be used directly to solve many of the questions associated with the distribution of human characteristics if these characteristics follow the normal curve.

The total area under the normal curve is directly proportional to the total number of observations (N) in a set of data which are normally distributed. We can use portions of the total area to yield information as to the number of individuals who are within varying Z score distances from the mean of the distribution.

The table of the normal curve is recorded in Z score form, and use of the table requires that the data be transformed into Z scores. Starting with the original scores, it is possible to determine the number of scores within any Z score distance of the mean, or between different Z score distances. It is also possible to start with the normal curve and find the score which deviates any given amount from the mean.

Since the area of the normal curve is directly proportional to the total number of observations, any subarea of the curve will yield information as to the proportion of the sample in that area, and this can be expressed as the probability of the occurrence of these events. Between the mean and one Z score will be found 34.13 percent of the total area under the normal curve. Between the mean and ±1.96 Z scores will be found 95 percent of the total area, and within ±2.58 Z scores of the mean will be found 99 percent of the total area.

The table of the normal curve can be used to establish an interval about a sample mean and to make probability statements as to the likelihood that a score will be within this interval.

Probability may be defined as the ratio of the number of ways an event can occur to the total of all events which might occur. The additive theorem of probability states: *The probability that any one of several mutually exclusive events will occur is the sum of their separate probabilities.*

The multiplicative theorem of probability states: *The probability that a combination of independent events will occur is the product of their separate probabilities.*

In working with two-choice situations in which the two choices are mutually exclusive, the binomial expansion may be used to solve probability problems. In the binomial expansion the exponents of a term identify the combination of events, and the coefficient of the term indicates the number of ways that combination of events can occur. If $p = q = .50$, then the ratio of the coefficient of the term divided by all events that could occur will give the exact probability of occurrence of that specific event. If $p \neq q$, substituting the values of p and q into the specific term in the binomial expansion will yield the exact probability of the occurrence of that event.

As the number of elements in a two-choice situation becomes large, the distribution of the coefficients of the binomial expansion approaches the normal curve. The mean event in the binomial is np, and the standard deviation is $\sqrt{npq}$; thus the normal curve can be used to solve discrete probability problems when n is reasonably large.

References

NORMAL-CURVE TABLE (MORE DETAIL)

QUINN MCNEMAR, *Psychological Statistics,* (4th ed.; New York: John Wiley & Sons, Inc., 1969), pp. 502–03.

POISSON DISTRIBUTION

W. J. DIXON and F. J. MASSEY, JR., *Introduction to Statistical Analysis.* (New York: McGraw-Hill Book Co., Inc., 1951), pp. 194–95. Tables for the Poisson distribution are also included in Dixon and Massey.

PROBABILITY

ALLEN, L. EDWARDS, *Probability and Statistics* (New York: Holt, Rinehart, and Winston, Inc., 1971).

QUINN MCNEMAR, *Psychological Statistics* (4th ed.; New York: John Wiley & Sons, Inc., 1969), pp. 41–48.

DAVID M. MESSICK (ed.), *Mathematical Thinking in Behavioral Sciences Readings from Scientific American* (San Francisco: W. H. Freeman and Company, 1968), pp. 4–32.

7

Sampling theory

To THIS POINT, we have been concerned with the *description* of an existing set of data. What is the central tendency or average of the data? How much variation in the data? What percentage of the scores are above a certain point? All of these questions involve the analysis and description of a set of data. But usually we collect data in order to make predictions or make inferences about a situation that we have not measured in full. What can a unit of government expect to collect in taxes during the next year? This kind of question requires making a prediction of a future condition based on our current knowledge. *Inferential* statistics is concerned with making predictions from one set of data to either a future condition or some larger body of data. One example of inferential statistics can be found in the polls that are used to predict the outcome of elections. The people who do election polling are attempting to do two things: (1) predict from a small sample how the population of voters would vote if the election were held at the time the poll was taken and (2) predict how the electorate will vote when the actual election is held. Both of these involve using a subset of the total electorate and, from this subset, predicting the behavior of the entire set or population.

SELECTING A SAMPLE FROM A POPULATION

On the night of a national election in the United States, the various television networks begin to "concede" various states to different candidates when only an extremely small percentage of the votes have been counted. In some cases, statements are made that candidate *A* has won the state when the election results at that time show candidate *B* to be ahead in the balloting! To one who is not familiar with modern sampling theory, this almost appears to be magic. Many individuals attribute the accuracy of these predictions to the use of computers. The computer is only a device to do high-speed calculations based on the information fed to the computer and the instructions given for the processing of this information. The success or failure of these predictions is determined by the utilization of the principles developed from sampling theory, and there is no magic involved—only the application of sampling theory.

DEFINING THE POPULATION.   One of the first problems to face anyone who wishes to make a prediction from a small sample of data is to determine the population involved in the prediction. If we are interested in voting behavior in the United States and we wish to draw a sample and then predict how the election will turn out, we had better not include anyone under 18 in the sample. The population of the United States is *not* the same as the population of voters in the United States. Even "all American citizens 18 years of age or older" is not an adequate definition of the population of voters. Many states have voter registration laws which require the individual to meet certain standards set by that state some weeks or even months before the election is held. Any individual who is not registered is not eligible to vote; hence, these people are not a part of the population of voters. For polling purposes the definition of a population may be a very expensive and time-consuming task. Once a population has been defined, it is necessary to obtain a sample of individuals that is representative of the entire population. It should be pointed out that a population does not have to be an extremely large group of individuals. A population can be large, such as all the individuals residing in the United States, or it may be as small as all the members of one family unit. Regardless of the size of the population, a sample is always a group of individuals from the specified population. The sample is usually much smaller than the population, but could be almost as large as the entire population.

STRATIFYING THE POPULATION.   After a population has been defined and a sample is to be selected from that population, we must be sure that the sample will be representative of the population. One approach is to use *stratified sampling*. In obtaining a stratified sample, we must categorize our population into meaningful subgroups. In political polling, this might involve such categories as race (Caucasian, Negroid, Oriental, etc.); religious preference (Catholic, Protestant, Jewish, etc.); area of residence (urban, suburban, rural); social-economic level, etc. We might have a category of Negro-Jewish-rural-female. After carefully subdividing the population into meaningful categories, we would then determine the percentage of the entire population which can be placed in each category. We would then make sure that our sampling was such that it reflected this same percentage of the total sample in each category as is true of the population. In this way the sample becomes a miniature population.

Failure to obtain a representative sample has been the cause of several spectacular failures in political polling. In the 1930s a popular magazine mailed postcards to several million potential voters. This was not a small sample. Based on the results of their poll, the magazine predicted that Franklin D. Roosevelt would *lose* the election! With such a large sample, how could it be so wrong? Postcards had been mailed to names secured from automobile registrations and telephone directories. During the depression of the 1930s, this oversampled the higher economic strata of the country.

RANDOM SELECTION.   Even after a refined stratified sampling plan has been developed, one additional step is essential if we wish to generalize from our sample to the entire population. This essential step is that the selection of the specific individuals for the sample *must* involve some type of *random* selection. Without random sampling, there is no basis for using probability theory or sampling theory for making predictions. Only with random selection can we place any confidence in the results of the sampling. There is probably no word which is more misused than the word *random*. In the most common usage, *random* seems to imply that no specific plan was used. We shall call this method of selecting subjects *haphazard* sampling. It definitely is not random sampling. For a sample to be random, two conditions must be met: (1) every individual in the population must have an equal opportunity to be selected in the sample (or more accurately, each individual must have a known probability of being selected); and (2) the selection of any one individual must not affect the selection of any other individual.

Some years ago a study was reported in which the researcher used grade school children as subjects. He reported that his subjects were a "random sample of grade school children." Let us examine this statement in detail and also the procedures which were used in actually selecting the children for inclusion in the study. The statement implies that the population was grade school children. Obviously, the research worker did not use grade school children from all parts of the world or the

United States. The children all came from the same community. Did all grade school children in this one community have an equal opportunity of being selected for the study? First, the selection of the specific subjects was left to the discretion of grade school teachers, and it was extremely clear that practically all students selected were of high intellectual ability. Secondly, the parents of the children selected had to agree to allow the children to participate; and the parents had to arrange transportation to and from the testing sessions, which occurred five days a week over a period of several weeks. This was described as a random sample of grade school children! A haphazard sample, yes; random, no.

Standing on the corner and interviewing people as they walk by is not taking a random sample of the community. Any inferences about a population which are based on haphazard sampling are in the same category as fortune telling. The fact that numbers are presented does not make the study better than going to a fortune teller and asking the crystal ball what the results will be. This kind of survey is the same as fortune telling unless the population has been carefully defined and the selection of the sample involves the principle of randomization.

MULTI-STAGE SAMPLING.   If the population from which a sample is to be drawn is large, it is often preferred to sample in two or more stages. The procedure currently in effect for selective service is one example of a multi-stage sample. All of the dates in the year were placed, one date per capsule, in a large drum which could be rotated. These were thoroughly mixed. The same procedure was followed for the numbers 1 through 365. A date was drawn at random from the one drum and a number was drawn at random from the second drum. The number was then attached to the date and became the order in which men of different birthdates were to be called for military service.

The same effect could have been obtained with one drawing—when the dates were drawn, the order in which they were drawn would be the order of being drafted; or the dates could have been numbered in advance from 1 to 365 and the numbers drawn. The first number drawn would indicate the birthday for the first men to be drafted. The use of two-stage sampling was unnecessary for purposes of randomization. It was probably necessary from the standpoint of public relations. The selective service officials were concerned that the draft lottery be random and that the public would accept the notion that the lottery was random. By including the two different random drawings they did not insure "more randomness," but they probably obtained more public acceptance of the procedures.

In most cases multi-stage sampling is more efficient than single-stage sampling. To draw a single-stage sample of people within the United States would be very expensive. One way to decrease this expense would be to organize the country into units that were much larger than individual people. The county could constitute such a first stage in multi-stage sampling. A random sample of counties could be drawn (counties could be stratified as to location, wealth, urban, suburban, rural, etc.). Within the counties which were selected, we could sample from the precinct within those counties as a second stage. A third stage might consist of areas of land such as city blocks. We would draw a random sample of blocks from the precincts which had been selected within the chosen counties. The last stage might consist of a random sample of individuals within the blocks.

With multi-stage sampling, the formulas presented in this book cannot be used directly. It is necessary to modify these formulas to take each sampling stage into consideration. Some of the stages may have very few elements in them. When this is the case it is necessary to adjust the various formulas to compensate for the small size of the population.

SAMPLING FROM A FINITE POPULATION. The formulas presented in this book assume that we are sampling from large populations. When this is not true, as is the case in many multi-stage-sampling plans, an adjustment must be made. This is related to the change that may occur in the probability of an element being selected as the random sampling occurs. This can be illustrated by using a deck of cards. What is the probability of selecting the ace of hearts in drawing five cards from a standard deck of playing cards? Since there are 52 cards in the deck and only 1 ace of hearts, on the first draw the ace of hearts has the probability of $1/52$ in being selected. If the first card is replaced, then the probability of selecting the ace of hearts on the second draw would again be $1/52$. This is referred to as sampling *with replacement*. If the first card drawn is not replaced, then the probability of selecting the ace of hearts on the second draw is $1/51$ (assuming that the ace was not drawn the first time). On the third draw the probability becomes $1/50$. This is called sampling *without replacement*. If the total population is large, the minor change which occurs in the probability of being selected as more cases are drawn is neglible. Sampling without replacement does present a serious problem when the population is small relative to the sample size. Information concerning the formulas to be used in multi-stage sampling and correction formulas for sampling from a finite population can be found in any standard treatment of survey sampling.

USE OF RANDOM NUMBERS.   To this point we have defined random
and we have referred to the use of a drum to physically draw samples
which would be random. However, the easiest way to draw random samples is to use a table of random numbers. Suppose that you wish to
survey the opinions of your fellow students on some issue or issues. You
decide to draw a random sample of these students rather than to contact
and interview every student. The first step, of course, is to define your
population. Who would you consider to be a student at your school?
Once you have defined the population, you must then list all the members.
Suppose that there are 8,000 students in the group that you have defined
as your population. You wish to draw a 10 percent sample, or 800 of
these students. Which ones do you select so that your sample will be
random and you can use sampling theory in making inferences about
the entire population? This is the type of situation in which a table of
random numbers is useful. A portion of table N from the Appendix
has been reproduced below as table 7–1. Notice that the random numbers

TABLE 7–1
Table of random numbers

	1	2	3	4	5	6	..	10	....	20
1	03	47	43	73	86	36	..	61	....	45
2	97	74	24	67	62	42	..	20	....	51
3	16	76	62	27	66	56	..	07	....	59
4	12	56	85	99	26	96	..	31	....	21
5	55	59	56	35	64	38	..	21	....	53
...	...	...	...	...	...	...	..	...	....	...
...	...	...	...	...	...	...	..	...	....	...
...	...	...	...	...	...	...	..	...	....	...
43	26	99	61	65	53	58	..	70	....	74

are arranged in columns with two-digit numbers, but if we number all
the students in the population from 1 to 8,000, we will need four digits.
We can use two adjacent columns which will yield four-digit numbers.
Where do we start in using the table? In one sense we could start any-
where but if we use the table many times we should start in a different
place each time to insure that each sample we draw is not systematically
like the last sample drawn. Once we have a starting point in the table
of random numbers, we move down the column and select the first 800
numbers. Any number above 8001 is inadmissible since there is no one

in the population with that number. Also numbers will repeat. Any repeat of a number is inadmissible, thus we will end up with 800 *different* numbers which are free to vary from 0001 to 8000. If we start at row 2, column 3 of table 7–1 we find the entry 24 but we need four digits. We can use both columns 2 and 3 or 3 and 4. Let us use 3 and 4. The first number, then, becomes 2467; the second person selected would be number 6227; the next number (8599) is inadmissible since no one in the population has that number assigned to him. The third person selected would be number 5635. This procedure would be continued until 800 cases had been selected.

A different situation arises when we have a group of 40 laboratory rats and we wish to conduct an experiment with 4 groups of 10 rats each. How do we assign the 40 animals to these groups in such a way as to have 4 *random* samples from this population of 40? Again we turn to the table of random numbers. Now we need to use only two columns of the numbers. We must number the animals from 1 to 40, then we select the first 10 *different* numbers that are between 01 and 40 in the column. We take the next ten numbers which are different from each other and do not include any of the ten numbers selected for the first group. This would be continued for the third group, and the fourth group would consist of those ten animals not selected in the first three groups.

An alternate system would be to assign the first random number to group 1, the second random number to group 2 and so on. The fifth random number would be assigned to group 1 and the cycle would continue through the 40 admissible numbers.

One question which arises is: Should sampling without replacement (as in the rat experiment) require that we use the corrections for sampling from a finite population? A discussion of the issue must be postponed until chapter 13.

DISTRIBUTION OF SAMPLE MEANS

Why is randomization so important? This question takes us into one of the most important characteristics of samples which are randomly obtained from a population. To illustrate this principle, we shall use a population of only eight individuals (remember that a population may be as large or small as we wish). The population of eight people have "scores" as shown in Table 7–2. The mean of the population can be obtained by adding all the scores and dividing by eight. Since this is

TABLE 7–2

Person	Score
A	6
B	4
C	6
D	2
E	10
F	4
G	6
H	2

the mean of the population, we shall use the Greek letter mu (μ) as the symbol for the mean:

$$\mu = \frac{\Sigma X}{N} = \frac{40}{8} = 5$$

We shall draw samples of 25 percent of the population (two individuals in each sample), and for each sample we shall compute the mean. Since there are eight individuals in the population and we are drawing samples of size 2, we can draw 28 different samples, which will pair each individual with every other individual in the population. These samples are given in table 7–3. Notice that if only sample *AB* had been drawn from

TABLE 7–3
All samples and sample means drawn from a population of eight individuals where the sample size is two individuals

Sample	Mean	Sample	Mean	Sample	Mean	Sample	Mea
$AB = 6 + 4$	5	$BC = 4 + 6$	5	$CE = 6 + 10$	8	$DH = 2 + 2$	2
$AC = 6 + 6$	6	$BD = 4 + 2$	3	$CF = 6 + 4$	5	$EF = 10 + 4$	7
$AD = 6 + 2$	4	$BE = 4 + 10$	7	$CG = 6 + 6$	6	$EG = 10 + 6$	8
$AE = 6 + 10$	8	$BF = 4 + 4$	4	$CH = 6 + 2$	4	$EH = 10 + 2$	6
$AF = 6 + 4$	5	$BG = 4 + 6$	5	$DE = 2 + 10$	6	$FG = 4 + 6$	5
$AG = 6 + 6$	6	$BH = 4 + 2$	3	$DF = 2 + 4$	3	$FH = 4 + 2$	3
$AH = 6 + 2$	4	$CD = 6 + 2$	4	$DG = 2 + 6$	4	$GH = 6 + 2$	4

the population, the mean of the sample would have been equal to the population mean. However, if we had happened to draw sample *AE*, the sample mean would have been considerably larger than the population mean, and sample *DH* would yield a sample mean considerably smaller than the population mean. These 28 different samples give 28 estimates of the population mean. Some estimates are too large, and some are too small. Table 7–4 shows a frequency distribution of these 28 sample means.

TABLE 7–4
Distribution of the 28 sample means drawn from a population of 8 individuals

Value of *sample mean*	*Frequency of* *sample mean*	$f\bar{X}$
8	3	24
7	2	14
6	5	30
5	6	30
4	7	28
3	4	12
2	1	2
Total	28	140

$$M_{\bar{X}} = \frac{\Sigma f\bar{X}}{N_{\bar{X}}} = \frac{140}{28} = 5$$

What is the "average" mean of these 28 samples? The mean of any set of numbers is the sum of the elements divided by the total number of elements in the set. Means are numbers like any other numbers, and so we can compute the mean of the means. The sum of the 28 means is 140; and when this is divided by 28, the total number of means, we find that the mean of the means is 5. But we found earlier that the mean of the population (μ) was also equal to five. *If we draw all possible samples of the same size from a population, the mean of all sample means will be exactly the same as the population mean,* or symbolically, $M_{\bar{X}} = \mu$.

Since the distribution of means is a distribution of numbers, we could describe these numbers in a great variety of ways. In addition to computing the mean of all these means, we could compute the median, the mode, and various measures of variability such as the standard deviation of the means, or the semi-interquartile range, or measures of kurtosis and skewness. If the distribution of means formed a normal distribution, we could compute the standard deviation of the means, compute Z scores, and answer such questions as: How many sample means will be at least three but no greater than seven? How large or how small must a sample mean be so that it will occur no more often than 5 times in 100? All we need do to find the mean of a population is to draw all of the possible random samples and compute the mean of all these means. Then we can describe this distribution of means by using the standard deviation of all these means.

We are somewhat in the position of the king with the magic suit of clothes, in the children's story. Everything went well until a child pointed out that the king was not wearing any clothes at all and the magic suit was nonexistent. To draw all possible samples from a population to determine the mean of that population may sound good, but it is not only impractical but far more labor than simply measuring the entire population. If we refer back to the population of eight individuals, we would need eight measurements to find the population mean. But if we selected all possible samples for two individuals in each sample, we would need 28 samples, or 56 different measurements, to arrive at the same place. What is needed is some method whereby we can draw one sample from a population and from this one sample estimate the population value.

THE STANDARD ERROR OF THE MEAN.   There is no direct way available to determine the mean of the population from a single sample. It is possible, however, to estimate the standard deviation of a distribution of means from a single sample. As was pointed out earlier, this standard deviation should be interpreted like any standard deviation. Just because it is a measure of the variability of a distribution of means does not make it different from any other standard deviation. This standard deviation has a special name which is sometimes misleading. The standard deviation of a distribution of means is called the *standard error of the mean*. Because the term *standard error* is used, some students tend to think of this statistic as being different from a standard deviation. Whenever the term *standard error* is used, it refers to a standard deviation, and the rest of the expression specifies that it is the standard deviation of some statistic. Thus the standard error of the mean is the *standard deviation* of a distribution of sample means; the standard error of a proportion is the *standard deviation* of sample proportions.

The standard deviation of a distribution of sample means (standard error of the mean) may be determined utilizing the following formula:

$$\sigma_{\bar{x}} = \frac{\sigma}{\sqrt{N}} \tag{7-1}$$

where

$\sigma_{\bar{x}}$ is the standard deviation of the distribution of means (standard error of the means)

σ is the standard deviation of the population from which these means were random samples

N is the size of the sample on which the means are based

Formula 7–1 is not directly useful to us, since it requires that we must know the standard deviation of the population before we can determine the standard error of the distribution of sample means. However, it is possible to estimate σ from a single sample. Formula 7–2 is used to obtain an *unbiased* estimate(s) of σ:

$$s = \sqrt{\frac{\Sigma x^2}{N-1}} \qquad (7\text{--}2)$$

In chapter 5 page 96, it was pointed out that the use of N as the divisor in computing the standard deviation will yield a biased estimate of the population value. The use of the degrees of freedom $(N-1)$ will yield an unbiased estimator. The standard error of the mean can be estimated as follows:

$$s_{\bar{x}} = \frac{s}{\sqrt{N-1}} = \frac{s}{\sqrt{N}} \qquad (7\text{--}3)$$

where

$s_{\bar{x}}$ is the standard error of the mean (unbiased estimate of $\sigma_{\bar{x}}$)

s is the standard deviation of the distribution of scores

s is an unbiased estimate of the population standard deviation (σ)

N is the sample size

$N-1$ is the number of degrees of freedom

On occasion, the standard deviation may have been computed using N in the denominator, and then it is desirable to compute the standard error of the mean. This can be accomplished by using $\sqrt{N-1}$ in the denominator of formula 7–3 rather than $\sqrt{N}$. Perhaps the easiest way to avoid errors in computing the standard error of the mean is to go to the sum of squared deviations and use the following formula which is an algebraic variation on formula 7–3.

$$s_{\bar{x}} = \sqrt{\frac{\Sigma x^2}{N(N-1)}} \qquad (7\text{--}4)$$

In reading research studies or textbooks, the student should examine the method of computing a standard deviation before deciding that it is an unbiased estimate of the population value. With a little care in reading, there is no problem in determining which statistic the author is dealing with.

CONFIDENCE INTERVALS. We are now in a position to estimate the standard deviation of a distribution of sample means from the standard deviation of one sample. We know that the population mean is equal to the mean of all the sample means that can be obtained from a population. We can estimate the standard deviation of these sample means. If the distribution of sample means is of the bell-shaped or normal-curve type, we can use our knowledge of the normal curve to estimate the location of the population mean.

If we knew the mean of all sample means (the population mean) and the standard deviation of these sample means (standard error of the mean), we could compute Z scores and determine the range within which we would expect any percentage of the sample means to be found. Between $-1Z$ and $+1Z$, we could expect to find 68 percent of all the sample means. Between $-1.96Z$ and $+1.96Z$, we could expect to find 95 percent of all the sample means; and between $-2.58Z$ and $+2.58Z$, we would expect to find 99 percent of all sample means. But of course, it is this mean of the population that is missing—it is the value we wish to estimate, and we want to make the estimate on the basis of only one sample.

We can utilize the normal curve to do this simply by reversing our logic! A sample mean which was $+1.96$ Z scores (or standard errors of the mean) above the population mean has a probability of occurrence of .025. Ninety-seven and a half percent of all sample means will be smaller than $+1.96$ standard errors of the mean. Now comes the reverse logic: If it is a rare event for a *sample* mean to be 1.96 or more standard errors of the mean *above* the *population* mean, it is just as rare an event that the *population* mean would be 1.96 standard errors of the mean *below* a given *sample* mean. But we don't know if our sample mean is larger or smaller than the true mean of the population. If we were to construct an interval of $-1.96s_{\bar{x}}$ to $+1.96s_{\bar{x}}$ about our sample mean, we could have great confidence that the population mean is located somewhere in that interval. Figure 7–1 illustrates this confidence interval. Note that we do not expect that the sample mean will in fact be as far as ± 1.96 standard errors of the mean away from the population mean, but we are confident that the population mean is no further than this from the sample mean.

FIGURE 7–1
Illustration of 95 percent confidence interval about a sample mean ($\overline{X}$),
showing two possible locations of the population mean and the associated
distributions of sample means

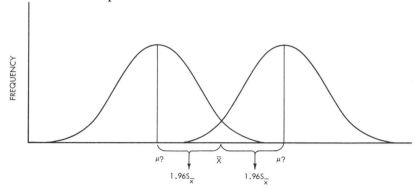

But we could be wrong. It is possible (not very probable) that the sample mean is more than 1.96 Z scores away from the population mean; but by using the normal curve, we can assign a probability statement to the likelihood of our being correct. If we construct an interval ±1.96 standard errors of the mean about our sample mean, we expect the population mean to be within this interval with 95 percent confidence. There is a 5 percent chance that we are wrong—the population mean is not within the interval. If we do not wish to run a 5 percent risk of being incorrect, we can use a different confidence interval. The chances that the population mean will be within +2.58 and −2.58 standard errors of the mean of the sample mean is 99 out of 100—this is the 99 percent confidence interval. The computation of the 95 percent confidence interval and the 99 percent confidence interval is shown in example 7–1.

Using the data in example 7–1, we would state that the mean of the population from which the sample was obtained is no less than 57.2 and no greater than 62.8, with confidence that in 95 out of every 100 similar situations we would be correct. If we are not willing to tolerate a 5 percent chance of being incorrect, we can use a confidence interval that is wider (less precision in our estimate), such as the 99 percent confidence interval; or we could use a confidence interval of ±4 standard errors of the mean, and then we would have 99.994 percent confidence in being correct—only .00006 probability of being incorrect that the population mean is within the interval from 54.28 to 65.72.

The width of the confidence interval around the sample mean is dependent upon three factors: (1) the variability within the sample, (2)

Example 7–1   Ninety-five percent and 99 percent confidence intervals for the population mean

$$\bar{X} = 60; \quad S = 10; \quad N = 50$$

Using equation 7–3:

$$s_{\bar{x}} = \frac{S}{\sqrt{N-1}} = \frac{s}{\sqrt{N}} = \frac{10}{\sqrt{49}} = \frac{10}{7} = 1.43$$

From table B of the Appendix, select that Z score appropriate to the middle 95 percent of the area of the normal curve:

$$Z = 1.96$$

The 95 percent confidence interval will be from $-1.96s_{\bar{x}}$ to $+1.96s_{\bar{x}}$ around the sample mean:

$$\text{CI}_{.95} = \bar{X} \pm 1.96s_{\bar{x}} = 60 \pm (1.96)(1.43) = 60 \pm 2.80$$
$$\text{CI}_{.95} = 57.20 \text{ to } 62.80$$

Determine the 99 percent confidence interval.

From table B of the Appendix, select that Z score appropriate to the middle 99 percent of the area of the normal curve:

$$Z = 2.58$$

The 99 percent confidence interval will be from $-2.58s_{\bar{x}}$ to $+2.58s_{\bar{x}}$ around the sample mean:

$$\text{CI}_{.99} = \bar{X} \pm 2.58s_{\bar{x}} = 60 \pm (2.58)(1.43) = 60 \pm 3.69$$
$$\text{CI}_{.99} = 56.31 \text{ to } 63.69$$

the sample size, and (3) the degree of confidence required by the research worker. The confidence interval can be made extremely narrow if the research worker is willing to run a large risk of being wrong. He could use an interval of $\pm.68s_{\bar{x}}$ and have only a 50 percent chance of being correct that the population mean is within the interval. In general, the best way to obtain a narrow confidence interval is to use a larger sample size. In practical situations the research worker must weigh the cost of obtaining additional subjects against the need for precision in estimating population values and, at the same time, minimize the risk of being incorrect in the estimate of the population value.

THE CENTRAL LIMIT THEOREM.   The use of the standard error of the mean in establishing limits within which we expect to find the population mean and the assignment of a probability statement to our chances of being correct (use of table B of the Appendix) assume that the distribution of the sample means closely approximates the normal curve. If the distribution of the underlying population is normal, then the distribution of samples from that population will tend to be normal (this is somewhat dependent on sample size). But what if the underlying distribution is nonnormal? We can still use normal-curve statistics if the sample size is large enough. The central-limit theorem may be stated as: *Regardless of the shape of the distribution of the population, as the sample size increases, the sampling distribution of the means approaches the distribution of the normal curve.* If the sample size is 100 or more, there is generally no question that the sampling distribution of the means will be normal. This allows us to compute the standard error of the mean and use the table of the normal curve to construct confidence intervals with assurance of the accuracy of our probability statements. Some writers indicate that a sample size of 50 may be used without any appreciable loss of accuracy in the probability statements; and if the underlying distribution of the population from which the samples are drawn is not badly "nonnormal," then a sample size of 30 may be sufficient for the central limit theorem to hold. Figure 7–2 shows the shape of different sampling

FIGURE 7–2
Distribution of estimates of average income from samples of various sizes drawn from a population of 12 million individuals

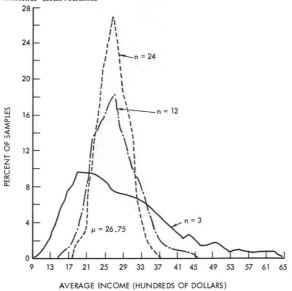

AVERAGE INCOME (HUNDREDS OF DOLLARS)

Adapted from M. H. Hansen, W. N. Hurwitz, and W. G. Madow, *Sample Survey Methods and Theory*, vol. I (New York: John Wiley & Sons, 1953), p. 28.

distributions of the mean based on different sample sizes, drawn from a positively skewed distribution of the population. Notice that with the extremely small sample size of only three cases per sample, the distribution of the means of these samples is positively skewed. With a sample size of 12 cases the distribution of the sample means is beginning to look quite symmetrical, and with a sample size of 24 cases the distribution of the means of these samples is very close to symmetrical and is approaching the normal curve (the leptokurtic appearance of the distribution of means based on samples of 24 is due to the scale used in drawing figure 7–2).

ESTABLISHING CONFIDENCE INTERVALS WITH NOMINAL DATA (STANDARD ERROR OF A PROPORTION)

To this point, we have discussed the establishment of confidence intervals around measurements which would generally assume interval or ratio scaling. Often the research worker is interested in drawing samples and simply counting the number of individuals in each category. With this type of observation we cannot compute a mean in the usual sense, but we can determine the proportion of the total sample which is found in each category. If we could count every individual in the population, we could compute the true proportion in each category. Since we are dealing with a sample from the population, we can consider the sample proportion as an estimate of the true proportion. Since the obtained proportion is a sample value, we would not expect it to be exactly the same as the true proportion; and if a second sample were drawn, we would expect this second estimate of the true proportion to be somewhat different than the estimate obtained from the first sample. If we could estimate the standard deviation of all possible sample proportions (standard error of a proportion), we could then construct confidence intervals about the sample proportion in the same manner that we used with the sample mean. The formula for the standard error of a proportion is:

$$\sigma_{prop} = \sqrt{\frac{PQ}{N}} \tag{7-5}$$

where

P is the true proportion in the population in one category
Q is $1 - P$ true proportion in the population not in the one category
N is the sample size

Note that P and Q are not sample statistics but rather true values based on the entire population (population parameters). This may create a problem, in that the size of the standard error of a proportion is directly dependent upon the true population values, which are not available. In many cases, this does not present a problem. In polling connected with elections, there is usually one true value which is important—namely, $P = .50$. The use of the standard error of a proportion using a political poll is given in example 7–2 for two different situations. In the first

---

Example 7–2   Use of the standard error of a proportion in political polling

A sample of 1,000 voters is asked how they intend to vote in the forthcoming election. Sixty-two percent indicate they will vote for candidate A, and 38 percent indicate they will vote for other candidates.

$$P = .50; \quad Q = 1 - P = .50; \quad N = 1,000$$

$$\sigma_{prop} = \sqrt{\frac{PQ}{N}} = \sqrt{\frac{(.50)(.50)}{1,000}} = \sqrt{\frac{.25}{1,000}} = \sqrt{.00025}$$

$$\sigma_{prop} = .0158$$

$$\text{Cl}_{.95} = P \pm 1.96\sigma_{prop} = .50 \pm (1.96)(.0158) = .50 \pm .03$$
$$\text{Cl}_{.95} = .47 \text{ to } .53$$

Thus, if the true proportion of the population who intended to vote for candidate *A* were .50, we would expect that 95 percent of all sample proportions would be within the limits of .47 to .53. Since our sample gave candidate *A* 62 percent of the votes, we decide that the true proportion in favor of candidate *A* must be more than .50; hence, candidate *A* will be elected.

An election is to be held for a school bond to construct a new building. Under the laws of the state a 60 percent favorable vote is required for passage. The school board has the right to call the election at any time. To insure passage of the bond issue, should the election be held as soon as possible, or should it be delayed in order to convince more voters of the need for the new building? A sample of 100 voters is selected, and 65 percent indicate that they are in favor of the bond issue.

$$P = .60; \quad Q = .40; \quad N = 100$$

$$\sigma_{prop} = \sqrt{\frac{PQ}{N}} = \sqrt{\frac{(.6)(.4)}{100}} = \sqrt{\frac{.24}{100}} = \sqrt{.0024} = .049$$

$$\text{Cl}_{.95} = P \pm 1.96\sigma_{prop} = .60 \pm (1.96)(.049) = .60 \pm .10$$
$$\text{Cl}_{.95} = .50 \text{ to } .70$$

If the true proportion of the population who intended to vote for the bond issue were .60, we would expect that 95 percent of all samples drawn from this population would have a proportion between .50 and .70. Since the obtained proportion is .65, it could easily have occurred as a sample from a population whose true value was .60. It might be advisable to postpone the election.

---

example, it would be highly unlikely for candidate *A* to be defeated (this assumes that the sample of 1,000 voters polled was, in fact, a random sample of the population of voters). In the second example the situation is not so clear-cut. Notice that the standard error of the proportion is relatively large (.05), and hence the 95 percent confidence interval is large. Since the standard error of a proportion varies with both the true proportions and the sample size, perhaps a larger sample should have been used to achieve a smaller confidence interval (increase the precision of the estimate).

A second approach to establishing the confidence interval about a sample proportion is to use the sample values themselves as direct estimates of the true values. This is not the preferred approach, but often there is no alternative. Example 7–3 shows the use of the standard error of

---

Example 7–3   Use of the standard error of a proportion based on the sample values

$p = .62$ = Proportion in sample selecting candidate A
$q = 1 - .62 = .38$ = Proportion in sample selecting candidates other than candidate A
$N = 1,000$

$$s_{prop} = \sqrt{\frac{pq}{N}} = \sqrt{\frac{(.62)(.38)}{1,000}} = \sqrt{\frac{.2356}{1,000}} = \sqrt{.0002356} = .0154$$

$CI_{.95} = p \pm 1.96 s_{prop} = .62 \pm (1.96)(.0154) = .62 \pm .03$
$CI_{.95} = .59$ to .65

We would conclude that candidate A will receive between 59 percent and 65 percent of the vote, with 95 percent confidence in the estimate (there is a 2.5 percent chance that the true proportion will be less than 59 percent and a 2.5 percent chance that the true proportion may be more than 65 percent).

$p = .65$ = Proportion of voters in sample favoring bond issue
$q = 1 - .65 = .35$
$N = 100$

$$s_{prop} = \sqrt{\frac{pq}{N}} = \sqrt{\frac{(.65)(.35)}{100}} = \sqrt{\frac{.2275}{100}} = \sqrt{.002275} = .0477$$

$CI_{.95} = p \pm 1.96 s_{prop} = .65 \pm (1.96)(.0477) = .65 \pm .09$
$CI_{.95} = .56$ to .74

We would conclude that the true proportion in the population is between .56 and .74, with 95 percent confidence in our estimate.

---

a proportion using the sample values as if they were the true proportions in the population.

In using the table of the normal curve in setting confidence intervals about a proportion, we are assuming that the sampling distribution of proportions approximates the normal curve. This assumption is tenable as long as the true proportion is in the middle range of possible proportions. If the true proportion were .99, sample values could only vary up to 1.00, but they would be free to vary below .99 to a considerable degree. This will result in a markedly skewed distribution of sample proportions. The degree of skewness in the distribution of sample proportions increases as the true value of the proportion approaches 1.00 (or 0.00). The degree of skewness is offset to some extent by using larger samples. In general, one should be cautious about using the normal curve to establish a confidence interval about a proportion when the true values appear to be about .80 and above (or .20 and below). In any case the normal curve should be used for establishing confidence intervals about a proportion only if $NP \geq 5$ and $NQ \geq 5$.

STANDARD ERROR AND SAMPLE SIZE. In determining the standard error of the mean or the standard error of a proportion, the size of the sample on which these estimates are based becomes extremely important. Formula 7–3 for the standard error of the mean and formula 7–5 for the standard error of a proportion are presented here to examine the different portions of these expressions.

$$s_{\bar{x}} = \frac{s}{\sqrt{N}} = \frac{S}{\sqrt{N-1}}$$

where

   s is the unbiased estimate of σ
   S is the standard deviation of the sample
   N is the sample size

$$\sigma_{prop} = \sqrt{\frac{PQ}{N}}$$

where

   P is the true proportion
   Q equals $1 - P$
   N is the sample size

In both equation 7–3 and equation 7–5 the sample size is in the denominator of the expression. In the case where the sample size is small (approaches one), the standard error of the mean approaches the standard deviation of the sample (or estimate of the population standard deviation); as the sample size approaches infinity, the standard error of the mean approaches zero. This means that with an infinitely large sample (the entire population) the sample mean is the population mean; hence, sampling error is nonexistent (zero). The same situation holds with the standard error of a proportion. As the sample size increases, the standard error of a proportion decreases. In order to increase the precision of an estimate of the true mean or true proportion in the population (decrease the width of the confidence interval) and still maintain the same degree of confidence in the estimate, the best procedure (usually) is to increase the size of the sample used in making the estimate.

Summary

The use of samples to make inferences concerning the population from which the sample was drawn requires a great deal of care and caution if the results obtained from the sample are to be valid for the population. One of the first requirements is to define clearly the population of interest. This must be done in such a manner as to identify clearly those elements which will be in the population and those elements which will not be within the population.

Once the population has been specified, the next step is to draw a sample that will be representative of the population. One approach is to draw a *stratified sample*. This is done by categorizing the population into meaningful subgroups. Each subgroup must be defined so that a member of the population is a member of that subgroup or is not a member of the subgroup. After defining each stratum to be used, it is necessary to specify the proportion of the population to be found within each stratum. After the stratification has been completed, it is *essential* that a random procedure be used to select the individuals from the population. *Random sampling* requires that each individual in the population must have a known probability of being selected and that the selection of any one individual will not affect the selection of any other individual.

If the population from which the sample is to be drawn is large, it is often preferable to use multi-stage sampling. This involves dividing the population into successively smaller units and using randomization at each stage to determine which units will be used in the study. Many different procedures may be used to draw a random sample, but one of the easiest and least-expensive methods is to use a table of random numbers.

If all possible samples of a given size are drawn from the same population, the mean of all the sample means will be identical to the population mean. This distribution of sample means will (under the central-limit theorem) be normally distributed for large samples. It is thus possible to determine the exact proportion of sample means which can be found any Z score distance from the mean of all sample means (population mean). Fortunately, the standard deviation of the distribution of sample means (standard error of the mean) can be estimated directly from one sample. Using the standard error of the mean and any Z score distance desired, it is possible to establish an interval about the obtained sample mean within which we would expect to find the true mean of the population. The particular Z score selected will give the probability of being correct in stating that the true mean of the population is within the interval constructed. An interval of 1.96 standard errors of the mean would allow us the probability of .95 that the true mean was within the limits set.

When the observations are nominal in nature, it is still possible to establish a confidence interval about the obtained proportion included in a category. It is necessary to dichotomize data into the proportion of cases found in the one category and the proportion of cases in all other categories combined. The standard error of a proportion is dependent upon the value of the true proportion in the population. This requires that in establishing a confidence interval about a sample proportion, the research worker should use a *hypothesized true value* in computing the standard error of a proportion.

The size of the interval within which one expects to find the true mean of the population (or the true proportion) is dependent upon both the degree of confidence required by the investigator and the sample size. For any given degree of confidence the interval will become smaller as the sample size increases. The smaller the confidence interval, the more precise is the estimate of the population value. The precision of an estimate is directly related to the size of the sample—the larger the sample, the greater the precision (the narrower the confidence interval). For any given sample size the confidence interval will be larger or smaller depending upon the degree of confidence required by the investigator. The interval for 99 percent confidence will be wider than required for 95 percent confidence. In planning an investigation, the research worker needs to determine the degree of confidence which is necessary for him, and the sample size which will permit the maximum precision possible.

References

SAMPLING THEORY AND PRACTICE

CHARLES H. BACKSTROM and GERALD D. HURSH, *Survey Research* (Evanston, Ill.: Northwestern University Press, 1963). A basic book in the area.

LESLIE KISH, "Selection of the Sample," in L. FESTINGER and D. KATZ (eds.), *Research Methods in the Behavioral Sciences* (New York: Dryden Press, 1953). More advanced than the Backstrom and Hursh treatment.

LESLIE KISH, *Survey Sampling* (New York: John Wiley & Sons, Inc., 1965).

RANDOMIZATION

D. R. COX, *Planning of Experiment* (New York: John Wiley & Sons, Inc., 1958), chap. v. Tables of random numbers and random permutations can also be found in Cox.

8

Correlation and regression

In our work to this point, we have been concerned with describing a single set of data. The description may be in terms specific to the one set of data, such as central tendency or variability; or the description may involve making inferences about a population from which the one set of data is a sample, such as establishing a confidence interval about a mean or determining the probability that a given sample deviates a certain distance from some true value. Although simple description of one set of data is an extremely important part of the work of the behavioral scientist, the examination of two or more sets of data at the same time has become even more important.

In general, most behavioral scientists prefer to introduce different treatments to different groups experimentally and test statistically to determine if the different treatments do produce differences in the groups (see chapter 10). There is, however, a large area of study in the behavioral sciences in which the researcher cannot control events but rather is searching for co-relationships among events that occur without experimental manipulation. We might observe that as readings on a barometer become lower (barometer falling), the weather tends to become stormy; and conversely, as the barometer readings rise, the weather tends to become clear and mild. We can say that changes in barometer readings are related to changes in the weather or that changes in the barometer readings and changes in the weather are correlated.

One of the difficulties with the correlational approach is the problem of deciding questions of cause-and-effect relationships. Certainly no one would say that a falling barometer causes storms. Nor would many people say that a storm has caused the barometer to fall; rather, in this particular case we might say that a third factor, air pressure, is related both to stormy weather and to the height of barometer readings. Correlation permits us to describe a covariation but does not permit statements concerning cause and effect.

There are many other examples of two variables tending to change together (concomitant variation). If we were to work only with men 20 years old and older, we would find that running speed would tend to decrease with the age of the men—younger men run faster than older men. This would be another example of concommitant variation (age and speed of running varying together). In this case the relationship is negative—small values of one variable (age) are associated with large values of the other variable (running speed). The younger a man is, the faster he can run; or the older a man is, the slower is his running speed.

We might also attempt to determine the relationship between the length of the big toe of the right foot and some measure of school success such as grade-point average. In this case we would find no co-relationship; that is, some people with long, big toes do well in school, but some individuals with long, big toes do poorly. Knowledge of the length of the big toe does not permit us to say anything about the person's grade-point average.

INDEX OF CO-RELATIONSHIP BETWEEN TWO VARIABLES

Thus far, we have mentioned degrees of relationship only in general terms—positive (as one variable increases, the other also increases); negative (as one variable increases, the second decreases); and no relationship (a change in one variable is not related to a systematic change in the other variable). What is needed is some precise index which will permit an unambiguous statement indicating the degree of relationship between two variables.

There are many different indexes of agreement that could be used; however, the specific index that is needed should have several properties. The index should be independent of the particular units of measurement that are used with the two variables, so that an index from one set of data could be directly compared with an index from a different set of data. When we mention comparing indexes from one set of data to another, this should suggest one of the properties of the Z score. Remember that the Z score is independent of the original measurement unit, and the use of Z scores allows us to compare deviations from different distributions. By comparing Z scores, we can say that a man is taller than he is heavy; that is, he deviates further from the mean height of his group in standard deviation units than he deviates from the mean weight. Thus, it would appear that an index based on Z scores would be independent of the specific measuring unit involved.

DIRECTION OF RELATIONSHIP AND THE SIGN OF THE INDEX.   An index of relationship will require two scores—one score on each of the two variables. If we call the first variable X and the second variable Y, each individual will have two scores—an X score and a Y score. Let age be the X variable and running speed the Y variable, and suppose that the relationship is perfect; that is, the youngest man runs the very fastest, the next youngest man runs next fastest, and so on to the oldest man in our group, who is the slowest runner. The youngest man will be below the mean in age for the group and therefore will have a minus Z score in X $(-Z_x)$. This same man will be the fastest runner; thus, his Z score in speed will be positive (Z_y). If we were to multiply these two Z scores together, we would have a negative product (a negative number times a positive number yields a negative product). If we converted each age score into a Z_x and each speed score into a Z_y and multiplied the Z_x and Z_y together for each person, all of these products of Z scores would be negative. The people who are older than the mean of their group $(+Z_x)$ run slower than the mean running speed for the group $(-Z_y)$, and thus the product of these Z scores for each individual would be negative $(-Z_xZ_y)$. If we sum these products of Z scores (ΣZ_xZ_y), the total will be negative. The *sign* of the sum of the cross products of these Z scores will indicate that the relationship is negative.

If the relation between two variables is positive, then individuals who are above the mean in X will also tend to be above the mean in Y. These individuals will have a positive Z_x and a positive Z_y, and the product of the Z_xZ_y will be positive. Consider those individuals below the mean in X and also below the mean in Y. The Z score in X for such an individual would be negative $(-Z_x)$, and the Z score in Y would also be negative $(-Z_y)$; but the product (Z_xZ_y) will be *positive,* and the sum of the cross products of Z scores (ΣZ_xZ_y) will be positive. This indicates that the relationship is positive (high values in X associated with high values in Y; low values of X with low values of Y). From these two illustrations the student should note the generalization that the *sign* of the sum of the cross products of Z scores will indicate the direction of the relationship.

MEAN OF THE CROSS PRODUCTS OF Z SCORES AS INDEX.   Although

$\Sigma Z_x Z_y$ will be independent of the original measuring units and the sign will indicate the direction of relationship, the size of the sum of the cross products will depend upon the number of sets of scores. A large sum could indicate either a high degree of relationship or a large number of scores. This problem can be overcome rather easily by dividing the sum of the cross products of Z scores by the number of pairs of scores (N). This will convert the index into a mean, and therefore the size of the index will be independent of the number of scores. Suppose that two different sets of cross products of Z scores have been obtained. In one case, $\Sigma Z_x Z_y = 25.26$; and in the other case, $\Sigma Z_x Z_y = 6.32$. Which indicates the higher degree of relationship? From the data given, we can't tell. But if we add the information that the first set is based on $N = 100$ and the second on $N = 10$, then we can state that there is a higher relationship in the second set of data! In the first case, 25.26 divided by 100 yields an index of .25. In the second case, 6.32 divided by 10 yields an index of .63.

In summary, the mean of the cross products of Z scores will have the following characteristics as an index of relationship between two variables: (1) it will be independent of the original measurement units; (2) it will be independent of the number of scores (N); (3) the sign of the index will indicate whether the relationship is positive or negative; and (4) the size of the index will indicate the strength of the relationship.

MAXIMUM VALUE OF THE MEAN OF THE CROSS PRODUCTS OF Z SCORES. If the size of the Z score in X and the Z score in Y are exactly the same for each individual ($Z_x = Z_y$), we shall obtain the highest possible sum of the cross products of the Z scores ($\Sigma Z_x Z_y = $ Maximum). If every individual in the group deviates as much from the mean in X as that individual deviates from the mean in Y ($Z_x = Z_y$ for each individual), a rather interesting situation is apparent. If for each individual $Z_x = Z_y$, then

$$Z_x Z_y = Z_x^2$$

and

$$\Sigma Z_x Z_y = \Sigma Z_x^2$$

Dividing both sides of the equation by N:

$$\frac{\Sigma Z_x Z_y}{N} = \frac{\Sigma Z_x^2}{N}$$

But

$$\frac{\Sigma Z_x^2}{N} = S_z^2$$

and

$$S_z{}^2 = 1$$

(See chapter 5 on the variance of any set of Z scores.) Is it possible for this index of co-relationship to be larger than one? The answer is no. Remember that the sum of a set of Z scores is always equal to zero. If any one Z score in a distribution were to change in value, at least one or more other Z scores in the distribution would also have to change in value. It will be left as an exercise for the student to construct a set of Z scores in X and a set of Z scores in Y. Construct these two distributions so that each Z_x is equal to a Z_y. The student should try pairing different Z scores in X and Y and determine for himself which type of pairing leads to the maximum value.

In addition to the characteristics of the index mentioned earlier, we may add an additional characteristic: Maximum relationship between the two variables will yield a numerical value of the index of one (the sign of the index will of course indicate the direction); minimum relationship will yield an index value of zero. This index, the mean of the cross products of Z scores, is called the Pearson product-moment correlation coefficient[1] and is designated by the symbol r.

USE OF r IN PREDICTION

Table 8–1 shows the scores of 20 freshman students on two different tests. The Scholastic Aptitude Test (SA) is designed to predict how well college students will do academically. The English Placement Test is designed to differentiate among students as to the knowledge and usage of the English language. The question is: Are these two tests correlated? If we know how well a student scores on the Scholastic Aptitude Test can we predict how well the student would score on the English Placement Test? The Pearson product-moment correlation coefficient will tell us the degree to which the tests are correlated. The correlation coefficient can then be used to make predictions from one set of scores to the other set.

If we were to draw a picture of the two distributions being correlated, we would place one variable along the abscissa and the other variable along the ordinate. Figure 8–1 shows such a graph. Scholastic Aptitude Test scores (SA) have been plotted along the abscissa and English Placement Test scores (EP) along the ordinate. The first student has a score

---

[1] Named for the English statistician Karl Pearson (1857–1936).

TABLE 8–1
Scores of twenty students on a Scholastic Aptitude
Test (SA) and an English Placement Test (EP)

Subjects	Scholastic Aptitude	English Placement
1	40	62
2	67	105
3	57	97
4	64	102
5	111	106
6	70	122
7	106	130
8	127	152
9	93	128
10	51	81
11	131	151
12	66	76
13	51	78
14	33	74
15	67	127
16	25	71
17	77	87
18	59	84
19	59	91
20	50	99

$$N = 20 \quad \begin{aligned} \Sigma X &= 1{,}404 & \Sigma Y &= 2{,}023 \\ \Sigma X^2 &= 114{,}982 & \Sigma Y^2 &= 217{,}765 \\ \bar{X} &= 70.2 & \bar{Y} &= 101.2 \\ S_x &= 28.65 & S_y &= 25.43 \end{aligned}$$

of 40 on the X variable (SA) and a score of 62 on the Y variable (EP). This individual's two scores have been plotted by placing a tally mark in the appropriate cell of the diagram (see circled tally in figure 8–1). After the tally marks have been placed in the appropriate cells, so that all individuals are represented in the diagram, a pattern is apparent. We note that the tallies seem to cluster together, running from the lower left corner of the scatter diagram to the upper right corner. In terms of scores, this shows that low scores in X are associated with low scores in Y; high X scores with high Y scores. Without computing r, it is apparent that the correlation will be positive.

If a line were drawn around the tally marks in the scatter diagram, the shape of this ellipse would also indicate, to some extent, the degree of relationship. A thin ellipse running from the lower left corner to the upper right corner would indicate a high positive relationship. A

FIGURE 8–1

Scatter diagram showing scores on an English Placement Test and a Scholastic Aptitude Test given to college freshmen

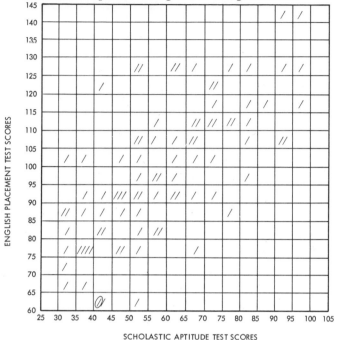

SCHOLASTIC APTITUDE TEST SCORES

long, thin ellipse running from the upper left corner down to the lower right corner would indicate a high but negative relationship. If the line around the tally marks were a circle, this would indicate no relationship. (See figure 8–2.)

If we wanted to make an estimate as to what score John Jones, new freshman, made on the English Placement Test but were unable to find Jones's test score, the best estimate that could be made would be that Jones scored at the mean of the group in the English test. By best estimate, we mean that the error in prediction for a large number of individuals will be least if the mean of the distribution is the predicted point. Remember that the mean is that point in the distribution about which the sum of the deviations is equal to a minimum (zero); thus the best point to predict for a large group is the mean. We shall be wrong by the least amount over many predictions.

From the scatter diagram, it is obvious that scores in the two tests tend to be related. Thus, we have a second bit of information. If we

FIGURE 8–2
Outline of scatterplots, showing various degrees of correlation

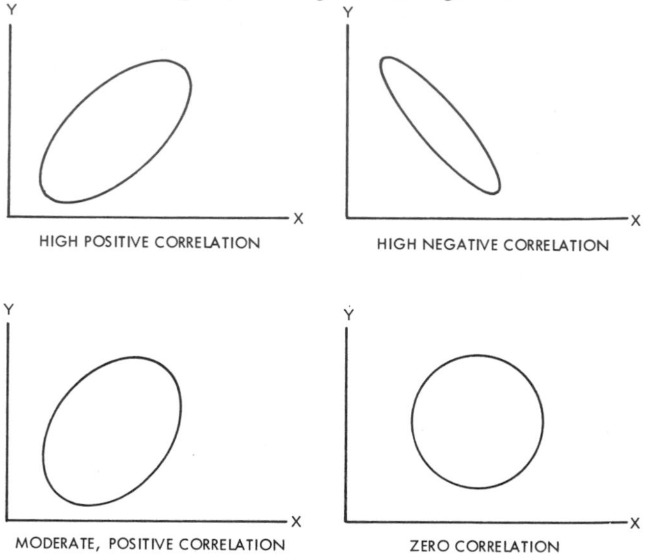

HIGH POSITIVE CORRELATION   HIGH NEGATIVE CORRELATION

MODERATE, POSITIVE CORRELATION   ZERO CORRELATION

know that John Jones scored 52 in the Scholastic Aptitude Test, we can make a better estimate of his score on the English test than the mean of all scores in English. Notice that the range of scores in English for all people who score 52 in scholastic aptitude is from 60 to 130. By inspection of the scatter diagram, it is apparent that the best estimate of Jones's score in English would be approximately 95. A better estimate would be the actual means of all scores in the particular array (column of the scatter diagram). The best estimate of the score on the English Placement Test, not only for John Jones but for anyone who scored 52 on the Scholastic Aptitude Test, would be the mean of the English scores in this particular array.

If we compute the mean for each array in X and plot these points (see figure 8–3), we notice that these dots (means of arrays) tend to fall in a line running from the lower left corner of the diagram to the upper right corner. For any given score group in scholastic aptitude, the best estimate of the English placement score will be the mean of English placement scores in that particular scholastic aptitude array. Instead of remembering both scores for each individual, knowledge of the mean of each array in X will permit a prediction for the Y scores with a minimum of error. But if the number of categories in X becomes large, it may be difficult to remember the mean for each of the different arrays.

FIGURE 8–3
Mean Y score (English Placement Test) for each array
in X (Scholastic Aptitude Test)

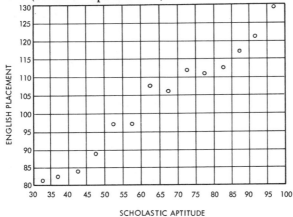

REGRESSION EQUATION. Suppose that we could draw a straight line through the series of means in such a way that the straight line would be the line of best fit to this particular set of data. (See figure 8–4.) We know from analytical geometry that the equation for any straight line may be written as $Y' = A + BX$, where Y' is a predicted score on the Y variable, X is any score on the X variable, A is called the *intercept point* (the point at which the line crosses the Y-axis), and B is a coefficient which

FIGURE 8–4
Straight line fitted to the mean Y score (English
Placement Test) for each array in X (Scholastic
Aptitude Test)

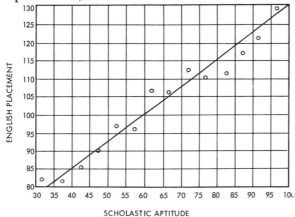

tells us the slope (steepness) of the line. If the equation for the straight line that best fits the data in the scatter diagram could be determined, it would no longer be necessary to remember the mean of each array; rather, only the equation for the straight line would be necessary. We could take any value of X, place it into the formula along with the slope and intercept point, and solve for Y'. This would yield a value that would be very close to the mean score in Y for that particular value of X. This equation for a straight line will allow us to predict not only the best estimate on the English Placement Test for Scholastic Aptitude Test scores of 52, but also for any value of X, such as 40.5, etc. It is beyond the scope of this course to show the precise relationship between the Pearson product-moment correlation coefficient and the slope of the straight line in a scatter diagram. Let us rather present the equation for this line (known as a *regression line*):

$$Y' = r_{xy}\frac{S_y}{S_x}X + \left(\bar{Y} - r_{xy}\frac{S_y}{S_x}\bar{X}\right) \tag{8-1}$$

where

Y' is the predicted score in the Y variable
r_{xy} is the Pearson product-moment correlation coefficient
S_y is the standard deviation of the Y variable
S_x is the standard deviation of the X variable
$\bar{Y}$ is the mean of the Y variable
$\bar{X}$ is the mean of the X variable

If we compare this equation to the equation for a straight line ($Y' = A + BX$), we note that the first term

$$\left(r_{xy}\frac{S_y}{S_x}\right)$$

is equivalent to the B in the formula for the straight line. This expression will yield the slope of the regression line for the original scores. The large, rather forbidding-looking expression on the right of the regression equation

$$\left(\bar{Y} - r_{xy}\frac{S_y}{S_x}\bar{X}\right)$$

is the intercept point—equivalent to A in the equation for a straight line. Thus the mean of the cross products of Z scores (r) is not only an excellent index of agreement which can be compared from one set of data to another; it is, in addition, related to the straight line fitted to the data in the scatter diagram. This means that once the value of *r* has been determined for a set of data, it is possible to predict values in one variable from knowledge of the second.

At this point the student may well ask: "If we already have scores on both tests for all the students, why not look at the student's English Placement Test score? That will certainly be more accurate than predicting his score from a regression line." Certainly this is correct. The most accurate estimate of the student's score on the English Placement Test would be his actual score on the test. Suppose we determine the correlation between some type of test score and success as a drill-press operator in a factory. If the original sample on which we computed the correlation coefficient is representative of the entire population, we can administer the test to people who apply for positions as drill-press operators and then hire only those for whom we would predict high success as drill-press operators. An inexpensive test, if it correlates highly with an expensive measurement or with a measurement that is difficult to obtain, may be used to predict the score on the expensive or difficult task much more economically. The prediction will not be as accurate as measuring the task in which we are interested, but it may be much less expensive. If the correlation were ± 1, prediction would be as accurate as measuring the task itself.

Without the knowledge of correlation, the best estimate of an individual's score is the mean score on that variable. Returning to John Jones, who scored 52 on the Scholastic Aptitude Test, we can improve our estimate of his score on the English Placement Test by using the regression equation. In a latter section of this chapter the *r* between SA and EP has been computed. The following information is taken from that section:

$$r_{xy} = .85; \; \bar{X} = 70.2; \; \bar{Y} = 101.2; \; S_x = 28.65; \; S_y = 25.43$$

Utilizing these numbers and equation 8–1 for the regression line:

$$Y' = r_{xy} \frac{S_y}{S_x} X + \left(\bar{Y} - r_{xy} \frac{S_y}{S_x} \bar{X} \right)$$

$$Y' = (.85) \left(\frac{25.43}{28.65} \right) (52) + \left[101.2 - (.85) \left(\frac{25.43}{28.65} \right) (70.2) \right] = 87.5$$

Inspection of Figure 8–1 shows clearly that the prediction of an English placement score of 87.5 for John Jones is a much better estimate of John's score than 101.2—the mean of the EP scores.

ACCURACY OF PREDICTION

The predicted Y value from a regression equation will be close to the mean value of Y associated with the particular X value, but notice that for any given individual we are going to be in error on the prediction. If we look at the scatterplot (figure 8–1) and regression line (figure 8–4), we immediately notice that very few scores in an array actually fall on the regression line. Some scores are above, some are below. If we could determine the standard deviation of all of the scores within each array, and then if these scores were normally distributed, we would be able to say that within ± 1 standard deviation of the scores in the array we would expect to find 68 percent of all the individuals with a given X score. Thus, if we find for all individuals who score 67 in SA that the standard deviation in EP scores within this array is 13.5, then we could say that between the limits of 91.5 and 118.5 (± 13.5), we would expect to find 68 percent of all the EP scores in this X array. The regression equation permits us to make a best estimate of what any one person's score may be; then, by using the standard deviation within an array, we can also specify the limits within which we expect the Y scores to cluster about the predicted point.

Fortunately, we do not have to compute the standard deviation for each of the arrays. We may use a term called the *standard error of estimate,* the formula for which is

$$S_{y.x} = S_y \sqrt{1 - r_{xy}^2} \qquad (8\text{-}2)$$

where

$S_{y.x}$ is the standard error of estimate
S_y is the standard deviation of the Y variable
r_{xy} is the correlation between the X and the Y variables

This statistic tells us the amount of variability in scores about the regression line. It is a standard deviation, but it is the standard deviation of the Y scores about the predicted point (regression line), and indicates the "error" in our predictions.

To illustrate the use of the standard error of estimate ($S_{y.x}$), we shall again call on John Jones, boy freshman. Inserting $S_y = 25.43$ and $r_{xy} = .85$ into the equation, we get

$$S_{y.x} = 25.43 \sqrt{1 - (.85)^2} = (25.43) \sqrt{1 - (.7225)} = (25.43) \sqrt{.225}$$
$$= (25.43)(.53) = 13.48$$

With this set of data, we can now state that within ±13.5 points of the predicted score we shall expect to find 68 percent of all the scores on the Y variable. If we take John's score of 52 on SA, what can we predict about his score on EP? From the equation for the regression line, we earlier determined that $Y' = 87.5$ for $X = 52$. The predicted point on EP is 87.5, but we can further state that the chances are two out of three (68 percent) that his score will be between 74 and 101 (87.5 ± 13.5). With the regression line, we can predict a better "best" estimate of an individual's EP score; with the use of the standard error of estimate, we can specify the limits of our prediction, as well as how much confidence we have in the prediction.

TESTING THE SIGNIFICANCE OF AN r

Since the researcher is seldom interested only in the sample upon which a given statistic is computed, some technique is necessary to allow inferences about the population from which the sample was drawn. If we were to draw a sample from a bivariate population (two variables—X and Y) and compute a correlation coefficient, and then draw a second sample from this same bivariate population and compute a second correlation coefficient, we would not expect both coefficients to be of exactly the same size. We would expect them to be of similar value and to be close to the value of the correlation coefficient which could be obtained by measuring everyone in the population (the true correlation). If it were possible to estimate the amount of variation to be expected among correlation coefficients obtained from different samples from the same population, it might be possible to construct a confidence interval within which we would expect to find the true correlation coefficient. In short, we need a statistic like the standard error of the mean, but to be used with correlation rather than means. Fortunately, the standard error of a Pearson product-moment correlation coefficient can be estimated by the following formula:

$$S_r = \frac{1}{\sqrt{N-1}} \qquad (8\text{--}3)$$

where

S_r is the standard error of a correlation coefficient
N is the number of subjects (pairs of scores in X and Y)

If sample r's were normally distributed, we could use the table of the normal curve to determine the various confidence intervals quite easily (95 percent, 99 percent, etc.). Unfortunately, as r gets larger and larger (approaches ± 1.00), the sampling distribution of sample r's becomes more and more skewed, and a straightforward approach to setting confidence intervals can be very misleading. It is possible to transform the r's and establish confidence intervals, and the techniques for this may be found in most intermediate or advanced textbooks on statistics. In many cases the researcher is more interested in the question whether this sample correlation indicates that there is in fact a real relationship between the two variables. Or stated a different way, is it likely that he could have obtained this large a correlation coefficient from the sample when the true correlation is zero ($\rho = 0$)? The sampling distribution about a true correlation of zero is normally distributed (if N is larger than about 30); thus, we can establish a confidence interval about a hypothesized true correlation of zero and determine if the sample value is within this interval. This technique is illustrated in example 8–1.

Example 8–1   Determination of a real relationship between two variables

$$r = .85; \quad N = 82$$

$$S_r = \frac{1}{\sqrt{N-1}} = \frac{1}{\sqrt{82-1}} = \frac{1}{\sqrt{81}} = \frac{1}{9} = .11$$

The 95 percent confidence interval about the true correlation of 0.00 would be:

$$\pm 1.96(S_r) = (1.96)(.11) = .2156 = .22$$
$$\rho \pm (1.96)(S_r) = 0 \pm .22 = -.22 \text{ to } +.22$$

We would expect 95 percent of all sample correlations to fall in the interval $-.22$ to $+.22$ if the true correlation were zero. Since our sample value is $r = .85$ and is outside this interval, it is extremely unlikely that the true correlation is zero.

We could turn the question around somewhat and ask: If the true correlation were zero ($\rho = 0$), what is the probability of a sample correlation this large or larger occurring by chance? If we think back to chapter 6 on the normal curve, this is one of the kinds of questions we were able to answer about test scores—assuming that the distribution is normal. Since the sampling distribution of r's about a true correlation of zero is normally distributed, perhaps we could set up a Z score formula similar to those used in chapter 6 but involving correlations rather than ordinary scores. The ordinary Z score formula is written as:

$$Z = \frac{X - \bar{X}}{S_x}$$

An analogous formula involving correlations would be:

$$Z = \frac{r - \bar{r}}{S_r}$$

This equation can be used as long as we use $\bar{r}$ to be a hypothesized true correlation equal to 0.00. The use of this technique is shown in example 8–2.

Example 8–2   Use of the Z score technique to determine the probability of a sample r occurring by chance if the true correlation is equal to zero ($\rho = 0.00$)

$$r = .85; \quad N = 82; \quad \rho = 0.00$$

$$Z = \frac{r - \rho}{S_r}$$

$$S_r = \frac{1}{\sqrt{N-1}} = .11$$

$$Z = \frac{.85 - 0.00}{.11} = \frac{.85}{.11} = 7.73$$

Using table B in the Appendix, a Z score of 7.73 is so large a deviation that it is not even recorded in the table. The probability of obtaining a Z score of 5.00 or larger (largest entry in table B) is $50 - 49.99997 = .00003$ percent, or $p = .0000003$. To obtain an $r = .85$ based on a sample of 82 subjects when $\rho = 0.00$ is an extremely unlikely event—perhaps equivalent to the sun rising in the West tomorrow morning. We would thus *reject* the hypothesis that $\rho = 0.00$.

Equation 8–3 shows that the size of the standard error of a Pearson product-moment correlation coefficient (S_r) is dependent upon the size of the sample (N). Because of this, it is simple to develop a table of different critical values of r to indicate the probability of the occurrence of these sample values if the true correlation is zero ($\rho = 0.0$). Table D in the Appendix shows these critical values of r. The first column in table D is labeled *degrees of freedom (df)*.

A full explanation of this concept must be postponed to chapter 10. However, for use with table D the degrees of freedom are equal to the total number of pairs of scores (N) minus two. Thus, $df = N - 2$. The column labeled *.05* indicates the critical Z score value such that the probability of the occurrence of a sample r of that size by chance, if the true correlation is zero, is .05. The same applies to the column labeled *.01*.

Suppose that we obtained a sample of $N = 23$ individuals and computed an $r = .492$ between two variables. What is the probability that this sample value could have been obtained if the true correlation in the population from which this sample was drawn were zero? Since $N = 23$, the *df* will be equal to 21 ($df = N - 2 = 23 - 2 = 21$). Entering table D in the column headed *degrees of freedom,* we find for $df = 21$ the entry .413 for the .05 level and .526 for the .01 level. This tells us that the probability of obtaining an $r = .492$ by chance—if the true correlation is zero—is less than .05 but greater than .01. We would probably conclude that the true correlation is not zero, but some value larger than zero. The two variables are related to each other.

Computation of the Pearson product-moment correlation coefficient

Z SCORE METHOD OF COMPUTING r

Since the Pearson product-moment correlation coefficient is simply the mean of the cross products of Z scores of the X and Y variables, a straightforward way to compute the correlation coefficient would be to compute $\bar{X}$, S_x, $\bar{Y}$, and S_y. Each score in X and each score in Y can be converted to a Z score. Next, multiply the two Z scores for each individual together, taking care to attach the proper sign to the product of Z scores; then, sum the cross products of Z scores for all individuals, and divide by the number of people. This will be the mean of the cross products of Z scores and, therefore, r. This method of computing r is shown in Example 8–3.

DEVIATION SCORE METHOD OF COMPUTING r

While the Z score technique can be used, it has distinct disadvantages. The first disadvantage is that we must subtract each score from the mean and then divide by the standard deviation. Aside from the labor involved, there is the possibility of introducing both arithmetic and rounding errors.

Example 8-3   Computation of r by the Z score method

Subjects	X	Y	x	y	Z_x	Z_y	Z_xZ_y
1	40	62	−30.2	−39.2	−1.05	−1.54	1.6170
2	67	105	−3.2	3.8	−.11	.15	−.0165
3	57	97	−13.2	−4.2	−.46	−.17	.0782
4	64	102	−6.2	.8	−.22	.03	−.0066
5	111	106	40.8	4.8	1.42	.19	.2698
6	70	122	−.2	20.8	−.01	.82	−.0082
7	106	130	35.8	28.8	1.25	1.13	1.4125
8	127	152	56.8	50.8	1.98	2.00	3.9600
9	93	128	22.8	26.8	.80	1.05	.8400
10	51	81	−19.2	−20.2	−.67	−.79	.5293
11	131	151	60.8	49.8	2.12	1.96	4.1552
12	66	76	−4.2	−25.2	−.15	−.99	.1485
13	51	78	−19.2	−23.2	−.67	−.91	.6097
14	33	74	−37.2	−27.2	−1.30	−1.07	1.3910
15	67	127	−3.2	25.8	−.11	1.01	−.1111
16	25	71	−45.2	−30.2	−1.58	−1.19	1.8802
17	77	87	6.8	−14.2	.24	−.56	−.1344
18	59	84	−11.2	−17.2	−.39	−.68	−.2652
19	59	91	−11.2	−10.2	−.39	−.40	.1560
20	50	99	−20.2	−2.2	−.70	−.09	.0630

$$\Sigma X = 1,404 \qquad \Sigma Y = 2,023 \qquad \Sigma Z_xZ_y = 17.0988$$

$$N = 20 \quad \begin{array}{l} \Sigma X^2 = 114,982 \quad \Sigma Y^2 = 217,765 \\ \bar{X} = 70.2 \qquad \bar{Y} = 101.2 \\ S_x = 28.65 \qquad S_y = 25.43 \end{array} \quad \begin{array}{l} r = \dfrac{\Sigma Z_xZ_y}{N} = \dfrac{17.0988}{20} \\ \\ r = .85 \end{array}$$

Through a bit of algebraic manipulation, it is possible to convert $\dfrac{\Sigma Z_xZ_y}{N}$ to a different formula:

$$r = \frac{\Sigma xy}{NS_xS_y} \tag{8-4}$$

This is known as the deviation-score formula for the Pearson product-moment r and is exactly equivalent to the mean of the cross products of Z scores.

To compute r using this latter formula, it is necessary to compute each deviation score in X (x) and similarly compute each deviation score in Y (y). Then, multiply the two deviation scores for each individual and add all of these together. We do not have to divide each of the deviation scores by the appropriate S. Thus, we save quite a bit of labor in using the deviation-score formula. This approach is shown in example 8–4.

Example 8–4   Computation of r by the deviation-score method

Subject	X	Y	x	y	xy
1	40	62	−30.2	−39.2	1,183.84
2	67	105	−3.2	3.8	−12.16
3	57	97	−13.2	−4.2	55.44
4	64	102	−6.2	.8	−4.96
5	111	106	40.8	4.8	195.84
6	70	122	−.2	20.8	−4.16
7	106	130	35.8	28.8	1,031.04
8	127	152	56.8	50.8	2,885.44
9	93	128	22.8	26.8	611.04
10	51	81	−19.2	−20.2	387.84
11	131	151	60.8	49.8	3,027.84
12	66	76	−4.2	−25.2	105.84
13	51	78	−19.2	−23.2	445.44
14	33	74	−37.2	−27.2	1,011.84
15	67	127	−3.2	25.8	−82.56
16	25	71	−45.2	−30.2	1,365.04
17	77	87	6.8	−14.2	−96.56
18	59	84	−11.2	−17.2	192.64
19	59	91	−11.2	−10.2	114.24
20	50	99	−20.2	−2.2	44.44

$$\Sigma X = 1{,}404 \quad \Sigma Y = 2{,}023 \quad \Sigma xy = 12{,}456.40$$

$$N = 20 \quad \begin{array}{l} \Sigma X^2 = 114{,}982 \quad \Sigma Y^2 = 217{,}765 \\ \bar{X} = 70.2 \qquad \bar{Y} = 101.2 \\ S_x = 28.65 \qquad S_y = 25.43 \end{array} \qquad r = \frac{\Sigma xy}{N S_x S_y} = \frac{12{,}456.40}{(20)(28.65)(25.43)} = .85$$

WHOLE-SCORE METHOD OF COMPUTING r

Through further algebraic manipulation, we can express the formula for r in terms of the whole scores:

$$r = \frac{N\Sigma XY - \Sigma X \Sigma Y}{\sqrt{N\Sigma X^2 - (\Sigma X)^2} \; \sqrt{N\Sigma Y^2 - (\Sigma Y)^2}} \qquad (8\text{–}5)$$

This permits the use of the whole scores in X and in Y directly in place of the deviation scores (x, y) or the Z scores (Z_x, Z_y); thus, r can be computed much more readily and will give the same numerical results as either of the two previous equations. Remember, in using this formula, that the same kind of procedure must be followed as with the two previous formulas. The whole score for any one individual in X must be multiplied by the whole score in Y for that individual; then sum all the XY products. This computing technique is illustrated in example 8–5.

Many other computational formulas for r are available. Most advanced textbooks will give other whole-score formulas which yield the same results as these particular formulas, but which may be more suitable for

Example 8-5   Computation of *r* by the whole-score method

Subject	X	Y	XY
1	40	62	2,480
2	67	105	7,035
3	57	97	5,529
4	64	102	6,528
5	111	106	11,766
6	70	122	8,540
7	106	130	13,780
8	127	152	19,304
9	93	128	11,904
10	51	81	4,131
11	131	151	19,781
12	66	76	5,016
13	51	78	3,978
14	33	74	2,442
15	67	127	8,509
16	25	71	1,775
17	77	87	6,699
18	59	84	4,956
19	59	91	5,369
20	50	99	4,950

$$\Sigma X = 1,404 \quad \Sigma Y = 2,023 \quad \Sigma XY = 154,472$$

$$N = 20 \quad \begin{array}{l} \Sigma X^2 = 114,982 \quad \Sigma Y^2 = 217,765 \\ \bar{X} = 70.2 \qquad \bar{Y} = 101.2 \\ S_x = 28.65 \qquad S_y = 25.43 \end{array} \quad r = \frac{N\Sigma XY - \Sigma X \Sigma Y}{\sqrt{N\Sigma X^2 - (\Sigma X)^2} \sqrt{N\Sigma Y^2 - (\Sigma Y)^2}}$$

$$r = \frac{(20)(154,472) - (1,404)(2,023)}{\sqrt{(20)(114,982) - (1,404)^2} \sqrt{(20)(217,765) - (2,023)^2}} = .85$$

certain types of data. The coding systems presented in chapters 4 and 5 for use in computing the mean and standard deviation can be utilized for the Pearson product-moment correlation coefficient. The wide availability of desk calculators and computers has made this approach much less useful than in past years. The student who is interested in these coding systems will find them in most statistics textbooks written prior to the middle of the 1960s.

ASSUMPTIONS INVOLVED IN THE INTERPRETATION OF *r*

The coefficient of correlation is one of the most complex of our statistics—complex in the sense that the assumptions involved in the use of *r* vary a great deal, depending upon the specific use of *r*. There are no special assumptions involved in utilizing *r* simply as the mean of the cross products of Z scores. In this case, *r* is simply an index with the limits of +1 or −1. However, in most work in the behavioral sciences, we wish to go further and use *r* for *predictions*. When we use *r* in this way, several assumptions are involved.

For most research, we must assume *linearity of regression*. This requires that the straight line fitted to the mean of our arrays is a reasonably good description of the best kind of line to fit. That is, the relationship can be described quite well as being linear, not curvilinear.

Although the best techniques for determining if the relationship between two variables is linear or curvilinear is beyond the scope of this course, we do have a straightforward technique available which will work if the relationship is decidedly curvilinear. We may (and should) construct a scatter diagram of the scores. An examination of this scatterplot will generally indicate whether the assumption of linearity is tenable. Figure 8–5 shows the outline of four scatterplots. One of the scatterplots indicates a linear relationship between the *X* and *Y* variables. The other three scatterplots show three different curvilinear relationships between the *X* and *Y* variables.

Another assumption is that the population standard deviation within each array of the scatterplot is equal to the population standard deviation within every other array. This assumption is called *homoscedasticity*.

FIGURE 8–5
Outline of scatterplots, showing one linear relationship and three different curvilinear relationships

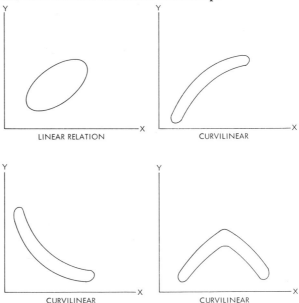

LINEAR RELATION   CURVILINEAR

CURVILINEAR   CURVILINEAR

If we use the standard error of estimate as a measure of the accuracy of predictions so that statements such as, "Within ±1.96 standard errors of estimate will be found 95 percent of all scores in that array," we have assumed that the distribution of scores within the array is normal. Thus the assumption of *normality within arrays* is necessary if we are to use the standard error of estimate to set limits on the accuracy of the predictions.

Indices of association with noninterval data

This entire chapter to this point has discussed the general question of correlation and, more specifically, the computation of the most widely used index, the Pearson product-moment correlation coefficient. The Pearson r is appropriate only when the data of both variables are reasonably close to an interval scale.

Suppose that at least one of the underlying categories of data fall into a nominal or ordinal scale. Or suppose that both variables are noninterval in nature. Is it still possible to have some type of index of association which would be applicable? Certainly the Pearson r cannot be used, since it is dependent upon interval-type data in both the X and Y variables. There are many techniques for computing an index of association between two variables when the underlying scale characteristics are nominal, ordinal, or a mixture of the various scales. Table 8–2 presents some of the different techniques that are available. It is not possible in an introductory textbook to discuss all the different indices; thus table 8–2 is intended as an aid for the student in using other references. Only three techniques presented in the table will be discussed: the Pearson product-moment correlation coefficient which has already been presented and is suitable when both variables form an interval or ratio scale; the Spearman rank-difference correlation coefficient which is useful when both variables form ordinal scales; and finally the contingency coefficient which can be used when both variables are in the form of nominal scales.

It should be emphasized that scales that are nominal or ordinal in nature do not have the amount of information that is contained in the numbers assigned to interval or ratio scales. It should be no surprise that correlation coefficients based on the association between noninterval scaled data will be less useful than the Pearson product-moment correlation coefficient. The Pearson r can be used in making predictions from one set of data to the other. It can also be used in setting limits on

TABLE 8–2
Some indices of association between variables organized by the scale characteristics

	Nominal	Ordinal	Interval and ratio
Nominal	Guttman's coefficient of predictability (λ) Contingency coefficient (C) Fourfold point correlation (ϕ)	Coefficient of differentiation (θ)	Correlation ratio (η)
Ordinal		Spearman rank-difference correlation (r_s) Kendall rank correlation (τ) Kendall coefficient of concordance (W)	Biserial correlation (r_b) Point biserial correlation (r_{pb}) Multiserial correlation (r_{ms})
Interval and ratio			Pearson product-moment correlation (r) Correlation ratio (η)

the accuracy of the predictions. The various coefficients presented in table 8–2 *cannot* be used in prediction equations. Their usefulness is limited to an index of association. A high number indicates a high degree of relationship, a low number a low degree of relationship.

SPEARMAN RANK-DIFFERENCE CORRELATION COEFFICIENT

There are several techniques for computing the correlation coefficient for ordinal-type data. The only one we shall be concerned with is called the *Spearman rank-difference correlation coefficient* and is symbolized by r_s.[2] The formula is as follows:

$$r_S = 1 - \frac{6\Sigma D^2}{N(N^2 - 1)} \tag{8–6}$$

---

[2] Older treatments of the Spearman rank-difference correlation coefficient often used the Greek letter rho (ρ) as the symbol for this coefficient. Since modern usage reserves Greek letters for population values, the symbol r_s is used for the Spearman rank-difference correlation coefficient, and ρ is used to represent the population value of the Pearson product-moment correlation coefficient. In reading older research articles, the student may find ρ used to indicate a Spearman coefficient. This possible source of confusion in symbols can be avoided by examining the procedures followed in computing the coefficent.

where

N   is the number of pairs of ranks
D   is the difference between a pair of ranks

Never compute D between scores, since with ordinal-type data the absolute numerical value does not indicate equal measurement units.

When assigning numbers to ordered data for computation of r_S, the highest score, by tradition, is assigned the rank of 1, with the lowest score assigned the rank of N. Occasionally, we find in ranking data, that two or more items are tied in their ordered position. When this occurs, the average of the ranks is assigned to the tied categories. Thus, if two items should both receive the ranks 8 and 9, we would assign the rank of 8.5 to each of the two items (the average of the ranks between 7 and 10). The computation of the rank-difference correlation is presented in example 8–6.

Example 8-6   Computation of Spearman rank-difference correlation coefficient

Subject	X	Rank in X	Y	Rank in Y	D	D^2
1	40	18	62	20	−2	4
2	67	8.5	105	8	.5	.25
3	57	14	97	11	3	9
4	64	11	102	9	2	4
5	111	3	106	7	−4	16
6	70	7	122	6	1	1
7	106	4	130	3	1	1
8	127	2	152	1	1	1
9	93	5	128	4	1	1
10	51	15.5	81	15	.5	.25
11	131	1	151	2	−1	1
12	66	10	76	17	−7	49
13	51	15.5	78	16	−.5	.25
14	33	19	74	18	1	1
15	67	8.5	127	5	3.5	12.25
16	25	20	71	19	1	1
17	77	6	87	13	−7	49
18	59	12.5	84	14	−1.5	2.25
19	59	12.5	91	12	.5	.25
20	50	17	99	10	7	49

$$N = 20; \quad \Sigma D^2 = 202.5$$

$$r_S = 1 - \frac{6\Sigma D^2}{N(N^2 - 1)} = 1 - \frac{(6)(202.5)}{20(400 - 1)} = 1 - \frac{1,215.0}{7,980} = 1 - .1523 = .8477 = .85$$

There are some very difinite limitations involved in the use of r_s The numerical size of r_s will indicate the degree of relationship between the *ranks* assigned to data, not the degree of relationship between the original scores. The r_s is a reasonable approximation to the value of r if the differences in ranks are approximately proportional to the differ-ences in the underlying scale values. The r_s cannot be used in the regres-sion equation, for the regression equation depends upon interval-type data. In view of the above, it should be apparent that r_s cannot be substi-tuted for r in the equation for the standard error of estimate. The r_s, then, is very limited in its usefulness. It can be used as an index of the degree of relationship of the ranks assigned to two variables, but it should not be used in predicting specific values of one variable or in setting limits to the predictions.

TESTING THE SIGNIFICANCE OF A RANK-DIFFERENCE CORRELATION COEFFICIENT.   Since the Spearman rank-difference correlation coefficient is based on ranked data, no standard-error term is available for establish-ing confidence intervals or for using a Z score approach to testing for the significance of a specific r_s. It is possible to determine the probability of an r_s occurring by chance if the true correlation is zero. The size of a rank-difference correlation needed for significance at the .05 and .01 levels of significance for different sample sizes is presented in table E of the Appendix.

In example 8–6, the rank-difference correlation coefficient for $N = 20$ was found to be .85. By using table E of the Appendix for $N = 20$, we find that an r_s of .450 is needed for significance at the .05 level and an r_s of .591 is significant at the .01 level of significance. Since the obtained value in example 8–6 was .85, we would reject the hypothesis that the true correlation between the two variables is zero. We would accept the hypothesis that the true correlation must be some value greater than zero.

CONTINGENCY COEFFICIENT

Often we have only categorical information concerning two variables in which we may be interested. In chapter 11, some hypothetical data are presented related to differences in library usage of biology majors and English majors. This information is presented in table 8–3.
According to this illustration, there are 105 biology majors and 101 English majors at the particular institution. During one 24-hour period, the number of biology majors and the number of English majors who used the library were counted. The resulting data were arranged in the form of a 2 $\times$ 2 table often called a contingency table. Is there a relation-

TABLE 8–3
Comparison of library usage of biology majors and English majors

MAJOR FIELD

	Biology	English	
Used library	47	62	109
Did not use library	58	39	97
	105	101	206

ship between major area and library usage? The contingency coefficient uses a statistic known as chi-square (χ^2) and is computed by:

$$C = \sqrt{\frac{\chi^2}{N + \chi^2}} \qquad (8\text{–}7)$$

where:

χ^2 is $\sum \dfrac{(O - E)^2}{E}$

N is the total number of individuals

An explanation of chi-square must be postponed until chapter 11, however a discussion of the contingency coefficient can be made without full knowledge of the chi-square statistic. In chapter 11 the χ^2 has been computed for these data and was found to be 5.12. Substituting this value in formula 8–7:

$$C = \sqrt{\frac{5.12}{206 + 5.12}} = \sqrt{\frac{5.12}{211.12}} = \sqrt{.0244} = .156$$

Unfortunately the magnitude of the contingency coefficient cannot be interpreted in the same way as the Pearson r or the Spearman r_s. The coefficient cannot be negative and the maximum size of the contingency coefficient is a function of the number of categories in the contingency table. Thus, for table 8–3, the maximum value the coefficient could take would be $\sqrt{\frac{1}{2}} = .707$. The maximum value of the contingency coefficient can only be computed for square tables and the maximum value will be:

$$C_{\max} = \sqrt{\frac{k-1}{k}} \qquad (8\text{--}8)$$

where:

k is the number of categories on one dimension of the table.

Unfortunately, the maximum value for nonsquare tables such as 2×3, 2×4, and 3×4 is not known. This makes the interpretation of the value of the contingency coefficient difficult.

If the data are such that a Pearson r could be computed, but the data are placed in categories and the contingency coefficient is computed, then for square tables, Stone and Skurdal[3] have shown that the ratio of the contingency coefficient to maximal contingency coefficient $(C/C_{\max})$ is close in value to the r computed from the same data. The reason for this relationship is not known, but the relationship does hold for a large number of situations. In the above example, the ratio .156/.707 is .22; thus, using the Stone and Skurdal work, we could interpret this coefficient as indicating a strength of association similar to a Pearson r of .22.

TESTING THE SIGNIFICANCE OF THE CONTINGENCY COEFFICIENT.   The chi-square statistic upon which the contingency coefficient is based is used to determine if a relationship greater than zero association does exist. The procedures for testing the significance of χ^2 are presented in chapter 11 along with the discussion of the statistic. In this particular situation, $\chi^2 = 5.76$ is statistically significant, and we would conclude that there is a relationship between library usage and the student's major of either English or biology.

Summary

The relationship between changes in two variables can be described in gross terms such as a high positive correlation, a high negative correlation, a moderate positive correlation, and no correlation. The mean of the cross products of Z scores is a precise index which yields a quantitative value indicating the degree of relationship between two variables. This

---

[3] LeRoy A. Stone and Marlo A. Skurdal, "Estimation of product-moment correlation coefficients through the use of the ratio of contingency coefficient to the maximal contingency coefficient," *Multiple Linear Regression Viewpoints*, vol. 1, no. 2 (October 1970), pp. 19–25.

index is *independent of the original measurement units*. The *sign* of the index will indicate whether the relationship is *negative* or *positive*. The *size* of the index is *independent* of the *number of observations*, and the *size* of the index will *indicate the strength of the relationship*. The mean of the cross products of Z scores is the Pearson product-moment correlation coefficient, and the value of this correlation coefficient has the range -1 to $+1$.

Not only is the Pearson product-moment correlation coefficient an excellent index of the degree of relationship between two variables; it is also related to the straight line of best fit between the two variables. Knowledge of the correlation between two variables allows us to write a *regression equation* to predict values in the one variable from our knowledge of the second variable. This regression line is the straight line of best fit for the data.

Without knowledge of the correlation between two variables, the best prediction that can be made for individuals is that each one will score at the mean of the group. With the regression equation, this prediction can be improved. However, the predictions will be to the regression line, and some predictions will be in error. The degree of error in the prediction can be ascertained by use of the standard error of estimate. The standard error of estimate is the standard deviation of the scores about the regression line. If the scores about the regression line are normally distributed, the standard error of estimate can be used to set a confidence interval about the score predicted from the regression equation.

The sampling distribution of the Pearson product-moment correlation coefficient is badly skewed for high values; thus, special techniques are necessary to establish confidence intervals about an obtained r. These techniques may be found in advanced texts. It is possible to establish a confidence interval about the hypothesized true value of zero by using the *standard error of r*. This same standard-error term can be used to establish a Z score and determine the probability of occurrence of a sample r if the population correlation is zero. Table D of the Appendix can also be used for testing the significance of a sample r.

The Z score approach in computing r is quite laborious and introduces increased opportunity for computational error. The whole-score method of computing r is generally favored especially if a desk calculator is available.

Although only the assumption of interval or ratio scaling is necessary to interpret the mean of the cross products of Z scores as an index of covariation, most uses of the Pearson *r* involve prediction, and this use requires that other more stringent assumptions be met. If the Pearson *r* is to be used in prediction, the relationship between the X and Y variables must be linear. To use the standard error of estimate requires that the assumption of homoscedasticity (equal population standard deviations in the arrays) be met. Finally, to establish confidence intervals about the scores predicted from the regression equation, the scores in each array must be normally distributed.

A large number of correlational techniques have been developed for special situations. The most widely used of these were organized into a table on the basis of the scalar characteristics of the variables being correlated. More detailed information concerning these special techniques can be found in the references given at the end of this chapter.

The *Spearman rank-difference correlation coefficient* can be used for ordinal data. This coefficient is computed on the *ranks* assigned to the observations. The rank-difference correlation coefficient is simply an index of correlation between the sets of ranks assigned to the two variables and should not be used in the regression equation. Table E of the Appendix provides critical values of r_s to test whether the sample correlation is significantly different from a true correlation of zero.

The *contingency coefficient* can be used with nominal data. This coefficient is based on the *chi-square* statistic. Unfortunately this coefficient does not have the same numerical meaning as the Pearson *r* or the Spearman r_s. The coefficient must always have a positive value, and the maximum size of the coefficient, in square tables, is a function of the number of categories. The maximum value for nonsquare tables is not known. Under special circumstances, the ratio of the contingency coefficient to the maximal contingency coefficient can be interpreted as showing the same strength of association as the Pearson *r*. To test if any association exists between the two sets of categories, the value of the χ^2 is tested for statistical significance.

References

CORRELATION AND PREDICTION

MORDECAI EZEKIEL and KARL A. FOX, *Methods of Correlation and Regression* (3d ed.; New York: John Wiley & Sons, Inc., 1959). An excellent basic text in this area.

SPECIAL CORRELATION TECHNIQUES

LINTON C. FREEMAN, *Elementary Applied Statistics* (New York: John Wiley & Sons, Inc., 1965), Chapters 7–12.

WILLIAM L. JEKINS, "Triserial *r*—A Neglected Statistic," *Journal of Applied Psychology,* Vol 40 (1956), p. 1.

QUINN McNEMAR, *Psychological Statistics* (4th ed.; New York: John Wiley & Sons, Inc., 1969), chap. xii.

SIDNEY SIEGEL, *Nonparametric Statistics for the Behavioral Sciences* (New York: McGraw-Hill Book Co., Inc., 1956), chap. ix. Contains nonparametric measures of correlation, including Kendall's coefficient of concordance.

MULTIPLE CORRELATION AND PREDICTION

QUINN McNEMAR, *Psychological Statistics* (4th ed.; New York: John Wiley & Sons, Inc., 1969), chap. xi.

9

Use of statistics in tests and measurements

As has been pointed out previously, a number by itself has little if any meaning. A score of 50 on a history examination could mean an *A*, an *F*, or any other grade, depending upon a great variety of factors. Thus, it is absolutely essential in reporting test scores to include other descriptive data and thereby give meaning to the numbers. If most data in the behavioral sciences dealt with ratio measurements, many problems of the relative meaning of scores would disappear. A score of 50 pounds has an absolute meaning, since we can specify an absolute zero point in weight, and weight is measured on an equal-interval scale. But unfortunately, many of the characteristics of human behavior are not measurable in this manner, What is zero knowledge of history? What is zero intelligence? Since the behavioral scientist cannot utilize ratio scales in many of his measurements, he must have additional information to give at least relative meaning to his observations.

Reporting test scores

The reporting of scores from tests involves the use of some one or more reference points. Intelligence test scores use the mean score of the standardization group as the major reference point. With most IQ tests, this mean score is arbitrarily given the value of 100. With this reference point, any "score" above 100 indicates that the individual is above the mean score on the test. This use of 100 for the mean score is utilized with several different types of tests.

A great variety of reference systems has been used in reporting test data. Some are used only in specialized situations, but two techniques and a variation of one of these are used most frequently.

REPORTING SCORES FOR NOMINAL DATA

If the measurements are only nominal in nature, little more can be done with the data than to report the number (or percentage) of cases in each category.

REPORTING SCORES FOR ORDINAL DATA

If the measurements meet the requirements for ordinal scaling, it is possible to make statements concerning the relative position of different scores. In chapter 2 the use of percentile scaling was discussed as a method of describing a distribution of scores. The use of the percentile scale is a special case of ranking or ordering the set of measurements. In this case the scores are ordered in terms of the *percentage* of the group who score at or below the particular value.

Once the percentile rank is determined, very little else can be done with the information. Although it is descriptive, it does not lend itself to further analysis. A danger present with percentiles is the tendency to attempt to compute averages from percentile ranks; and since the percentile ranks are a form of ordinal scale, such averages are not only meaningless but may convey incorrect information. An even greater limitation in the use of percentile ranks is inherent in any ordinal scale. The ranks cannot tell us how far the ranked objects are apart—only that one is greater than another. Thus, to say that an individual is at the 53d percentile on both a history and an English test does not tell us that he necessarily did equally well on both tests. It merely tells us that his relative rank in each test is the same—the same percentage of people scored at or below his score on each test.

REPORTING SCORES FOR INTERVAL OR RATIO DATA

When an equal-interval unit of measurement has been obtained, many different techniques can be used to report the results of measurements.

The most widely used procedure involves a variation of the use of the Z score.

THE Z SCORE

In chapter 5, it was pointed out that the Z score permits one to compare scores obtained from different measuring instruments in units which are independent of the original measuring units. Thus the Z score, in addition to its use with the normal curve, may be used as a simple, descriptive technique of reporting test results. This is a technique which permits us to compare an individual's score on one test with his score on a second test—in effect, to state that a given individual is heavier than he is tall.

The major limitations of the use of the Z score are threefold: (1) the test data should meet the requirements of an interval or ratio scale; (2) both plus and minus signs are used, which may lead to errors; and (3) the use of a decimal may also cause confusion and result in errors in reporting and interpreting test data. To overcome these latter two problems, many tests report normative data as "derived" scores.

THE DERIVED SCORE

Actually, there are many different types of derived scores. A discussion of these various derived scores more properly belongs in a course in tests and measurements. We shall discuss only one procedure—the procedure which results in the use of only positive whole numbers and which retains the advantages of Z score scaling. This form of derived scale utilizes two characteristics of numbers: (1) the multiplication of all numbers in a distribution by a constant does not change the relationship between the various numbers except to spread them out in a uniform way, and (2) adding a constant to all numbers in a distribution does not change the relationship between the numbers but merely shifts them all by the same amount.

The derived score is obtained by multiplying all Z scores in a distribution by a constant. This constant is selected so that no decimal need be used. The net effect is to change the standard deviation of the distribution from one (remember that the standard deviation of any set of Z scores is one) to whatever constant value is selected. The next step is to add a constant to all scores. The effect of this is to shift the mean of the distribution from zero (the mean of any set of Z scores is always zero) to the value of the constant that is added.

The advantages of the derived score can be illustrated by the Army

General Classification Test (AGCT) used during and after World War II. The constant of 20 was selected as the multiplier for all the Z scores. This had the effect of making the standard deviation of the distribution 20. Next, the constant of 100 was added to each score, making the mean of the distribution 100. To illustrate, a Z score of +1.6 would become a derived score of 132.

$$D = (20)(Z) + 100 \qquad\qquad D = (20)(1.6) + 100 = 32 + 100 = 132$$

This type of transformation of scores is especially useful with tests where the "test" consists of a battery of tests. If ten tests are combined into a single battery, it can become confusing for a person to attempt to remember the means and standard deviations of all the subtests. The conversion to Z scores makes comparisons much easier and converting to derived scores makes the process even less confusing. Thus, all tests have a mean of 100 and a standard deviation of 20. A score of 112 on two different tests could be interpreted in a very similar manner.

A similar procedure was used by the Navy during World War II; but instead of a standard deviation of 20 and a mean of 100, the Navy elected to use a derived SD of 10 and a mean of 50. Thus the distribution of scores for the various subtests in the Navy's battery could be set up by using the formula

$$D = (10)(Z) + 50$$

Any combination of constants can be used as long as they serve the purpose of establishing comparable score units from test to test (the advantage of the Z score scaling) and the absence of negative scores and decimals.

Figure 9–1 shows the relationship among the various methods of re-

FIGURE 9–1
Relationship in a normal distribution among area under the curve, original scores, percentile ranks, Z scores, and derived scores (M = 100, S = 20)

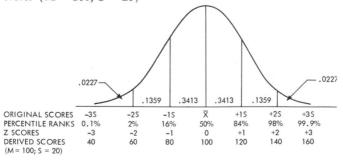

ORIGINAL SCORES	−3S	−2S	−1S	X̄	+1S	+2S	+3S
PERCENTILE RANKS	0.1%	2%	16%	50%	84%	98%	99.9%
Z SCORES	−3	−2	−1	0	+1	+2	+3
DERIVED SCORES (M = 100; S = 20)	40	60	80	100	120	140	160

porting test scores *when the underlying distribution of scores is normal.* Original scores, Z scores, and derived scores will maintain the same relationship with each other regardless of the shape of the underlying distribution. Putting this another way, the Z score- and derived-score transformations do not change the distribution in any way—the transformation *does not* turn a nonnormal distribution into a normal distribution. The area under different segments of the curve and the percent of area (percentile rank) are, of course, entirely dependent upon the shape of the underlying distribution.

The use of correlation in the evaluation of tests

The widest application of the techniques of correlation in the behavioral sciences has been in the area of tests and measurements. With most tests, there will be several correlation coefficients expressed which describe different aspects of the test.

RELIABILITY OF TESTS

The reliability of a test may be defined as the stability or consistency of measurement. Will the test yield the same score on an individual if two or more measurements are made? The test user should be reasonably sure that whenever a particular test is administered to an individual, the score for the individual will not change appreciably from one testing to a different testing. This, briefly, is what is meant by the *reliability of a test.*

Since reliability involves the idea of consistency, consistency itself would indicate that we need at least two observations on an individual in order to determine whether the measuring instrument is doing the same thing each time. That is, is it consistent (reliable)?

Whenever there are two measurements of each individual in a group, the correlation coefficient may be used to indicate the degree of relationship (agreement) between the two sets of measurements. Just as the Pearson product-moment r is an index of agreement between two variables, X and Y, so also the r can be used as an index of agreement between two measurements with the same test. Thus the coefficient of reliability is simply the correlation between two measurements of the same thing. There are several different methods of determining the reliability of a test. Three are used more often than others, and these three all utilize *r* as the measure of reliability.

TEST-RETEST RELIABILITY. In utilizing this technique, a test is administered to a group of individuals, and scores are obtained for each individual. Later the same test is administered a second time to the same group of people, and a second set of scores is obtained. The Pearson product-moment correlation coefficient (r_{xx}) between these two sets of scores would be the test-retest reliability. It indicates how consistently the test is measuring from one testing period to the second.

Since test-retest reliability requires the administration of the test at two different periods of time, the length of time between testings may become extremely important. A rather ridiculous example of this problem can be illustrated by a test-retest of children's height. Suppose we measure the height of a group of three-year-old children. Then we decide to determine the reliability of our test (the ruler), so we again measure the height of the same children one year later. The two measures of height will not agree—not because the ruler is not a reliable measuring instrument, but because the thing being measured has changed between the two measurements. The same type of problem is present with most tests.

A second difficulty with test-retest determination of the reliability of tests is that the first application of the test may make the test unsuitable for further use with the same subjects. Suppose the class repeated the same course examination. Certainly the students' answers would change. Some eager-beaver students might have looked up specific items of information, and most students would probably do better on problem and essay items. With many tests, there is an ever-present danger that *the act of measurement will change the thing being measured.*

The test-retest method for assessing the reliability of a test is most suitable when the characteristic being measured does not change in the time interval between testings and the act of measurement does not destroy the usefulness of the test with the same individuals.

COMPARABLE FORM RELIABILITY. Comparable form reliability is determined by administering one form of a test to a group of individuals and then administering a second form to the same group. The correlation is then computed between the two sets of scores obtained with these comparable measuring instruments.

Comparable form reliability overcomes one of the limitations of the test-retest technique. Any specific knowledge gained about items on the one test will not be of direct importance in determining the score on the comparable second test. However, comparable form reliability does not solve the problems of change in the thing being measured—change due to the activities in the time interval between administration of the comparable forms or change due to the first measurement. If the comparable forms are administered fairly closely in time, then the problem of change is usually minimized.

An extremely important consideration in this form of assessment of reliability resides in the definition of comparable forms. Constructing comparable forms involves equating the content and statistical characteristics of each item on one form of the test with each item on the other form. Because of the labor involved in equating specific items on the two forms, comparable form reliability is not used as often as it might be. A good example of this form of reliability is with Forms L and M of the 1937 revision of the Stanford-Binet Intelligence Test.

SPLIT-HALF RELIABILITY.   Both the test-retest and comparable form techniques for assessing the reliability of measurement involve a time lag between the first and second applications of the measuring instrument(s). As was pointed out in the preceding sections, the thing being measured may change during this passage of time. The split-half technique was devised to overcome this problem of time lag between successive measurements. The basic idea is to think of a test not as a single test, but as many different tests. One way to do this is to let all even-numbered items in the test be one form and all odd-numbered items be the second form. Under these circumstances, it would be possible to administer the two comparable forms of the test during one testing period. There would be no intervening time period between the administration of the two forms, and it would be possible to assume that any change brought about by the act of measurement would affect both forms of the test in the same manner.

The correlation between two halves of a single test is the basic statistic necessary for computing split-half reliability. Actually, any division of the test into two halves could be used; however, for most purposes, it is more satisfactory to consider even-numbered items as one half and odd-numbered items as the other half. The obtained correlation is between comparable forms, each of which is only one half as long as the full test. It can be shown that the magnitude of the reliability coefficient will change with the length of the test. Thus the r obtained with the two "half" tests will be smaller than an r obtained between the two tests each as long as the original full test. Traditionally, the obtained correlation between the two half tests is "corrected" for the length of the test. This correction is accomplished through the Spearman-Brown prophecy formula:

$$r_{xx} = \frac{2r_{12}}{1 + r_{12}} \qquad (9\text{--}1)$$

where

r_{xx} is the reliability
r_{12} is the correlation between the two halves of the test

The reliability computed from the split-half technique will generally be higher than the r_{xx} determined through test-retest or comparable form techniques. This increase is considered to be spurious; i.e., the obtained r_{xx} is higher than the true reliability of the test. In deciding which one of several available tests should be used, high reliability would be one desirable characteristic, but a test with a higher reliability coefficient determined by the split-half technique is not necessarily more reliable than a test with a lower reliability coefficient determined by the test-retest or comparable form technique.

STANDARD ERROR OF MEASUREMENT. To individuals who have worked with tests, test results, and correlation coefficients, the size of r_{xx} takes on some rather definite meanings. For the beginner in this area the interpretation of r_{xx} remains a bit of a mystery. The question of what is a "high" coefficient of reliability or how large r_{xx} must be for a test to be useful is difficult to answer in any general way. One statistic which is based on the reliability coefficient is rather easy to interpret. This is known as the *standard error of measurement* (S_e) and is given by the following formula:

$$S_e = S_x \sqrt{1 - r_{xx}} \qquad (9\text{--}2)$$

where

S_x is the standard deviation of the measurements
r_{xx} is the reliability of the measuring instrument

The interpretation of the standard error of measurement involves only one new idea. This is the idea of a "true" score. Think of any score (X) obtained from any individual as consisting of two parts: (1) the "true" value of the thing being measured and (2) an error of measurement. Thus, if we measure the height (or intelligence) of an individual, we can think of the measurement as being a combination of the true height (or intelligence) plus some error in the measurement. The standard error of measurement is nothing more or less than the standard deviation of these errors of measurement over a larger number of people. The standard error of measurement can be interpreted in exactly the same way as any standard deviation. If we can assume normality of errors (and this is generally a safe assumption), then we can use S_e to establish confidence intervals within which we would expect the true score to fall with a given degree of confidence. Thus, taking the test score X, we can say that we would expect the true score for the individual to lie within $\pm 1.96S_e$ of X, 95 percent of the time; within $\pm 2.58S_e$, 99 percent of the time.

By using the standard error of measurement in this way, the test interpreter can think in terms of a region of scores on a test as the best estimate of the individual's score, and the size of the region within which we expect to find the true score will depend upon the size of r_{xx}. If r_{xx} equals one, the test is perfectly reliable, and the score (X) will be the true score for that individual.

Example 9–1   Determination of the 95 percent confidence interval for a test score

$$X = 81; \; S_x = 16; \; r_{xx} = .91; \; S_e = S_x \sqrt{1 - r_{xx}}$$
$$S_e = 16 \sqrt{1 - .91} = 16 \sqrt{.09} = (16)(.3) = 4.8$$

The standard error of measurement is 4.8. The establishment of the 95 percent confidence interval for this test score is the same as establishing any confidence interval—we need to go out 1.96 standard deviations from the mean in each direction.

$$CI_{95} = 81 \pm 1.96S_e = 81 \pm (1.96)(4.8) = 81 \pm 8.408 \text{ (rounding to 8.4)}$$
$$CI_{95} = 72.6 - 89.4$$

This 95 percent confidence interval means that if the same test were administered many times to the individual who scored 81 on the first administration, we would expect that 95 percent of the retest scores would be in the interval between 72.6 and 89.4. An alternate way of stating this is to say that we would expect the true score of the individual who obtained the score of 81 to be no lower than 72.6 and no higher than 89.4, with 95 percent confidence in the statement.

As was stated earlier, the standard error of measurement (S_e) is the *standard deviation* of the distribution of errors of measurement. As a standard deviation, it can be used in the same manner as any standard deviation; and specifically, it is used as an estimate of the variability to be expected on repeated measurements.

Many standardized tests for use in industry and schools are now using this idea in reporting norm data. After the score has been determined for a student and one wishes to look up the percentile rank of that score, a band of ranks is presented—i.e., 64 percent to 73 percent. This reflects the use of the standard error of measurement in reporting the normative data.

VALIDITY OF TESTS

Validity of a test may be defined as how well the test measures whatever it is that the test constructor says it measures. Determining the validity of a test is often a complex and difficult task. A special committee of the American Psychological Association has recommended that the concept of validity be subdivided into three different types of validity. These have been called *content, construct,* and *criterion-related validity.* We shall confine this discussion to those forms of validity (criterion-related) which utilize the Pearson product-moment correlation coefficient as the coefficient of validity.

If we state that a test is measuring that which it purports to measure, we are saying that we have two different measurements—one set of measures from the test (test scores) and one set of measures from what the test is measuring (criterion scores). We can compute a correlation between these two sets of numbers, and the correlation coefficient (r_{xy}) is the coefficient of validity. Note that the only basic difference in the use of correlation in determining reliability and determining validity involves the kinds of scores being correlated. In determining reliability, we are correlating scores on a test with scores on the same test or some comparable form of that test. In determining validity, we are correlating scores on a test with some independent measure or criterion.

A test may have high reliability (consistency) without validity. Suppose someone were to propose a new measure of height. As his measuring instrument, this individual proposes to use a bathroom scale. Repeated measurements with this measuring device (test-retest) would yield highly comparable "scores" for each individual. The measuring device would be highly reliable. However, few people would accept this as a valid measure of height.

Whenever the criterion-related validity of a test is to be determined, the first problem is to select a criterion. Consider the use of a test to predict success in college. What is an adequate criterion for success in college? Certainly there would be much disagreement and discussion as to what is an adequate criterion. In the last analysis, however, the question of a suitable criterion becomes a subjective question—the researcher or a group of individuals decide what to use as the criterion. In the case of success in college, we could probably get agreement among people that the grade-point average would be one satisfactory criterion. Next, we must develop a measuring instrument that will correlate highly with grade-point average.

PREDICTIVE VALIDITY.   To illustrate this form of validity, let us continue with the test to predict college success (grade-point average). We could develop a test and administer it to a large number of students prior to their entrance in college. Let them all go to college and then determine how well the scores on the test correlate with grade-point average in college. The correlation between test scores obtained *before* entering college and grade-point average earned *later* would indicate the predictive validity. The essential characteristic of predictive validity is that the test scores are obtained *before* the individuals are permitted to have experience on the criterion task (taking college work and thereby obtaining a grade-point average).

CONCURRENT VALIDITY.   If we were to administer the test to a group of seniors ready to graduate from college and correlate their test scores with grade-point average, the correlation between test scores and grade-point average would be the index of concurrent validity. The essential characteristic of concurrent validity is that the test scores are obtained from individuals who are already performing on the criterion task.

One difficulty in this approach should be immediately apparent. The only individuals being measured are those who have obtained some degree of success on the criterion measure. Those individuals who did not maintain a certain minimum grade-point average would not be tested. This truncation (cutting down on the range of possible scores on both the test and the criterion) would have the effect of lowering the r_{xy} between test scores and grade-point average. In general, predictive validity is to be preferred to concurrent validity. Unfortunately, predictive validity is often more expensive and time-consuming to obtain. Suppose that some organization wishes to use a test to predict success in a complex job which requires several months (or even years) of training. Suppose further that the failure rate among the people training for this job is quite high and the organization desires to reduce this as much as possible. To determine the predictive validity of a given test would take a great deal of time, and it might turn out that the correlation between the criterion and test scores would be too low for practical use. A great deal of time, effort, and money might be expended with little or no benefit to the organization. In this situation, concurrent validity is generally used. People who are already successful on the job are tested, and a correlation is obtained between test scores and the criterion. If the test proves to be inadequate, other tests can be tried with little additional effort. When a suitable test (high r_{xy}) is found, it can be used with applicants for the position.

Criterion-related validity (either predictive or concurrent) is always specific. The validity of a test is how well the test correlates with a specific criterion. A test may be reliable when used in many different situations, but validity must always be stated in terms of the specific criterion with which the test scores are correlated. A test would rarely if ever be considered valid for many different purposes. The validity would be expressed as the correlation between the test scores and the *specific* criterion.

FACTORS WHICH AFFECT THE MAGNITUDE OF r_{xy}. The correlation coefficient is an extremely powerful tool in constructing and using measuring instruments. But as with any tool, there are cautions and safeguards which must be observed if the tool is to be used or interpreted correctly. We have already discussed many of these factors—i.e., the assumptions underlying the use of r; r used in reliability, in validity, etc. Two additional factors which affect the size of the correlation coefficient will be discussed in this section.

RANGE OF TALENT. In chapter 8, in the discussion of correlation, a line was drawn around the tally marks in the scatterplot. The shape of the figure (circle, ellipse) was a rough indication of the degree of correlation between two variables. If the line drawn around the tally marks tends to be a circle, we know that there is little if any relationship between the two variables. If, however, the line drawn around the tally marks tends to be very narrow and elongated, we know that the relationship is high, either positive or negative, depending upon the direction of the ellipse.

Suppose that we were to administer two tests—a reading test and an arithmetic computation test—to all children in an elementary school. We would not expect children in the first grade to do well on either the reading test or the arithmetic computation test. The scores of these first graders would tend to be very low and would appear in the lower left corner of the scatter diagram. We would expect most of the eighth graders to score quite well on both tests, and we would find most of the tally marks for these students in the upper right corner of the scatterplot. The tally marks for each grade group—first grade through eighth—would be progressively higher as we move from the lower left corner of the scatterplot to the upper right corner. Drawing an ellipse around these scores would make a long, thin, narrow type of figure; and by this, we could judge that the correlation between reading scores and arithmetic computation scores is high and positive. The situation is shown in figure 9–2.

FIGURE 9–2
Scatterplot showing how the correlations between two variables may be low for any restricted group, but high over a wider distribution of the variables

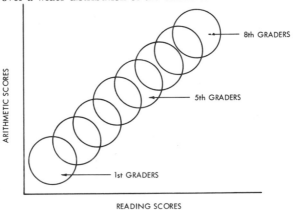

Examine the tally scores of just the students in the fifth grade. If we draw a line around the fifth grade scores, we see that this figure is very nearly a circle. There seems to be little correlation between the two tests for fifth graders. The same relationship holds true for any single-grade group. The correlation between the two tests for any single grade is low, yet the overall correlation between the two tests is high when we consider all students in all grades. The size of the correlation coefficient is directly related to the range of scores in both the X and the Y distributions. If we take a narrow segment of a given population and compute a correlation between two measures taken from this narrow segment, the correlation will be lower than if the same measures are correlated over a large range.

Whenever a correlation coefficient is reported, there should be some statement describing the group on which the r was computed, for the magnitude of the correlation coefficient itself will be dependent upon the spread or range of scores in both the X and the Y variables.

In determining reliability (r_{xx}) using the split-half technique, it was necessary to "correct" the correlation between the two halves for use with the full test. The reasons for this correction should now be clear. The scores earned on a half test consisting of 50 items can only range from zero to 50. Thus, we would expect a higher reliability from a 100-item test than from a 50-item test. The r_{xx} for the 50-item half test is an underestimate of the reliability of a 100-item test.

EFFECTS OF UNRELIABILITY ON VALIDITY

A validity coefficient will of necessity require that two scores be obtained for each individual—a test score and a criterion score. The measuring instruments used to obtain these two scores may vary with respect to reliability. It has often been said that the first requisite of any test is reliability. Whatever the test is measuring, the test must be able to measure it consistently before that test can be utilized to predict a second variable.

Let us go back and use the idea of a "true" score and again think of any measurement or score as consisting of a true score and an error of measurement. Thus, we could state that $x = t_x + e$. By the proper algebraic manipulation, the following equation can be derived:

$$r_{xy} = r_{tt} \sqrt{r_{xx}} \sqrt{r_{yy}} \qquad (9\text{-}3)$$

where

r_{xy} is the obtained correlation between variables X and Y
r_{tt} is the correlation between the true score in X and the true score in Y
r_{xx} is the reliability of the X measurements
r_{yy} is the reliability of the Y measurements

Notice from this formula that if the correlation between the true scores of the two measures is ± 1 ($r_{tt} = \pm 1$), the size of the obtained correlation (r_{xy}) can be one only if the two measures, X and Y, are perfectly reliable ($r_{xx} = 1$ and $r_{yy} = 1$). If the reliability of either of the two measurements were zero, then even if the correlation between the true scores of the two measures were one, the obtained correlation would be zero.

This probably illustrates more clearly than any other single thing that if we do not have high reliability in our basic measuring devices, we cannot expect to obtain stable estimates of the correlation between two different variables. Although we are interested in the relationship between variables, the first concern must always be with the reliability or consistency of the measurement of these individual variables. After the reliability of our measuring devices has been established, we can turn to the problem of correlation between variables.

Summary

The procedures used in reporting the results of measuring some characteristics will depend in part upon the scaling characteristics of the objects being measured. If the data are nominal in nature, very little can be done in reporting the results other than reporting frequency or proportions in each of the categories of the nominal scale. The percentile scale is most often used for reporting data which represent an ordinal scale of measurement. The percentile scale indicates the percentage at or below each point, but does not allow for statements as to the distance between measured objects.

If interval scaling has been achieved, the most often used technique for reporting test results is a variation on the Z score. The Z score is independent of the original measuring unit used. Since the mean of any set of Z scores is always zero and the standard deviation is always one, the Z score can be used for direct comparison of scores obtained with two different measuring instruments. The disadvantage of the Z score is the use of decimals and negative numbers.

Derived scores can be established which overcome the problem of using decimals and negative numbers. A convenient constant is selected to *multiply* all Z scores in the distribution. This has the effect of changing the standard deviation of the distribution from one (standard deviation of the Z scores) to the value of the constant. A second constant is selected to be *added* to all scores in the distribution. This has the effect of changing the mean of the distribution from zero (mean of a set of Z scores) to the value of this second constant. The examinations developed by the Educational Testing Service use 100 as the multiplier constant and 500 as the additive constant. Thus, examinations such as the College Boards and the Graduate Record Examination have a mean of 500 and a standard deviation of 100.

The idea of the reliability of measurement involves some type of stability or consistency. The Pearson product-moment correlation coefficient can be used to indicate the degree of agreement between two sets of measurements on the same individual. Three types of reliability are used most commonly. *Test-retest* reliability involves testing the same individual twice with the same measuring instrument. Two cautions need to be pointed out in assessing reliability in this manner. First, the characteristic being measured may change in the interval between measurements; and second, the use of the measuring instrument may destroy its usefulness for a second administration to the same individuals. *Comparable form* reliability utilizes two measuring instruments which are comparable to each other. This minimizes the problem of the destruction of the usefulness of the instrument after it has been used once, but does not solve the problem of a possible change in the characteristic being measured in the interval between measurements. The *split-half* technique uses the idea of one measuring instrument being composed of two or more instruments. The correlation is obtained between the scores made on each of two halves of the test. This obtained correlation is "corrected" for the length of the test by using the Spearman-Brown prophecy formula.

The standard error of measurement (S_e) is an extremely useful statistic for determining the stability of an obtained measurement. This is the standard deviation of the "errors of measurement" as indicated by the reliability of the measuring instrument. A confidence interval can be constructed around the obtained score with any specified degree of confidence that repeated measurements on the same individual with comparable measuring instruments will yield scores within the interval.

The validity of a test is a statement as to how well the test measures whatever it is said to measure. Only those forms of validity which involved the correlation of the test scores with a criterion were discussed (criterion-related validity). *Predictive validity* is the correlation of a set of test scores obtained before the individuals were allowed to have experience with the criterion. *Concurrent validity* is the correlation of test scores obtained from individuals who are already performing on the criterion task.

The size of the correlation between a set of measurements and a criterion will also be influenced by the spread or range of scores on both the test and the criterion. Severe restrictions on the range of observation (truncation) will reduce the obtained correlation between the two sets of measurements.

If either the test or the criterion is unreliable, one cannot expect to obtain any sizable correlation between the two measures, regardless of the size of the correlation between the true scores of the two sets of observations. The first requisite in attempting to determine the relationship between two variables is to develop reliable measures of each of the variables. Without reliability, there cannot be validity.

References

SPECIFIC RECOMMENDATIONS ON RELIABILITY AND VALIDITY

AMERICAN PSYCHOLOGICAL ASSOCIATION, COMMITTEE ON TEST STANDARDS, "Technical Recommendations for Psychological Tests and Diagnostic Techniques," *Psychological Bulletin,* Vol. 51 (1954), No. 2, Part 2.

ANNE ANASTASI, *Psychological Testing* (3d ed.; New York: Macmillan Co., 1968).

GENERAL REFERENCES ON TESTS AND MEASUREMENTS

L. CRONBACH, *Essentials of Psychological Testing* (2d ed.; New York: Harper and Row, 1960).

G. C. HELMSTADTER, *Principles of Psychological Measurement* (New York: Appleton-Century-Crofts, Inc., 1964.

J. C. NUNNALLY, JR., *Tests and Measurements* (New York: McGraw-Hill Book Co., Inc., 1959).

10

Significance of differences between two groups with interval or ratio scaling

IN CHAPTER 7, we explored the use of the standard error of the mean to establish a confidence interval about a sample mean. This technique allows us to make probability statements concerning the location of the true mean of the population from which the sample was drawn. In chapter 8 this idea was extended to the establishment of a confidence interval around a value in Y predicted from our knowledge of X and the correlation between X and Y. In chapter 9 the idea of a confidence interval was applied to the stability or reliability of a test score. The logic used in establishing a confidence interval about a sample mean, a regression prediction, or the reliability of a score, can be extended to the situation where the research worker is concerned with the *difference* between two groups.

Establishing a confidence interval around the difference in means

Rather than drawing one random sample from a population, we shall draw two random samples. We can compute the mean of each sample, and we can also compute the difference between the two means. We can draw two more samples and again compute the mean of each sample and the difference in the means. This procedure can be continued until we have an extremely large number of differences between sample means.

This distribution of differences in sample means can be described by determining the central tendency of the distribution and the variability of the distribution. If each sample were a random sample from the same population, we would expect most of the differences in sample means to be small—near zero. Suppose that we obtained one difference between sample means that was extremely large. We could ask: What is the probability of a difference this large occurring by chance? If we could determine the standard deviation of the differences in means, it would be a relatively easy matter to convert the obtained large difference in sample means into a Z score. Using the table of the normal curve, we could determine the probability of this large a difference occurring by chance.

THE STANDARD ERROR OF THE DIFFERENCE IN MEANS

The catch in all this is the problem of determining the standard deviation of the differences in means. Fortunately, this can be estimated from the two samples, and we do not need to obtain an infinite number of differences in means to determine the standard deviation of these differences. It can be shown that in drawing samples from the same population with equal sample sizes ($N_1 = N_2$), the standard error of the differences in mean is

$$ s_{D_{\bar{x}}} = \sqrt{s_{\bar{x}_1}{}^2 + s_{\bar{x}_2}{}^2} = \sqrt{\frac{s_1{}^2}{N_1} + \frac{s_2{}^2}{N_2}} \tag{10-1} $$

where

$s_{D_{\bar{x}}}$ is the standard error of the difference in means
$s_{\bar{x}_1}$ is the standard error of the mean of the first sample
$s_{\bar{x}_2}$ is the standard error of the mean of the second sample

In formula 10–1, notice that s was used rather than S. This indicates that these are unbiased estimates of the standard error of the mean. (See chapter 7, p. 141). This particular formula is useful when the sample sizes are equal, but often the researcher is working in a situation in which $N_1 \neq N_2$. If the two N's are unequal, then equation 10–1 should be avoided. It is necessary to utilize an expression which will take into account that the two different estimates of the $s_{\bar{x}}$ are based on unequal numbers of observations. Since $s_{\bar{x}_1}$ and $s_{\bar{x}_2}$ are both estimates of the same standard deviation of a distribution of means (they are based on two different samples from the same population), it is possible to "pool" these two estimates into a single "better" estimate. The student will recall (chap. 7, p. 141) that the standard error of the mean is based upon an unbiased estimate of the standard deviation of the population:

$$s_{\bar{x}} = \frac{s}{\sqrt{N}} = \frac{s}{\sqrt{N-1}}$$

We form this better estimate by returning to the sums of square deviations used to determine the standard deviations. Thus the pooled estimate of the standard deviation can be obtained by use of the following equation:

$$s = \sqrt{\frac{\Sigma x_1^2 + \Sigma x_2^2}{(N_1 - 1) + (N_2 - 1)}} \tag{10-2}$$

where

   s is an unbiased estimate of the population standard deviation(s)
   Σx_1^2 is the sum of squared deviations based on the first sample
   Σx_2^2 is the sum of squared deviations based on the second sample

Thus the standard error of the difference in means becomes

$$s_{D\bar{x}} = \sqrt{\frac{s^2}{N_1} + \frac{s^2}{N_2}} \tag{10-3}$$

Combining formulas 10–1 and 10–2 the standard error of the differences in means becomes

$$s_{D\bar{x}} = \sqrt{\left(\frac{\Sigma x_1^2 + \Sigma x_2^2}{N_1 + N_2 - 2}\right)\left(\frac{1}{N_1} + \frac{1}{N_2}\right)} \tag{10-4}$$

If the original scores are used to compute the standard error of the differences in means directly, formula 10–4 becomes

$$s_{D\bar{x}} = \sqrt{\left(\frac{\Sigma X_1^2 + \Sigma X_2^2 - (N_1\bar{X}_1^2 + N_2\bar{X}_2^2)}{N_1 + N_2 - 2}\right)\left(\frac{1}{N_1} + \frac{1}{N_2}\right)} \tag{10-5}$$

If $N_1 = N_2 = N$, formula 10–4 simplifies to

$$s_{D\bar{x}} = \sqrt{\frac{\Sigma x_1^2 + \Sigma x_2^2}{N(N-1)}} \tag{10-6}$$

With original scores, formula 10–6 becomes

$$s_{D\bar{x}} = \sqrt{\frac{\Sigma X_1^2 + \Sigma X_2^2 - N(\bar{X}_1^2 + \bar{X}_2^2)}{N(N-1)}} \tag{10-7}$$

This standard error of the differences in means is a standard deviation and can be interpreted like any standard deviation. We can now use this standard deviation of the differences in means as the basis for establishing a confidence interval about an obtained difference. This procedure is shown in example 10–1.

The establishment of a confidence interval around the obtained difference in means allows us to state that we would expect that the differences between the true means of the two populations are within the interval we have constructed. It also allows us to state that it is unlikely that the true difference is larger than that indicated by the confidence interval or smaller than the differences within the confidence interval. In the case of the data given in example 10–1, we believe that the difference between the true means of the two distributions is no smaller than 2.29 but no larger than 13.51, with 95 percent confidence in the statement. We would

---

Example 10–1   Illustration of the establishment of a confidence interval about an obtained difference in two sample means

$$\bar{X}_1 = 73.6 \qquad\qquad \bar{X}_2 = 65.7$$
$$S_1 = 8.32 \qquad\qquad S_2 = 9.14$$
$$N_1 = 20 \qquad\qquad N_2 = 20$$

$$s_{D\bar{x}} = \sqrt{\frac{s^2}{N_1} + \frac{s^2}{N_2}} \qquad D_{\bar{x}} = \bar{X}_1 - \bar{X}_2 = 73.6 - 65.7 = 7.9$$

Note that the expression for the standard error of the differences in means requires the unbiased estimate of the population variance (s^2). The data given indicate the sample standard deviations (S). We must convert the data given above into sums of squared deviations (see equation 10–2). Since

$$S^2 = \frac{\Sigma x^2}{N}$$

then

$$\Sigma x^2 = NS^2$$

Thus the Σx^2 for sample 1 is equal to $N_1 S_1^2$, or

$$\Sigma x_1^2 = (20)(8.32)^2 = 1{,}444.45$$

and for sample 2:

$$\Sigma x_2^2 = N_2 S_2^2 = (20)(9.14)^2 = 1{,}670.79$$

$$s^2 = \frac{\Sigma x_1^2 + \Sigma x_2^2}{(N_1 - 1) + (N_2 - 1)} = \frac{1{,}444.45 + 1{,}670.79}{(20 - 1) + (20 - 1)} = \frac{3{,}115.24}{38} = 81.98$$

$$s_{D\bar{x}} = \sqrt{\frac{s^2}{N_1} + \frac{s^2}{N_2}} = \sqrt{\frac{81.98}{20} + \frac{81.98}{20}} = \sqrt{4.10 + 4.10} = \sqrt{8.20} = 2.86$$

$$CI._{95} = D_{\bar{x}} \pm (1.96)(s_{D\bar{x}}) = 7.9 \pm (1.96)(2.86) = 7.9 \pm 5.61$$

Thus, we would expect to find the true difference between the means or the difference between the true means of the two populations ($\mu_1 - \mu_2$) to be at least 2.29 but no more than 13.51, with 95 percent confidence in the statement.

---

also be able to state that there is a small probability that the true difference could be as large as 15.

Notice that in establishing this confidence interval, we combined the sums of squared deviations from the two different samples and the sample sizes to obtain one estimate of the population standard deviation(s). This pooling of the two sets of data has the implicit assumption that s_1^2 and s_2^2 are unbiased estimates of the *same population* variance, (σ^2). This is known as the assumption of *homogeneity of variance*.

Testing specific hypotheses about differences in means

The confidence-interval approach allows us to establish a region or interval within which we would expect the difference between the true means of the two populations to fall. There are many occasions when the research worker is more interested in making a direct test to see if the true difference is more (or less) than some specific amount. The testing of the hypothesis of a specific difference is of importance in many areas in which a decision to adopt a new technique must be weighed against the cost of installing the technique. Suppose that an agricultural experiment station is conducting experiments on the uses of certain fertilizers on grain crop production. The evidence seems to indicate that the use of a given fertilizer will increase the yield of the grain, but fertilizer is costly to buy and apply. It would not be economically feasible to use the fertilizer unless it would increase the yield by more than the costs of application. Suppose that an economist determines that the yield must be increased by 6 bushels per acre for the farmer to break even in using the fertilizer. The researcher is now faced with a very specific question: Will the true difference in yield per acre be at least 6 bushels? To test this specific hypothesis, we shall need to go back to the Z score.

From the table of the normal curve, it is a simple matter to determine the probability of the occurrence of a Z score of any given magnitude or larger. A Z score of $+1.00$ or larger will occur 16 times out of 100 (50 percent of the scores will be below the mean, and 34 percent of the scores will be between the mean and a Z score of $+1.00$). But the Z score we use will be a Z score based on the distribution of differences. The usual formula for the Z score is

$$Z = \frac{X - \bar{X}}{S}$$

We shall need to use a similar formula, but one based on the distribution of differences in means:

$$Z_{D_{\bar{X}}} = \frac{D_{\bar{X}} - \bar{D}_{\bar{X}}}{s_{D_{\bar{X}}}} \qquad (10\text{–}8)$$

where

$D_{\bar{X}}$ is the difference in sample means $(\bar{X}_1 - \bar{X}_2)$

$\bar{D}_{\bar{X}}$ is the mean of all the possible differences in means and is equivalent to the difference between the true means of the two populations $(\mu_1 - \mu_2)$

$s_{D_{\bar{X}}}$ is an unbiased estimate of the standard deviation of the sampling distribution of differences in means

The research worker could do an experiment growing some grain with fertilizer and some without the fertilizer. The mean yield per acre of each grain crop could be determined. Using formulas 10–2 and 10–3, he could compute the standard error of the difference in means and would be ready to test if the difference in yield is large enough to warrant using fertilizer. For this, he would use formula 10–8. What value should be used as the mean of all possible differences in means (the estimate of the true difference between μ_1 and μ_2)? He should use the value of 6 bushels, since he is trying to determine if the obtained difference is sufficiently larger than the break-even point to be worth using. By using formula 10–8 and the table of the normal curve, he can determine the probability of the occurrence of a difference this large or larger, if the true difference is 6 bushels. Example 10–2 illustrates this use of the Z score formula in testing for a specific hypothesis.

The Z score used in this way is often referred to as the *critical ratio* (CR). Based on the size of the critical ratio (its probability of occurrence by chance), the research worker makes a decision to reject the hypothesis that the true difference was the specific stated amount (if the probability of the occurrence of the observed difference is small), or he decides that there is not sufficient evidence to reject the specific hypothesis. In the case of the fertilizer experiment the research worker would have to decide whether he believed that the true difference was greater than 6 bushels per acre.

THE NULL HYPOTHESIS

In testing a specific hypothesis, the research worker will deal with a Z score equation with three elements: (1) the observed difference between the

---

Example 10–2 Illustration of the testing of a specific hypothesis about the difference in true means of two different populations

$$\bar{X}_1 = 18.63 \qquad \bar{X}_2 = 10.52$$
$$s_1{}^2 = 23.05 \qquad s_2{}^2 = 21.49$$
$$N_1 = 120 \qquad N_2 = 130$$

Hypothesis: $\mu_1 - \mu_2 = 6.00$

Using formula 10–3:

$$s_{D\bar{x}} = \sqrt{.3574} = .598$$

$$\text{CR (or Z score)} = \frac{(\bar{X}_1 - \bar{X}_2) - \bar{D}}{s_{D\bar{x}}} = \frac{(18.63 - 10.52) - 6.00}{.598}$$

$$\text{CR} = \frac{8.11 - 6.00}{.598} = \frac{2.11}{.598} = 3.53$$

Using table B in the Appendix, we find that a Z score of $+3.50$ or larger will occur by chance less than 2 times in 10,000. Thus, $p < .0002$. We would reject the hypothesis that the difference between the means of the two populations is six. It must be some value larger than six.

---

means of the two groups $(\bar{X}_1 - \bar{X}_2 = D\bar{x})$; (2) the hypothesized true difference in population means $(\mu_1 - \mu_2 = \bar{D}\bar{x})$; and (3) the standard error of the differences in means $(s_{D\bar{x}})$. The research worker is in effect saying, "If the hypothesized true difference is correct, then the observed difference should be of about the same size as the hypothesized true difference or $D\bar{x} = \bar{D}\bar{x} = 0$. This relationship is called the *null hypothesis*. No matter what the value of the hypothesized true difference is, the null hypothesis refers to the difference between the observed difference and the hypothesized true difference. In example 10–2 the hypothesized true difference was 6. The null hypothesis states, in this case, that the observed difference, $\bar{X}_1 - \bar{X}_2 = D\bar{x} = 6$, or $D\bar{x} - \bar{D}\bar{x} = 6 - 6 = 0$. In example 10–2, the observed difference in means was 8.11 and the CR was 3.53. The conclusion was that it is extremely unlikely $(p < .0002)$ that this large a difference would have been observed if the true difference were 6.00. The research worker would *reject* the null hypothesis and instead accept the idea that the true difference must be some value greater than 6.00.

In much of the research in the behavioral sciences, we have not progressed to the point where we are concerned about the *amount* of an effect of the variable; rather, we are attempting to determine if a given variable is having *any* effect. The sociologist may be concerned with the question: Does the social-economic level of a teen-ager have an effect on the amount of education the individual will receive? The psychologist interested in the learning process might ask: Does the level of motivation of the learner have an effect on the amount learned? Research workers from several different branches of the behavioral sciences might be interested in the question: Does church attendance have *any* effect on attitudes toward current social and political issues? In the situations mentioned above, the specific hypothesis under test would be $(\mu_1 - \mu_2) = \bar{D}_{\bar{x}} = 0$. The null hypothesis, in this case, would be $D_{\bar{x}} - \bar{D}_{\bar{x}} = 0 - 0 = 0$. The null hypothesis refers to an expected difference of zero between the *observed* value and the *hypothesized* value. The special situation where the hypothesized true difference is zero is not, itself, the null hypothesis. Generally, research is done because the scientist believes (or has a hunch) that variable *B does* have an effect on variable *C*. Thus the psychologist might state his research hypothesis as: "I believe that differences in motivation will lead to differences in the amount learned." But now we come to the curious logic of the null hypothesis. This is the hypothesis of *no* difference. The research worker will not directly test *his* hypothesis (the research hypothesis). Instead, he will test the hypothesis of *no* difference. If he finds that the difference obtained from the research is too large to have occurred by chance if the true difference was zero, then he can *reject* the null hypothesis (no difference) in favor of his research hypothesis (there is a difference).

MAKING DECISIONS CONCERNING THE NULL HYPOTHESIS

The use of the table of the normal curve with a given critical ratio will indicate the probability that a difference this large or larger could have occurred by chance if the true difference was zero. The investigator must decide if the obtained difference was a chance difference (a rare sampling error) or if it reflects a true difference (the null hypothesis is incorrect). The research worker must make a decision—to reject the null hypothesis or not to reject it.

The different combinations of decisions and consequences can be diagrammed in a 2×2 table showing four different consequences. These are given in table 10–1. Along the top of the table are presented the two possible conditions for the null hypothesis—it is true (there is no true difference between the observed difference and the hypothesized difference), or it is false (there is a true difference). Along the side are the two decisions that the research worker can make. He can reject the null hypothesis (decide that there is a true difference), or he can fail to reject the null hypothesis.

ERROR OF THE FIRST TYPE. For convenience, the cells in table 10–1 have been labeled *a*, *b*, *c*, and *d*. We shall first examine the situation in

TABLE 10-1
Different decisions concerning the null hypothesis

		The null hypothesis is	
		True	False
Experimenter's decision is	Reject H_0	*a* Type I error $p = \alpha$	*b* Power $1 - \beta$
	Do not reject H_0	*c*	*d* Type II error $p = \beta$

which the null hypothesis is true (cells *a* and *c*). Based on the size of the critical ratio he obtained from his data, the experimenter's decision may be to reject the null hypothesis (cell *a*). If this is his decision, he has made an error. By rejecting the null hypothesis, he is stating that he believes, based on his research evidence, that there is a true difference. But he is wrong—the difference he observed was a very unlikely chance difference. This is called an *error of the first type,* or Type I error. How often will this type of error occur? What is the probability of the occurrence of a Type I error? If the obtained critical ratio were 1.96, we could say that the probability of occurrence of a difference this large or larger by chance was 5 percent. Thus, he made an error which will occur only 5 times in 100 with a critical ratio of this magnitude. The lowercase Greek letter alpha (α) is used to indicate the probability of a Type I error. In this case, $\alpha = .05$. If the experimenter is not willing to run that big a risk of making a Type I error, he can use some other value of the critical ratio to make his decision to reject the null hypothesis. If he wants more certainty that he is not rejecting a true hypothesis, he may decide to use a critical ratio of 2.58, in which case $\alpha = .01$. Figure 10-1 shows the probability of a Type I error under the null hypothesis for two decision points.

The probability of the occurrence of a Type I error is completely in the control of the research worker, but he should make his decision as to what probability of Type I error is acceptable to him *prior to the collection of his data.* The probability of the occurrence of a Type I error is dependent only upon the size of the critical ratio the research worker will use to reject the null hypothesis.

FIGURE 10–1
Distribution of differences under the null
hypothesis (H_0) showing the probability of a
Type I error (α) at the .05 level (fig. a) and
the .01 level (fig. b).

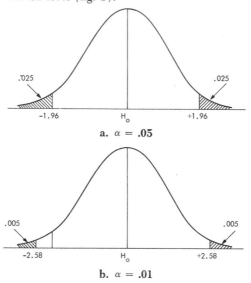

NONREJECTION OF THE NULL HYPOTHESIS. If the null hypothesis
is true and the research worker decides *not* to reject it, (cell *c* of table
10–1), we are in an interesting position. This does *not* mean that the
researcher has *proved* that no difference exists. An adequate discussion
of this topic would take us too far into the realm of philosophy, but
the outline of the argument is as follows: We can prove a universal
statement to be true only if we examine all possible cases. In a sampling
sense, we can never hope to prove that something is universally true—we
would have to examine the entire population. We can prove that some
proposition is not universally true if we can find one case that does not
support the proposition. Thus, if we start with the proposition that there
is no true difference in the means of the two groups, we can reject that
universal statement (the null hypothesis) if we can find one case which
indicates that there is a difference. On the other hand, if we draw a
sample from a population and, basing our conclusion on the one sample,
do not find a difference large enough to reject the null hypothesis, we
have not proved the null hypothesis to be correct. We can only say that
we don't have enough evidence to show that the hypothesis is wrong.

In one sense, cell *c* of table 10–1 is of little use to us. Failure to reject the null hypothesis does not mean that the null hypothesis is true; it means that we don't have the evidence to show that the hypothesis is false.

ERROR OF THE SECOND TYPE. Suppose the null hypothesis is false (cells *b* and *d* of table 10–1). The research worker, on the basis of his evidence, can still make only one of two decisions: (1) he may reject the hypothesis, or (2) he may fail to reject the hypothesis. Let us examine the latter of these two situations first. If the null hypothesis is false but the evidence is such that the research worker does not reject the hypothesis, then he has made an error. This is called an *error of the second type,* or Type II error. The lowercase Greek letter beta (β) is used as the symbol for the probability of the occurrence of a Type II error. The probability of Type I error (α) is set directly by the experimenter but the probability of Type II error (β) is determined by several factors, some of which are not under the direct control of the experimenter. Notice that if the experimenter is very concerned about the probability of the occurrence of a Type I error and hence sets α at a low level (e.g., $\alpha = .001$), this will make the probability of a Type II error larger (β will be larger). Decreasing the probability of a Type I error makes it more difficult to reject the null hypothesis (we need a larger critical ratio). But if the null hypothesis is false, then demanding a larger critical ratio makes it more likely that we shall fail to reject the null hypothesis and thus commit a Type II error. Obviously, if the null hypothesis is false, we wish to be able to reject it. This brings us to cell *b* in table 10–1.

POWER OF A STATISTIC. The ability of a statistic to reject the null hypothesis when it is false is called the *power of the statistic.* Since most research is done because the research worker believes that a difference does exist, much effort should be directed toward the rejection of a false null hypothesis. The power of a statistic is given by $1 - \beta$. This brings us right back to cell *d*. In order to have the research as powerful as desired, we must keep low the probability of a Type II error. One way of doing this is to let α become large, but increasing the probability of a Type I error may be a very poor solution if the null hypothesis is true. Figure 10–2 shows the relationship between the Type I error and the Type II error. Note that a Type I error can occur only if the null hypothesis is *true.* The distribution to the left in figure 10–2 shows the sampling distribution of the *CR*s if the null hypothesis is true. If the decision point to reject H_0 is set at $\alpha = .05$, then a Type I error

FIGURE 10–2
Relationship between the probability of a Type I error and the probability of a Type II error, with two different decision points.

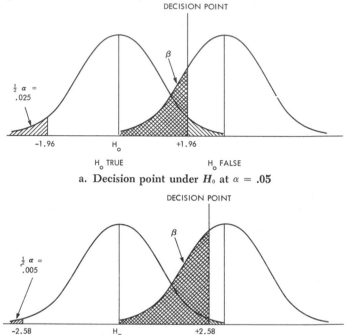

a. Decision point under H_0 at $\alpha = .05$

b. Decision point under H_0 at $\alpha = .01$

can occur only 5 percent of the time if the null hypothesis is true. This is shown by the shaded area under the curve to the left in figure 10–2. *If the null hypothesis is false, a Type I error cannot occur.*

If the null hypothesis is false then we either reject H_0 (the correct decision) or we can fail to reject H_0 (a Type II error). The distribution to the right in figure 10–2 shows the sampling distribution of the *CRs* if the null hypothesis is false. The decision point should always be set before the data is collected and is based on the hypothesis that H_0 is true. The crosshatched area in figure 10–2 shows the area under the curve and to the left of the decision point, when H_0 is false. Note that this crosshatched area indicates the situation in which the investigator fails to reject H_0 when it is false. This area indicates the probability of the occurrence of a Type II error (β). The same situation is shown in the *b* portion of figure 10–2, with one important change. The probability of the Type I error has been changed from $\alpha = .05$ to $\alpha = .01$. That is, the decision point under the null hypothesis has been changed from a *CR* of ± 1.96 to a *CR* of ± 2.58. The decision to use 2.58 rather than 1.96 for rejection of the null hypothesis is excellent *if the null hypothesis is true*. But notice what has happened to the probability of the occurrence of the Type II error! All other factors equal, a decrease in α will produce an increase in β. An increase in β means a decrease in $1 - \beta$, the power of the statistical test!

REDUCING THE PROBABILITY OF THE TYPE II ERROR

Research costs a great deal. Not just the direct expenditure of money but also the time and energy of the research workers. While the null hypothesis is always the hypothesis under test for statistical purposes, the research worker usually has a positive hypothesis which he or she is investigating. The research is being conducted because the investigator believes that the differential treatment of the subjects will produce a difference. Within the framework of decision theory, the null hypothesis is usually designated by the symbol H_0. The research hypothesis is designated by H_1 and if there is more than one research hypothesis the others would be H_2, H_3 $\cdot$ $\cdot$ $\cdot$ H_k. The logic of the situation requires the investigator to reject H_0 which will then allow the acceptance of H_1.

To conserve resources, the investigator should make the research as powerful as possible (maximize $1 - \beta$). This means that the probability of the occurrence of a Type II error (β) must be minimized. Obviously, the investigator should do everything possible to increase the power of the statistical test. On the other hand, if the null hypothesis is true, the investigator does not wish to declare a difference to exist when in fact there is no difference (Type I error). The very essence of research is to make a "fair" test, and this includes minimizing the probability of making a Type II error as well as minimizing the probability of making a Type I error. The Type I error is controlled by setting the decision point. The Type II error is much more difficult to control. We shall use the Z score called the critical ratio to illustrate the control of Type II errors. The principles discussed are also relevant for other statistical techniques which will be introduced later in this textbook.

$$CR = \frac{(\bar{X}_1 + \bar{X}_2) - \bar{D}_{\bar{X}}}{\sqrt{\dfrac{s^2}{N_1} + \dfrac{s^2}{N_2}}}$$

Arithmetically, there are two ways in which we can increase the size of *CR*. We may either increase the absolute size of the numerator or decrease the size of the denominator.

INCREASING THE OBSERVED DIFFERENCE.   Let us examine the first of these two possibilities. The only way in which we can increase the size of the difference between means (other than introducing bias into the experiment) is by careful selection of different values of the independent variable. Suppose that we suspected that students who score high on the Taylor Scale of Manifest Anxiety will learn nonsense syllables faster than students who score low in this scale. It would be a waste of time, effort, and money to select two groups so that they differed by only one or two points on the scale. Far better to select the top and bottom 10 percent on this anxiety scale and then let both groups learn the list of nonsense material. If the experiment is well designed and controlled, it is more likely that we shall detect a difference between the two samples if a true difference does exist in the population. The effect of increasing the treatment effect on Type I and Type II errors is shown in figure 10–3. Note that the Type I error is influenced only by the decision point set by the investigator.

Notice in figure 10–3, that if the null hypothesis is false but the difference is small, the probability of a Type II error is large, as shown in the *a* portion of the figure. The *b* portion of the figure shows that when the null hypothesis is false and the difference is large, the probability of a Type II error becomes almost negligible.

INCREASING PRECISION.   A decrease in the denominator of the *CR* can be accomplished by decreasing s^2 or by increasing N. Either one of these will lead to a decrease in the standard error of the difference in means and thus an increase in the precision of the research. An increase in N will serve to decrease the $s_{D_{\bar{x}}}$ and consequently increase the size of *CR*. It certainly would be foolish to use only one or two subjects per group in most research, but might it not be almost as foolish to use 1,000 subjects per group? Increasing the N is one way to increase precision; but it can also be quite expensive in terms of man-hours, money, and research effort.

ONE-TAILED AND TWO-TAILED TESTS OF SIGNIFICANCE.   In most cases, the research worker is interested in any departure from the null hypothesis. A situation where the observed difference is larger than the hypothesized difference is important, but so is the situation when the observed difference is smaller than the hypothesized difference. In the case where the hypothesized true difference is zero, the investigator is interested in $\mu_1 \neq \mu_2$. If he decides to reject the hypothesis of no difference, it can be because

FIGURE 10–3

Type I and Type II errors as a function of an increase in the treatment effect

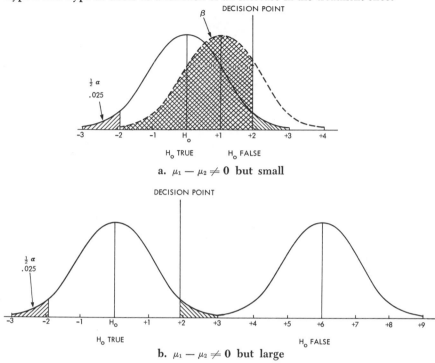

a. $\mu_1 - \mu_2 \neq 0$ **but small**

b. $\mu_1 - \mu_2 \neq 0$ **but large**

$\mu_1 > \mu_2$ or because $\mu_2 > \mu_1$. Thus, he is making a "two-tailed" test of significance. If he is willing to accept the probability of a Type I error at $\alpha = .05$, then he is selecting a CR such that 2.5 percent of the area in the distribution is beyond a $\pm CR$ of the critical value. In a few situations the research worker, *before collecting his data,* decides that he is interested in one specific directional hypothesis, such as $\mu_1 > \mu_2$. In this case a one-tailed test of significance would be indicated. The one-tailed test is more powerful than the two-tailed test as can be seen in figure 10–4.

While the one-tailed test is more powerful than the two-tailed test in rejecting a false null hypothesis, it does present a danger. The decision as to the α level to be used and the decision to make a one-tailed or two-tailed significance test should be made *before the data is collected.* A one-tailed test implies that only differences in the specified direction will be considered. Any difference in the nonspecified direction *no matter how large* would have to be declared as nonsignificant if the one-tailed

FIGURE 10–4

Comparison of the power of the one-tailed test of significance versus the two-tailed test

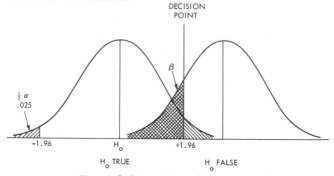

a. Two-tailed test of significance ($\alpha = .05$)

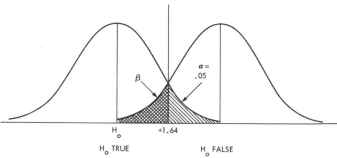

b. One-tailed test of significance ($\alpha = .05$)

test had been used. In most situations the research worker should be making a two-tailed test.

A discussion of the factors involved in the decision to use a one-tailed test of significance rather than the more usual two-tailed test would take more space than is warranted. The reader can find excellent discussions in most advanced texts in statistics.

THE STATISTIC USED TO ANALYZE THE DATA.   In general, those statistics which require interval or ratio scaling will be able to detect a true difference more readily than will statistics based on ordinal or nominal scaling.

To summarize, the probability of a Type II error (β) is affected by:

1.   *The α level set by the research worker for rejecting the null hypothesis.* If the research worker sets a very small value for α and thus makes it more difficult to reject the null hypothesis, this will increase the probability of the occurrence of a Type II error (β).

2. *The size of the true difference between population means* $(\mu_1 - \mu_2)$. Since a Type II error can occur only if the null hypothesis is false (a true difference does exist), the probability of the occurrence of a Type II error will depend in part on the magnitude of the true difference. Careful selection of treatment conditions will make this difference large, if a true difference does exist. Small differences are difficult to detect; large differences are relatively easy to detect. The research should be planned to maximize any possible difference.

3. *The amount of variability(ies) in the research material.* The denominator of most statistics is based on some estimate of the variability of the research material. Any technique which will reduce variability will reduce the probability of a Type II error and thus increase the power of the study.

4. *The sample size.* The larger the sample on which the study is based, the lower will be the probability of a Type II error.

5. *The statistic used to analyze the results of the study.* In general, those statistics which require interval or ratio scaling will be able to detect a true difference more readily than will statistics based on ordinal or nominal scaling. In addition a one-tailed test of significance will be more powerful than a two-tailed test.

Up to this point, we have used the critical ratio and the table of the normal curve to make a decision to reject the null hypothesis or not to reject the hypothesis. This assumes that the distribution of the differences in means when divided by the standard error of the difference in means is normally distributed. This assumption is certainly tenable if the sample size is large enough (see chap. 7, p. 144). If the sample size is small, the distribution of the critical ratios will not be normal; hence, we cannot use the table of the normal curve to obtain probability statements concerning the obtained differences. The use of the critical ratio also assumes that the numbers used have met the requirements of interval or ratio scaling.

The null hypothesis can be tested with a great variety of different statistics. The remaining sections of chapter 10 discuss hypothesis testing with interval or ratio scales for comparing two groups. Chapter 11 presents a variety of different tests which are suitable for use with nominal or ordinal scales. Chapter 12 is an introduction to the analysis of variance with interval or ratio scaling and in addition presents multi-group statistics for use with nominal and ordinal scales. Since the reader is just beginning a journey through a variety of new statistical tests,

TABLE 10–2
Statistical tests of H_0

Level of measure-ment	One-sample case	Two-sample case		k-sample case	
		Independent samples	Correlated samples	Independent samples	Correlated samples
Nominal	Binomial test p. 233 χ^2 one-way classification p. 235	χ^2 test for two-way classification p. 237 Fisher-Yates test p. 243	McNemar test of change p. 241	χ^2 test p. 280	Cochran Q test p. 286
Ordinal		Median test p. 246 Mann-Whitney U test p. 248	Sign test p. 252 Wilcoxon matched pairs signed-ranks test p. 254	Extension of median test p. 282 Kruskal-Wallis one-way analysis of variance for ranks p. 283	Friedman analysis of variance by ranks p. 289
Interval or ratio	Student's t test p. 216	Student's t test p. 216	Student's t test p. 221 Sandler's A p. 226	Analysis of variance p. 262	

a "road map" has been provided in table 10–2. This guide has three dimensions: (1) the scalar characteristics of the measurement, (2) the number of groups in the study, and (3) the independence or noninde-pendence of the samples. This same table, with minor modification, is also presented in chapter 13 as a summary.

STUDENT'S t STATISTIC

In 1908 an article appeared addressed to the problem of the sampling distribution of means based on small samples. The author, W. S. Gosset, published the article under the pseudonym "Student" and thus the term *Student's t* has been associated with the small-sample approach to the sampling distribution of differences in means.

Interestingly, Student's t is not different from the usual Z score formula. The difference is in the shape of the sampling distribution of the t statistic as the sample size changes.

$$t = \frac{D_{\bar{X}} - \mu}{s_{D_{\bar{X}}}} \tag{10-9}$$

where

$D_{\bar{X}}$ is the difference between two sample means $(\bar{X}_1 - \bar{X}_2)$
μ is the true difference in means in the populations
$s_{D_{\bar{X}}}$ is the standard error of the difference in means

The computation of the t statistic is exactly the same as the computation of the critical ratio. In fact, the term *critical ratio* has dropped out of modern usage and has been replaced by the student's t to indicate any Z score arrangement to test for the differences in means. The difference between the two techniques is in the sampling distribution for the t statistics for large samples and for small samples. For large samples the table of the normal curve can be used. For small samples a special table of the t statistic is needed.

SAMPLING DISTRIBUTION OF THE t STATISTIC. As Gosset pointed out, there is no one single distribution of the t statistic. The sampling distribution of the t statistic is different for each different number of observations. Figure 10–5 shows the sampling distribution for three different numbers of observations.

FIGURE 10–5
Normal distribution compared with t distribution
for $N = 4$ and $N = 8$

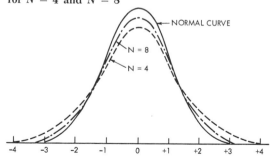

As the number of observations increases, the sampling distribution of the t statistic changes from a leptokurtic distribution to the normal distribution. With an infinitely large sample size, the distribution of the t statistic becomes the normal curve. As a practical matter, as the sample size becomes larger than 30, the sampling distribution of the *t* statistic is close enough to the normal curve so that almost no error is introduced by using the normal curve for the decision to reject or not to reject the null hypothesis. In order to use the *t* statistic, we would need at least 30 different tables, all similar in format to the normal curve table, but each table different in the amount of area within given *Z* or *t* distances from the mean. Since the *t* statistic is used primarily in making decisions concerning hypotheses under test, only a few values in each of the different tables are of interest to us. The values of *t* for significance at the 10 percent, 5 percent, 1 percent, and 0.1 percent levels, for the different number of observations, have been placed together in one table and are presented in table G of the Appendix. A portion of table G is reproduced as table 10–3.

TABLE 10–3
Distribution of *t*

Degrees of freedom (*df*)	*p* = .10	*p* = .05	*p* = .01	*p* = .001
1	6.314	12.706	63.657	636.619
...	.....	......	......	....
10	1.812	2.228	3.169	4.587
...	.....	......	......	....
30	1.697	2.042	2.750	3.646
...	.....	......	......	....
∞	1.645	1.960	2.576	3.291

The left-hand column of table 10–3 is headed *degrees of freedom* (*df*). This is a term which we shall return to in the next paragraph. For the moment, let us examine the use of table 10–3. If we have only one degree of freedom, it will require a *t* statistic of 12.706 or larger to reject the null hypothesis at the .05 level of significance. With 10 degrees of freedom, a *t* of 2.228 or larger is sufficient to reject the hypothesis at the 5 percent level. If, with 30 degrees of freedom, we obtained a *t* of 3.46, by examination of the table we find that we could reject the null hypothesis at the .01 level of significance; but if we had required that the obtained *t* had to reach the .001 level, we would not be able to reject the hypothesis. Remember that the research worker should make his decision as to the required level of significance for rejection of the null hypothesis *before* examining his data. The bottom entry in the table indicates an infinite number of degrees of freedom. Notice that the entry for the 5 percent significance level is the old familiar 1.96 from the normal curve. Similarly, the entry for the 1 percent level is 2.576, or the same value of the normal curve which indicates the middle 99 percent of the area:

DEGREES OF FREEDOM (df). Obviously, the idea of degrees of freedom is related to the total number of observations in the data, but it is not exactly the same. In computing statistics from samples, we are trying to estimate certain values. The sample mean is an estimate of the population mean. The standard deviation is an estimate of the population standard deviation (biased estimate). When we estimate the population mean from the sample, we sum all the scores and divide by the number of scores; but when we determine the standard deviation, we must find the deviation of each score about the sample mean. Unfortunately, the sample mean is not the population mean (it is an unbiased estimate of the population mean). In computing the standard deviation of the sample, we are placing a restriction on the deviations—we are requiring that the sum of the deviations about the mean must add to zero. Placing this restriction on the numbers interferes with their freedom to vary. Suppose that we have five numbers: 2, 6, 4, 5, 3. The sum of these numbers is 20, and the mean is 4. As long as the problem is to determine the mean of five numbers, the numbers can be of any value. If we have five numbers but require that the mean must be four, we have interfered with the freedom of these numbers to vary. How many of the five numbers are free to vary and still meet the condition that the mean must be four? We shall illustrate this in table 10–4. If four of the numbers are 3, 8, 2, 5,

TABLE 10–4

2	3
6	8
4	2
5	5
3	?
$\Sigma = 20$	$\Sigma = 20$
$\bar{X} = 4$	$\bar{X} = 4$

and we require that the mean of the five numbers must be four, then the fifth number (indicated by the question mark) *must be two*. No other number will fulfill the requirements. Any four of the five numbers are free to vary; but by the requirement that the mean shall be a specified value, the fifth number is no longer free to vary. Thus, we would have four degrees of freedom associated with these five observations. If we draw a sample of N observations from a population, how many degrees of freedom do we have in estimating the population mean? The answer is N, since we are not placing any restrictions on the sample mean. It is the sum of all the observations divided by the degrees of freedom (N). If we estimate the standard deviation of the population (σ) from the sample, how many degrees of freedom do we have making this estimate? $N - 1$. To determine a standard deviation, we must use an estimate (the mean of the sample) in the computations. We have placed one restriction on the freedom of the deviations to vary. The sum of the deviations must add to zero about the sample mean (an estimate). Thus the unbiased estimate of the population standard deviation (s) will be found by summing all of the squared deviations (Σx^2) and dividing by the degrees of freedom ($N - 1$).

Returning to the use of table G in determining the significance of the t statistic, we must determine the number of degrees of freedom for the particular set of data. The unbiased estimate of the population standard deviation based on *both* samples was given in formula 10–2, which is repeated here:

$$s = \sqrt{\frac{\Sigma x_1^2 + \Sigma x_2^2}{(N_1 - 1) + (N_2 - 1)}}$$

Note that each sample contributes observations to the determination of the population standard deviation. There are N_1 observations from sample 1; but in computing the sum of squared deviations, the sample mean had to be used, and thus there are only $N_1 - 1$ degrees of freedom for the estimation of the population standard deviation from the first sample. Sample 2 contains N_2 observations, or $N_2 - 1$ degrees of freedom for the estimate of the population standard deviation. What are the degrees of freedom to be used in entering table G? $N_1 - 1$ or $N_2 - 1$? The sum of the two sets of degrees of freedom. We have $N_1 + N_2$ observations, but each set of observations has had a restriction placed on the freedom of the deviations to vary, and thus we have lost two degrees of freedom. The total degrees of freedom for entering the table of the t statistic will be

$$(N_1 - 1) + (N_2 - 1) = N_1 + N_2 - 2$$

The computation of the t statistic is shown in example 10–3.

Example 10–3   Illustration of the computation of Student's t with unequal and small samples

Scores for group I		Scores for group II	
X	X^2	X	X^2
58	3,364	94	8,836
51	2,601	82	6,724
41	1,681	73	5,329
40	1,600	67	4,489
33	1,089	66	4,356
27	729	62	3,844
20	400	60	3,600
18	324	$\Sigma X_2 = 504$	$\Sigma X_2^2 = 37,178$
$\Sigma X_1 = 288$	$\Sigma X_1^2 = 11,788$		

$$N_1 = 8 \qquad N_2 = 7$$
$$\bar{X}_1 = 36.0 \qquad \bar{X}_2 = 72.0$$

Using formula 10–5 for the standard error of the difference in means

$$s_{D\bar{x}} = \sqrt{\left(\frac{\Sigma X_1^2 + \Sigma X_2^2 - (N_1\bar{X}_1^2 + N_2\bar{X}_2^2)}{N_1 + N_2 - 2}\right)\left(\frac{1}{N_1} + \frac{1}{N_2}\right)}$$

$$s_{D\bar{x}} = \sqrt{\left(\frac{11,788 + 37,178 - [8(36)^2 + 7(72)^2]}{8 + 7 - 2}\right)\left(\frac{1}{8} + \frac{1}{7}\right)}$$

$$s_{D\bar{x}} = \sqrt{\left(\frac{48,966 - (10,368 + 36,288)}{13}\right)(125 + .143)}$$

$$s_{D\bar{x}} = \sqrt{\left(\frac{48,966 - 46,656}{13}\right)(.268)} = \sqrt{\left(\frac{2,310}{13}\right)(.268)}$$

$$s_{D\bar{x}} = \sqrt{(177.692)(.268)} = \sqrt{47.621456} = 6.90$$

$$t = \frac{D\bar{x} - 0}{s_{D\bar{x}}} = \frac{\bar{X}_1 + \bar{X}_2}{s_{D\bar{x}}} = \frac{36.0 - 72.0}{6.90} = \frac{-36.0}{6.90} = 5.22$$

(the sign of t does not make a difference)

$$t = 5.22; \quad df = (N_1 - 1) + (N_2 - 1) = (8 - 1) + (7 - 1) = 13$$

Using table G of the Appendix, we find that $df = 13$, $p < .001$, for a $t = 5.22$.

MATCHED SAMPLES

To this point we have used two different samples which were selected *independently* of each other. In many research situations it is desirable to use the same sample of individuals in two or more conditions of the same experiment. Hence these different conditions, because they utilize the same subjects, will not be independent of each other. In other cases, different subjects may be used in the different conditions of the study but it is desirable to pair similar subjects together—match subjects.

Perhaps this can be illustrated by a hypothetical experiment to find which of two methods of teaching a course in American government might be better. The dependent variable could be the scores earned on the tests during the course, and the independent variable could be the techniques of teaching. Obviously, it would be foolish to have a group of students take the course twice—once under each teaching method. In addition, there certainly would be some transfer from the first time to the second. We would have to use two different groups of subjects. One variable that we cannot eliminate from the study is scholastic aptitude. Students who score high on scholastic aptitude tests tend to do better in college courses than students who obtain low scores on these tests. It would never do to have more students with high-aptitude scores in one group. To equalize this effect, we would simply make sure that the two groups are equal with respect to aptitude test scores. The most straightforward way to accomplish this is through matching the two groups. Thus, we would examine the aptitude scores of all the students and pair the students by these scores. One high scorer would be assigned to one section, and the other member of the pair would be assigned to the second section. We would continue this procedure for each of the pairs of matched students (the decision as to which member of the pair should be assigned to each section should be a matter of chance— some randomization technique would be used).

After completion of the study, we could use a t test to test the significance of the difference in means between the two groups. We might (incorrectly) use formula 10–9 for t. It was pointed out that each of the two samples must be random samples and must be obtained *independently of each other*. When each of the two samples is random, then equation 10–3 for the standard error of the difference in means is correct for use in the t test. For convenience, equation 10–3 is repeated below:

$$s_{D\bar{x}} = \sqrt{\frac{s^2}{N_1} + \frac{s^2}{N_2}}$$

STANDARD ERROR OF THE DIFFERENCE IN MEANS FOR MATCHED SAMPLES.   In the present situation the samples are not independent of each other; in fact, they have been carefully matched, so that each group has within it a student with the same aptitude score as a student in the other group. The two groups are *correlated* and not independent. It can be shown that the standard error of the difference in means for correlated groups may be obtained from the following equation:

$$s_{D_{\bar{x}}} = \sqrt{\frac{s^2}{N} + \frac{s^2}{N} - 2r\frac{s^2}{N}} \qquad (10\text{--}10)$$

where

s^2 is the unbiased estimate of the population variance based on both samples
N is the sample size for each of the matched groups
r is the product-moment correlation between the groups

Notice that the difference between this expression and the usual denominator of the t ratio is in the term to the right of the negative sign. If the groups are correlated, we must subtract a given amount from the usual denominator for t. The only new term in the expression is the correlation coefficient (r). This is the correlation between the *dependent* variable and the variable used to *match* the subjects. The effect of this correlation will be to reduce the size of the denominator of the t ratio. A reduction of the denominator will of course result in a larger t (provided the difference in means remains the same). Decreasing the denominator of the t ratio results in an increase in the *precision* of the experiment. By increasing the precision of the experiment, we are more apt to detect a difference between the two samples if a true difference does in fact exist between the populations from which the samples were drawn.

Suppose that we were to match subjects in an experiment and then discover that no correlation exists between the matching variable and the dependent variable. Notice that the negative term in the equation for the standard error of the difference in means becomes zero and the denominator reduces to the usual expression for the denominator of t with random groups. Actually, this term for use with random samples is a special case of the general equation for the standard error of the difference in means—a special case where the correlation is zero. (Remember that since the groups are selected at random, they are independent of each other; hence, no correlation exists between the two.)

DEGREES OF FREEDOM FOR MATCHED SAMPLES.   At first glance, it would appear that we have everything to gain and nothing to lose by matching the two groups. At best, we may gain in precision; at worst, the study will have the same precision that would have been achieved by selecting the two samples independently of each other. Any slight correlation would increase precision. This is true insofar as the size of the t ratio is concerned, but consider one other factor—the degrees of freedom which are needed to determine the significance of the t. For two random samples the degrees of freedom will be equal to $(N_1 - 1) + (N_2 - 1) = N_1 + N_2 - 2$. What about the case of matched samples? Remember that the concept of degrees of freedom involves the idea of the independence of the scores—freedom to vary.

Let us examine one pair of subjects that have been matched for the study on methods of teaching a course on American government. If we examine the two students with the highest aptitude scores, what would be the prediction about their scores on the course examinations? Since we have stated that there is a high correlation between aptitude test scores and college grades, we would predict that both students would do well in the study of American government. The freedom of the two scores to vary has been restricted by the matching procedure. The freedom of variation comes only in the selection of the *pair* of subjects. When one subject is chosen, this automatically fixes the second subject in the pair as being one with the same score on the matching variable and hence a very similar score on the dependent variable. The sampling unit becomes the pair of subjects—not the individual subject. The total degrees of freedom becomes $N - 1$, where N is the number of *pairs* of subjects. Matching has resulted in the number of degrees of freedom being cut in half.

Returning to the situation where the r between the matching variable and the dependent variable is zero, we find that we have not increased the precision of the study, but we shall have reduced the degrees of freedom by one half. The t ratio obtained would be the same size as that which we would have obtained by using two random samples, but the degrees of freedom would be smaller. For a t ratio of a given size, we would be less likely to detect a true difference if such a difference did exist, since the degrees of freedom available would have been reduced drastically. The probability of a Type II error has been increased.

Matching of subjects in an experiment is an extremely useful and powerful tool, but only if there is a substantial correlation between the dependent variable and the matching variable.

COMPUTATION OF t WITH MATCHED GROUPS. To illustrate the computation of the t statistic with matched groups, we shall use some fictitious data on the learning of nonsense material under two different degrees of distributed practice. Let us assume that all subjects first learned a verbal maze, and that it was possible to match the subjects in terms of their performance with verbal maze learning. After the pairing of scores, one subject from each pair was assigned randomly to the zero-second-rest group, and the other member was assigned to the group receiving a 30-second rest. Both groups then learned the same list of nonsense syllables to the criterion of two consecutive trials without error. The number of trials to reach the criterion was determined for each subject. The data are presented in example 10–4.

Example 10–4 The t statistic for matched groups

Number of trials to learn a list of nonsense syllables by a zero-second-rest and a 30-second-rest group—groups were matched on ability to learn a verbal maze.

| Pair | Delay period | | Difference in scores |
	0 seconds	30 seconds	
1	31	8	23
2	37	13	24
3	45	22	23
4	28	25	3
5	50	29	21
6	37	31	6
7	49	35	14
8	25	38	−13
9	36	42	−6
10	69	52	17
	$\bar{X} = 40.7$	29.5	11.2
	$S = 12.35$	12.56	12.47
	$s^2 = 172.25$		

The correlation between these two sets of scores is $r = .50$. Using the equation for t with matched groups:

$$t = \frac{D_{\bar{x}} - 0}{s_{D\bar{x}}} = \frac{\bar{X}_1 - \bar{X}_2}{\sqrt{\dfrac{s^2}{N} + \dfrac{s^2}{N} - 2r\dfrac{s^2}{N}}}$$

$$t = \frac{40.7 - 29.5}{\sqrt{\dfrac{172.25}{10} + \dfrac{172.25}{10} - (2)(.50)\dfrac{172.50}{10}}} = \frac{11.2}{\sqrt{34.45 - 17.22}}$$

$$t = \frac{11.2}{\sqrt{17.32}} = \frac{11.2}{4.15} = 2.70$$

Since there are ten *pairs* of subjects, the number of degrees of freedom is equal to nine. Entering the table of the t statistic with $df = 9$, we find that a t of 2.70 is significant beyond the .05 level.

Now, suppose that the subjects had been matched but the experimenter (incorrectly) used the equation for the t statistic with random groups (unmatched):

$$t = \frac{\bar{X}_1 - \bar{X}_2}{\sqrt{\dfrac{s^2}{N} + \dfrac{s^2}{N}}} = \frac{40.7 - 29.5}{\sqrt{\dfrac{172.25}{10} + \dfrac{172.25}{10}}} = \frac{11.2}{\sqrt{43.45}}$$

$$t = \frac{11.2}{5.87} = 1.91$$

Entering the table of the t statistic with $df = 9$, a t of 1.91 is not significant at the .05 level. In this case, by using the inappropriate equation for the t, the research worker would fail to reject the null hypothesis and could well be committing a Type II error.

THE DIFFERENCE METHOD OF COMPUTING t.   The procedure for computing t from matched groups can be simplified so that the computations are more direct and do not require the computation of the correlation coefficient. It can be shown that the difference in means is equal to the mean of the differences of the paired scores ($D_{\bar{x}} = \bar{X}_D$). Referring to example 10–4, the mean of the column labeled *differences in scores* is 11.2. This is equal to 40.7 − 29.5, or the differences in means. Similarly, it can be shown that the standard deviation of the difference in mean ($s_{D_{\bar{x}}}$) is equal to the standard deviation of the mean of the set of differences ($s_{\bar{x}_D}$). To compute t by the direct-difference method, simply find the difference between each pair of matched scores and compute the standard deviation of those differences, and then find the standard error of the difference:

$$s_{\bar{x}_D} = \frac{S_D}{\sqrt{N - 1}}$$

and the t by the difference method would be

$$t = \frac{\bar{X}_D}{s_{\bar{x}_D}} = \frac{11.2}{\dfrac{12.47}{\sqrt{10 - 1}}} = \frac{11.2}{4.16} = 2.70$$

This is the same result as the longer technique which requires the computation of the r between the scores.

SANDLER'S A STATISTIC FOR MATCHED SAMPLES.   For matched (correlated) samples in which the hypothesized true difference is zero ($\mu_1 - \mu_2 = 0$), Sandler[1] has derived a variation of the direct-difference

---

[1] J. Sandler, "A Test of the Significance of the Difference between the Means of Correlated Measures, Based on a Simplification of Student's t," *British Journal of Psychology* 46(1955):225–26.

method of computing t which is simple to use. A special table must be used, but the probability values obtained are identical to those obtained with the t statistic. Sandler's A statistic is found by

$$A = \frac{\Sigma D^2}{(\Sigma D)^2} \qquad (10\text{-}11)$$

where

ΣD^2 is the sum of the squared differences between the matched individuals
$(\Sigma D)^2$ is the square of the sum of the differences between the matched individuals

Table H of the appendix is used to determine the probability of the occurrence of an A of this value or *smaller* if the true difference is zero. Table H presents both one-tailed and two-tailed probabilities. All that is needed to use the table are the degrees of freedom $(N-1)$ and the value of A as obtained from formula 10–11. Example 10–5 shows the computation of Sandler's A.

Example 10–5   Sandler's A for matched groups

The data for this example is the same as that in example 10–4.

Pair	Group I	Group II	Difference (D) in scores	D²
1	31	8	23	529
2	37	13	24	576
3	45	22	23	529
4	28	25	3	9
5	50	29	21	441
6	37	31	6	36
7	49	35	14	196
8	25	38	−13	169
9	36	42	−6	36
10	69	52	17	289
			$\Sigma D = 112$	$\Sigma D^2 = 2{,}810$

$$A = \frac{\Sigma D^2}{(\Sigma D)^2} = \frac{2{,}810}{12{,}544} = .224$$

Table H with $df = 9$ shows a value of .276 for the .05 level two-tailed. Since .224 is less than .276, the null hypothesis would be rejected ($\alpha = .05$) which is the same result as in example 10–4 with this same data.

ASSUMPTIONS IN THE USE OF STUDENT'S t.   The t statistic, like any other statistic, is a mathematical model. Certain conditions or requirements are necessary to derive that particular statistic. The statistic is useful to us in our research to the extent that the data we obtain come reasonably close to meeting the requirements or assumptions of the model. In the case of the t statistic, there are four of these conditions or assumptions which are fairly important:

1. The measurements should be reasonably close to interval or ratio. Since the t statistic utilizes the means and standard deviations, this assumes that the operations of addition may be carried out on the numbers.

2. The samples used in computing the statistics are random samples from their respective populations.

3. The populations from which these samples are drawn are normally distributed. Notice that the t statistic was developed primarily for use with small samples. If the sample size is large, then the central-limit theorem would hold, and the underlying distribution of the population would not be of any marked importance. The t statistic for small samples does require that the underlying population(s) be normally distributed.

4. The variances of the two populations are equal. In the equation for the standard error of the difference in means the standard deviations based on each sample are combined to form a single, unbiased estimate of one population standard deviation. (See equation 10–2.)

ROBUSTNESS OF THE t STATISTIC.   The validity of the probability statements obtained from hypothesis testing would appear to be dependent upon the degree to which the obtained data meet the requirements of the mathematical model. Since many data of the behavioral sciences are measured in a crude fashion, a good deal of effort has been directed toward determining how critical these assumptions are. Can one use the t statistic for testing for differences in IQ scores between two different groups? Certainly IQ scores (and most testing data) do not meet the requirement of interval scales. If the scaling assumptions are met but the underlying population distributions are not normally distributed, what effect does this have on the use of the t statistic? With the development of high-speed computers, these assumptions have been tested empirically. The usual procedure is to construct a population of scores with a given degree of nonnormality or with different population variances. The computer is then programmed to draw small samples from these special populations. A t statistic is computed, and another sample is drawn and a second t computed. Thousands of t statistics are computed based on random samples, all drawn from these nonnormal distributions. The frequency distribution of these thousands of t statistics is compared with the theoretical distribution of t's. The results of such empirical research have been reassuring. It would appear that the assumptions of the *t* statistic can be violated (within limits) without producing a great difference in the accuracy of the probability statements. This characteristic of a statistic to yield reasonably accurate results even though the basic assumptions have been violated is referred to as *robustness of the statistic*. The Student's *t* is a highly robust test.

Summary

The idea of estimating the true mean of a distribution through use of a confidence interval can be extended to estimating the true *difference between the means* of two different distributions. The *standard error of the difference in means* can be estimated from the two samples and this standard deviation then used to establish a confidence interval about the observed difference between the samples. This assumes that each sample variance is an unbiased estimate of the same population variance.

If the research worker is interested in testing a hypothesis that the true difference between two populations is some specific value, a Z score approach is used. If the Z score is larger than a predetermined value, the research worker will reject the idea that the true difference was the specific amount hypothesized. Although it is possible to use the Z score or critical-ratio approach to test any hypothesized true difference, the hypothesis of no difference is most commonly used in the behavioral sciences.

The decision to reject the null hypothesis or not to reject it is based on samples drawn from the two populations. Thus, it is possible that an incorrect decision concerning the null hypothesis may have been made. Rejecting the null hypothesis when it is true is referred to as a *Type I error* (stating that a difference does exist when there is no true difference). The Greek letter alpha (α) is used to symbolize the probability of a Type I error. Alpha is completely under the control of the research worker. Whatever critical ratio or Z score is selected in advance for rejection of the null hypothesis sets the probability of the Type I error.

Failure to reject the null hypothesis when it is false is also an incorrect decision. This is called a *Type II error*. The probability of a Type II error is symbolized by the Greek letter beta (β).

On a sampling basis, it is not possible to "prove" the null hypothesis. Failure to reject the null hypothesis simply means that there is not sufficient evidence to indicate that the hypothesis is false.

The rejection of the null hypothesis when it is false is referred to as the *power of the statistic*. The power of a statistic is $1 - \beta$; thus the factors which influence the probability of a Type II error are the same that lead to the correct decision—rejection of the null hypothesis when it is false.

The probability of the occurrence of a Type II error is governed by the following factors:

1.   The α level set by the research worker for rejecting the null hypothesis
2.   The size of the true difference between the population means
3.   The amount of variability in the research material
4.   The sample size
5.   The statistic used to analyze the data

The Student's t statistic for small samples is computationally the same as the familiar Z score. The sampling distribution of the t statistic changes as a function of the size of the sample from a leptokurtic distribution for very small samples to the normal curve for large samples. The table of the t statistic is actually many tables combined into one. Since the sampling distribution for each sample size is different, only certain frequently used points in the distribution are included in the table of the t statistic. To use the table of the t statistic, it is necessary to determine the number of *degrees of freedom* (df). In general, the number of degrees of freedom is the total number of independent observations less the number of population values which must be estimated from the samples.

Most tests of hypotheses in the behavioral sciences are two-tailed. That is, the research worker is attempting to find if groups differ without predicting which group will have the larger mean. The table of the t statistic is set up for this two-tailed type of test. On rare occasions the research worker may predict in advance the direction of differences between two groups, in which case a one-tailed test of significance should be used.

It is sometimes desirable to use the same individuals in more than one condition of a study or to match different individuals on some variable which is related to the dependent variable. When the groups are *not independent* (they are correlated), the standard error of the difference in means becomes smaller by a factor related to the magnitude of the correlation between the groups. The degrees of freedom for correlated groups is $N - 1$ where N is the number of matched *pairs*. While the t test for correlated groups can be computed by using the correlational factor directly in the computations, the difference technique is much simpler. The preferred technique for matched samples is to use Sandler's

A statistic which is a variation of the difference method of computing a correlated *t* test. The Sandler *A* test requires the use of table H in the Appendix, rather than the table of the *t* statistic.

The assumptions necessary for interpretation of the *t* statistic are: (1) the measurements are interval or ratio, (2) the samples are random samples from their respective populations, (3) the populations from which the samples are drawn are normally distributed, and (4) the populations have equal variances. A statistic is said to be robust if departure from the assumptions does not have a severe effect on the accuracy of probability statements derived from the statistical results. The *t* statistic is highly robust; thus, failure to meet the assumptions of the *t* statistic does not destroy its usefulness in testing hypotheses.

References

HYPOTHESIS TESTING

PAUL BLOMMERS and E. F. LINDQUIST, *Elementary Statistical Methods in Psychology and Education* (2d ed.; Boston: Houghton Mifflin Co., 1960), chap. x.

SIDNEY SIEGEL, *Nonparametric Statistics for the Behavioral Sciences* (New York: McGraw-Hill Book Co., Inc., 1956), chap. ii.

ASSUMPTIONS OF THE *t* STATISTICS

C. A. BONEAU, "The Effects of Violations of Assumptions Underlying the 't' Test," *Psychological Bulletin,* Vol. 57 (1960), pp. 49–64.

11

Significance of differences between groups with nominal or ordinal scaling (nonparametric statistics)

Since the late 1940s, there has been an "explosion" in the number of different statistical tests developed for specialized situations in which the t statistic is not appropriate. The term *nonparametric* or *distribution-free statistics* has been applied to these tests. These nonparametric statistics are designed to analyze data for two types of situations: (1) measurements that are ordinal or nominal in nature, and (2) distributions whose shapes are either nonnormal or unknown.

Since the nonparametric statistics do not utilize all the information that is contained in numbers of an interval or ratio scale, they are not as powerful as the t statistic. That is, there will be a larger probability of committing a Type II error with a nonparametric statistic than with the t test. This is not a serious problem since one way to guard against the Type II error is to increase the sample size. If a larger N is used, then in many cases, the nonparametric statistic will have adequate power for the situation. Direct comparison of the nonparametric statistics with the t test is perhaps not too important. When possible, the investigator will use the most powerful statistic which is suitable for the particular research situation. The nonparametric statistics are used when it is not possible to use the more powerful parametric techniques.

It is not possible to include all these different tests in an introductory book. However, the tests presented in this chapter are included for two reasons: (1) to acquaint the student with this general area and (2) because they are the tests most likely to be useful for the beginning student.

Testing for significance of differences with nominal data

Nominal scaling involves assigning cases to categories. Thus a significance test using nominal data must involve two questions: (1) What frequency do we expect in each category? (2) What frequency have we observed in each category? A significance test will involve a comparison of the *expected frequency* in each category with the *observed frequency* in the same categories.

If, in tossing a coin with a "friend" to determine who buys coffee, the friend wins ten times in a row, we might become a little skeptical. Our suspicions would be greater if the friend always used his own coin and always called heads. At this point, we might ask to examine the coin. There is only one way that ten heads in a row can occur in tossing a coin ten times (if it is an honest coin). There are 1,024 ways in which the pattern of heads and tails can be arranged in tossing the coin 10 times (see chap. 6, pp. 122–25). The probability that the friend is using an honest coin is 1/1,024—a very rare occurrence for an honest coin! The student will recognize the use of the binomial expansion for the solution of this problem. Whenever events can be divided into two mutually exclusive categories and we can count the frequency of cases in each category, the binomial expansion can be used.

THE BINOMIAL TEST

Many binomial problems are concerned with the special situation of

$$p = .50 \text{ and } q = 1 - p = .50$$

Since this situation occurs fairly often, a special table has been constructed (table I of the Appendix) for ease in determining the probability of occurrence by chance. A portion of table I is reproduced below as table 11–1. Suppose that we are tossing eight coins. What is the probability of all eight coins being heads? Three heads and five tails? These problems can be answered by direct use of the binomial expansion; but since only two categories are involved and $p = q = .50$, table I can be used directly. Using table 11–1 for these specific questions, find in the first column (labeled N) the number of elements (coins) in the problem. In the first row of the table (labeled x), we must locate the frequency of cases observed. In table 11–1, x is the smaller frequency of the two categories. In the cell for $N = 8$ and $x = 0$ (no tails in tossing eight coins), we

TABLE 11–1
Probabilities associated with values as small as observed
values of x in the binomial test

x / N	0	...	3	...	6	...	9	...	12
5	031		812						
...	...		...						
8	004		363		965				
...	...		...		...				
20		...	001		058		412		868
...	...		...		...		...		...

read .004, which is the probability of this event occurring by chance if the expected outcome is four heads and four tails. The probability of *three or fewer* heads is found in the cell $N = 8$ and $x = 3$ and is .363.

The general term for the expansion of the binomial was presented on page 123 (formula 6–1). Although this can be used to determine the coefficients of the binomial for any combination of events, it is rather laborious to compute.

Table C of the Appendix gives the coefficients for all expansions up to and including $n = 20$. If p and q are reasonably equal to .50, the distribution of the coefficients of the binomial is approximately the normal curve for n greater than 20. In such situations the normal curve is used to determine the probabilities rather than the direct solution of the binominal expansion (see chap. 6, pp. 125–29).

THE CHI-SQUARE TEST

The statistic χ^2 has been developed for use with data that are not expressed in measurements, but rather in terms of the number of individuals (or objects) in each of several categories. This type of data is often referred to as "count" data—that is, we count the number of objects in the category, we do not measure these objects. If some set of categories is established, we can count the number of individuals in each category, and we have a set of observed frequencies. If these individuals were a random sample from a population, we could ask if the observed frequencies deviated markedly from the number of cases we expected in each category. The categories may be nominal in nature, or they may be ordered, or the categories could be based on interval or ratio scales. The important point is that the recorded data are in the form of frequency (counts) within each category. Two items of information are necessary to compute a χ^2: (1) the actual number of cases *observed* in each category and (2) the number of cases *expected* in each category. The formula for the computation of the χ^2 is

$$\chi^2 = \sum \frac{(O - E)^2}{E} \qquad (11\text{--}1)$$

where

O is the number of cases observed in one category
E is the number of cases expected in the same category

The hypothesis we are testing is that the observed frequencies do not differ from the frequencies that are expected in each category (null hypothesis). An alternate way of expressing the same idea is that the· differences between the observed frequency and the expected frequency are due to chance factors. If the χ^2 is too large, then we would reject the null hypothesis and conclude that the observed frequencies differ significantly from the expected frequencies.

We stated that if the χ^2 were too large, we would reject the null hypothesis. How large is too large? As in the case of the t statistic, we must determine the probability of the occurrence of a χ^2 of this size or larger. Just as we needed to know the sampling distribution of t, we must determine the sampling distribution of chi-square.[1] As in the case of the t statistic for small samples, there are many different sampling distributions of chi-square. The sampling distribution of chi-square is dependent upon the number of degrees of freedom. Sampling distributions for chi-square with differing degrees of freedom are shown in figure 11–1.

Figure 11–1 indicates the marked shift in the shape of the sampling distribution of chi-square as the number of degrees of freedom increases. As the number of degrees of freedom approaches infinity, the sampling distribution of chi-square approaches the normal curve.

ONE-WAY CLASSIFICATION.   The computation of χ^2 depends upon the observed frequencies and the expected frequencies. To determine if a specific χ^2 is statistically significant, we also need to determine the number of degrees of freedom associated with the χ^2. The degrees of free-

---

[1] The symbol χ^2 will be used to refer to the statistic computed from the data. The term *chi-square* will be used to refer to the theoretical distribution.

FIGURE 11–1
The chi-square distribution for various *df*'s

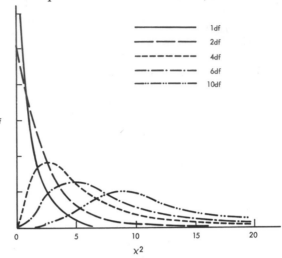

dom for a χ^2 *are* *not* related to the number of individuals in the sample. The degrees of freedom *are* dependent upon the number of *independent* differences between observed and expected frequencies. The computation and determination of degrees of freedom are illustrated in example 11–1.

To determine if the special group of students is significantly different from the national norms, we would turn to table J in the Appendix. Entering the table with $df = 4$, we find that the χ^2 of 112.93 will occur by chance less than 1 time in 100 (112.93 > 13.28); hence, we can reject the null hypothesis and state that the special sample is different from the national norm group.

Why four degrees of freedom? Remember that the number of degrees of freedom are dependent upon the number of independent $O - E$ deviations. There are five $O - E$ deviations, but only four of them are free to vary. The sum of $O - E$ must equal zero; thus, with five categories, only four are free to vary and still meet the requirement that the sum of $O - E$ equals zero.

The situation illustrated in example 11–1 is a one-way classification of the data. All of the data can be classified into categories lying along one dimension. The expected values were determined from outside the data—the proportion of individuals within the national norm group in each category.

Example 11-1 Computation of χ^2 from reading test scores of a sample of freshman college students

A reading examination was administered to a special group of 942 college freshman. Do these freshman differ significantly in reading ability from the group which was used to establish norms on the test? Since the information about the test scores was reported in percentile form (ordinal data), it did not seem reasonable to use the t test. The test scores were divided into quintiles.

Percentiles based on national norms	Observed frequency	Expected frequency	$O - E$	$(O - E)^2$	$\dfrac{(O - E)^2}{E}$
80–99	95	188.4	−93.4	8,723.56	46.30
60–79	184	188.4	−4.4	19.36	.10
40–59	213	188.4	24.6	605.16	3.21
20–39	293	188.4	104.6	10,941.16	58.07
0–19	157	188.4	−31.4	985.16	5.23
	$\Sigma = 942$	$\Sigma = 942$	$\Sigma = 000.0$		$\Sigma = 112.92$

Since the data are placed in quintiles based on the national norms, we would *expect* that 20 percent of the students should be in each quintile. Multiplying the total number of students (942) by 20 percent will yield an *expected* frequency of 188.4 students in each category, if the special sample is no different than the national group.

$$\chi^2 = \sum \frac{(O - E)^2}{E} = 112.92; \text{ degrees of freedom } (df) = 4$$

TWO-WAY CLASSIFICATION (INDEPENDENT SAMPLES). Oftentimes, we are interested in classifying individuals in two ways to determine if membership in one classification implies membership in some other classification. Suppose we were interested in the use of the college library by biology majors as compared to English majors. We have the one set of categories, biology majors and English majors. The other categories might be *use* and *nonuse* as defined in a particular way. Carrying this example a bit further, suppose that we have identified 206 students who are either English majors or biology majors (105 in biology and 101 in English). We decide that during one 24-hour period we are going to identify the students who used the library and then determine if there is a difference between the number of students in these two major fields in their library usage on that specific day. The computation is shown in example 11–2.

In example 11–2 the expected value for the biology students who used the library (cell *a*) is given as 55.6. Where did we obtain this value? We do not have any outside information on which to base the expected value in each cell. This value will have to be determined from

Example 11–2 Comparison of the library usage of biology majors and English majors

MAJOR FIELD

	Biology	English	
Used library	*a* 47 (55.6)	*b* 62 (53.4)	109
Did not use library	*c* 58 (49.4)	*d* 39 (47.6)	97
	105	101	206

For convenience, the four cells in the table have been labeled *a*, *b*, *c*, and *d*. The observed frequencies are entered in each cell, and the expected frequencies have been placed in parentheses in each cell.

$$x^2 = \sum \frac{(|O - E| - 0.5)^2}{E} = \frac{(|47 - 55.6| - 0.5)^2}{55.6} + \frac{(|62 - 53.4| - 0.5)^2}{53.4}$$
$$+ \frac{(|58 - 49.4| - 0.5)^2}{49.4} + \frac{(|39 - 47.6| - 0.5)^2}{47.6}$$

$$x^2 = \frac{(-8.1)^2}{55.6} + \frac{(8.1)^2}{53.4} + \frac{(8.1)^2}{49.4} + \frac{(-8.1)^2}{47.6} = 1.18 + 1.23 + 1.33 + 1.38$$

$$x^2 = 5.12; \quad df = 1; \quad p < .05$$

the data themselves. Since there were 109 students out of the total of 206 who used the library, we can state that 53 percent of the 206 students used the library. If there were no differences in library usage between biology majors and English majors (null hypothesis), we would expect that 53 percent of each group would use the library. Since there are 105 biology majors, under the null hypothesis we would expect 53 percent of 105 (55.6) of the biology majors to use the library on that day. Similarly, the expected frequency of English majors who used the library (cell *b*) should be 53 percent of 101, or 53.4. The expected frequency for cell *c* would be found by dividing 97 (number of students not using the library) by 206 (total number of students in the group), and multiplying this by 105, or $(97/206) \times 105 = 49.4$. The same procedure would be followed to find the expected frequency for cell *d*.

In example 11–1, there were five categories and four degrees of freedom. In example 11–2, there are four categories, but the degrees of freedom are given as only one. At first glance, it would appear that

there should be three degrees of freedom associated with these four categories. Why only one? One of the conditions for the use of chi-square
is that the sum of the expected frequencies must equal the sum of the
observed frequencies. In the table in example 11–2, 109 students used
the library; thus, when we determine that the expected frequency in cell
a is 55.6, this "sets" the frequency in cell *b* at 53.4, so that the total
is 109. Only one cell frequency is free to vary. Although one can always
determine the number of degrees of freedom by inserting different numbers and finding how many are free to vary, a simpler procedure is available. In any two-way table for χ^2, the number of degrees of freedom
will be the product of one less than the number of rows times one less
than the number of columns.

$$df = (r - 1)(c - 1) \tag{11-2}$$

where

  r is the number of rows in the table
  c is the number of columns in the table

Notice that the formula used in the computation of example 11–2
was not the same as the formula used in example 11–1. The derivation
of the chi-square distribution assumes that chi-square is continuous and
not discrete. It also assumes that the expected values will tend to be
large. Many writers warn against using the test when the degrees of
freedom are one, or when the expected frequencies are less than five.
These restrictions can be overcome by the use of *Yates' correction*. This
is accomplished by subtracting 0.5 from the positive $(O - E)$ discrepancies and by adding 0.5 to the negative $(O - E)$ discrepancies before
these values are squared. This is shown in formula 11–3.

$$\chi^2 = \sum \frac{(|O - E| - 0.5)^2}{E} \tag{11-3}$$

where

$|O - E|$ is the absolute value of the discrepancy

In working with a 2×2 table for χ^2, a convenient computing arrangement has been developed by Quinn McNemar. This is especially convenient
if a desk calculator is available. If the letters *A*, *B*, *C*, and *D* represent
the frequencies in the four cells of a 2×2 table, then the marginal

frequencies will be $A + B$, $C + D$, etc. These are shown in the following diagram.

A	B	$A + B$
C	D	$C + D$

$$A + C \qquad B + D \qquad A + B + C + D$$

The expected frequency for the cell with the frequency of A will be

$$E_a = \left(\frac{A + B}{A + B + C + D}\right)(A + C)$$

The expected frequency for cell d will be

$$E_d = \left(\frac{C + D}{A + B + C + D}\right)(B + D)$$

By a little algebraic manipulation, it can be shown that in a 2×2 table

$$\chi^2 = \frac{N(|AD - BC| - .5N)^2}{(A + B)(C + D)(A + C)(B + D)} \qquad (11\text{–}4)$$

The computation of χ^2 using formula 11–4 is shown in example 11–3.

In the two-way classification with one degree of freedom, certain cautions should be observed. The Yates correction for continuity must be

---

Example 11–3   Computation of χ^2 for a 2×2 table

MAJOR FIELD

	Biology	English	
Used library	47	62	109
Did not use library	58	39	97
	105	101	206

Using formula 11–4:

$$\chi^2 = \frac{(206)[|(47)(39) - (62)(58)| - (.5)(206)]^2}{(109)(97)(105)(101)} = \frac{(206)(|1{,}833 - 3{,}596| - 103)^2}{112{,}126{,}665}$$

$$\chi^2 = \frac{567{,}653{,}600}{112{,}126{,}665} = 5.06; \quad df = 1; \quad p < .05$$

---

used—i.e., equations 11–3 or 11–4. If the expected frequency in any cell is less than five the chi-square distribution should not be used. The Fisher-Yates exact-probability test which is presented later in this chapter, should be used when the expected frequency in any cell is less than five.

THE MCNEMAR TEST OF CHANGE (CORRELATED SAMPLES). It is often possible to divide individuals into two categories before the administration of some experimental procedure and then determine the proportion in each of the two categories after the experiment. The research worker is interested in the number of individuals who "changed" and the number who did not change. A political scientist might be interested in the impact of college education upon the party identification of students. He might ask new freshman to identify themselves as Republicans or Democrats. (In this situation, only two categories can be used; thus, third-party identification would have to be discarded. Or the categories could be *Republican* and *Other*.) Four years later, these same individuals would again be asked to identify themselves as to their political party. A 2×2 table could be constructed as follows:

After college

		Democrat	Republican
Before college	Republican	A	B
	Democrat	C	D

Since the political scientist is interested in those individuals who have changed, cells A and D are the two of interest. McNemar has shown that in this type of situation a χ^2 with one degree of freedom can be computed, using only those individuals who changed:

$$\chi^2 = \frac{(|A - D| - 1)^2}{A + D} \tag{11-5}$$

where

A is the frequency of cases that changed in one direction
D is the frequency of cases that changed in the other direction

SIGNIFICANCE OF χ^2 WHEN $df > 30$. As the number of degrees of freedom increases, the sampling distribution of chi-square approaches the normal curve. With degrees of freedom equal to or larger than 30, the normal curve can be used for testing for the significance of any obtained χ^2. To use the table of the normal curve, we must be able to cast the data into Z score form. The χ^2 can be converted into a Z score by use of the following formula:

$$Z = \sqrt{2\chi^2} - \sqrt{2df - 1} \qquad (11\text{--}6)$$

where

χ^2 is computed from the data
df is the degrees of freedom for the χ^2

ASSUMPTIONS FOR THE USE OF THE CHI-SQUARE DISTRIBUTION. The computation of a χ^2 and the use of the table of the chi-square distribution are not different in principle from the computation and use of other statistics. The chi-square is a theoretical or mathematical model. Since it is a mathematical model, certain conditions are assumed to be true in the development of the theoretical distribution. If these same conditions hold true for the data we are analyzing, then the conclusion (probability statements) of the model will hold true for the data. If the data do not meet the requirements of the mathematical model, the conclusions we draw may not be warranted. It is the responsibility of the research worker to understand the assumptions of the statistic he wishes to use and then make a judgment as to whether his data come reasonably close to meeting the requirements (assumptions) of the model. For use of the chi-square distribution to determine the probability of occurrence of a χ^2, the following conditions must be met:

1. *The data must be of the count or frequency type.*
2. *Each observation must be independent of every other observation.*
3. *The sampling distribution of the observed frequencies around the expected frequency should be normal.* If the expected frequency in any cell is small, it would not be possible for samples to vary much below the expected frequency. Hence the sampling distribution of observed frequencies would be positively skewed. If the expected value in any cell is less than five, the chi-square distribution should not be used.
4. *The sum of the observed frequencies must always equal the sum of the expected frequencies.*
5. *The categories used to classify the data should be established before the data are collected.*

THE FISHER-YATES EXACT PROBABILITY TEST

When the expected frequency in any cell of a 2×2 table is less than five, the Fisher-Yates exact-probability test may be used. If the two classifications are independent of each other and if the marginal frequencies are considered to be fixed, the exact probability of the occurrence of the frequencies A, B, C, and D can be determined from the following formula:

$$p = \frac{(A+B)!(C+D)!(A+C)!(B+D)!}{A!B!C!D!N!} \tag{11-7}$$

where the frequencies are arranged as follows

A	B	$A + B$
C	D	$C + D$
$A + C$	$B + D$	N

and

! is the factorial of the number

If one of the cell frequencies is zero, then the Fisher-Yates test is not too laborious to compute. The values are placed directly in formula 11–7 and the exact probability can be computed. If none of the cell frequencies is zero, then the test must be computed on the frequencies as observed. The cell with the lowest frequencies is reduced by one (keeping the marginal totals the same) and the test recomputed. This procedure is followed until the cell frequency in this cell has been reduced to zero. The sum of each of the separately computed probabilities is determined, and this is the one-tailed probability of occurrence of the observed set of frequencies or more remote frequencies. If a two-tailed probability is desired, this number is doubled.

The computation of the Fisher-Yates test is shown in example 11–4.

From example 11–4, it is readily apparent that the computation of the Fisher-Yates test can become quite laborious if the smallest cell frequency is of any appreciable size. If the smallest value were four, then five different probabilities would have to be computed and summed together. If the exact probability is not needed but significance levels will suffice, table K of the Appendix may be used.

Example 11-4   Computation of the Fisher-Yates exact-probability test

With 20 observations arranged in the following 2×2 table, the expected frequencies in cells B and D would be less than five.

Using formula 11-7

A	B	
8	1	9
C	D	
5	6	11
13	7	20

$$p_1 = \frac{9!11!13!7!}{8!1!5!6!20!} = .0536$$

Reducing cell B by one yields the following table (marginal totals the same)

A	B	
9	0	9
C	D	
4	7	11
13	7	20

$$p_0 = \frac{9!11!13!7!}{9!0!4!7!20!} = .0042$$

$$p = p_1 + p_0 = .0536 + .0042 = .0578$$

The probability computed is a one-tailed probability. If the two-tailed probability is desired, the obtained number is doubled. In this case the two-tailed probability would be .1156.

Dodge and Kolstoe[2] report a study on the use of the Shaw-Matthew Pseudoneurologic Scale (a subset of items from the Minnesota Multiphasic Personality Inventory) in differentiating between very early stages of multiple sclerosis and conversion hysteria. Twenty-seven patients were located who had a confused early diagnosis between multiple sclerosis and conversion hysteria. The MMPI had been administered at that early point in time. Later, the progress of the cases allowed a clear diagnosis as to either multiple sclerosis or conversion hysteria. The results of this double classification are shown in table 11-2.

A small portion of table K of the Appendix has been reproduced below as table 11-3. The Dodge and Kolstoe data is used to illustrate the use of table K with the Fisher-Yates exact-probability test.

The following steps should be followed in using table K of the Appendix:

1. *Compute the values of* $(A + B)$ *and of* $(C + D)$ *from the* 2×2 *table.* In the example $A + B = 14$ and $C + D = 13$.

[2] Gordon R. Dodge and Ralph H. Kolstoe. "The MMPI in differentiating early multiple sclerosis and conversion hysteria," *Psychological Reports* 29(1971):155-59.

TABLE 11–2
Distribution of patients on the Shaw and Matthew P–N scale

Group	Neurologic	Pseudoneurologic	Total
Multiple sclerosis	A 10	B 4	$A + B$ 14
Conversion hysteria	C 2	D 11	$C + D$ 13
Total	$A + C$ 12	$B + D$ 15	$N = 27$

TABLE 11–3
Portion of table of critical values of D (or C) in the Fisher-Yates test

Totals in right margin	B (or A)†	Level of significance .05	.025	.01	.005
$A + B = 14$ $C + D = 13$	14	9	8	7	6
	13	7	6	5	5
	12	6	5	4	3
	11	5	4	3	2
	10	4	3	2	2
	9	3	2	1	1
	8	2	1	1	0
	7	1	1	0	0
	6	1	0	—	—
	5	0	0	—	—
$C + D = 12$	14	8	7	6	6
	13	6	6	5	4

2. In table K under "Totals in Right Margin," *locate the observed value of $A + B$. $A + B = 14$.*
3. In the same section of table K, also under the heading "Totals in Right Margin," *locate the observed value of $C + D$. $C + D = 13$.*
4. For the observed value of $C + D$, several possible values of B are listed in the table. *Locate the observed value of B among those listed. If the observed value of B is not listed, use the observed value of A instead. $C + D = 13$; $B = 4$ and is not listed in the table. We must use $A = 10$.*

5. *Locate the observed value of D in the table. If the observed value of D is equal to or less than the value in the table under a particular significance level, then the null hypothesis can be rejected at that significance level. If A is used instead of B in step 4, then C is used instead of D in this step.* Since A was used in step 4, C is used instead of D. $C = 2$. The number 2 is located under the level of significance of .01 and also .005. Since $C = 2$ is equal to or less than the tabled value, the null hypothesis can be rejected at $p < .005$. We would conclude that the Shaw-Matthew PN scale can reliably differentiate between early multiple sclerosis and conversion hysteria with these patients.

Testing for the significance of differences with ordinal data

Many of the data in the behavioral sciences are ordinal in nature and do not appear to have sufficient interval characteristics to permit use of the t statistic or others that require interval or ratio measurement. We can reliably place individuals in categories; but if we have more information than implied by nominal scaling, we are willing to say that one individual possesses more of a given quality than another. The categories are ordered. To use statistics which do not utilize the ordered characteristics of the data is to waste a great deal of information.

THE MEDIAN TEST (INDEPENDENT SAMPLES)

The median test can be used with different individuals in each group. In recent years, there have been a great deal of interest in leadership training. Training programs have been established by industry, the military, and schools and colleges. To illustrate the use of the median test, we shall construct a hypothetical research situation in a college leadership training workshop.

A sociologist is interested in the effectiveness of a workshop at his college. Invitations are sent to students who have shown evidence of leadership by the fact that they hold some office on the campus. Thirty-nine students indicate an interest in participating in the leadership workshop. Due to the organization of the workshop, it is only possible to accept 23 of the applicants. A table of random numbers is used to select 23 students for the program. Sixteen students are not permitted to participate. Two years later, all 39 students are rated as to their effectiveness as student leaders. The research question would be: Are those students who participated in the workshop more effective leaders than those students who did not participate? The null hypothesis would be that participation in the workshop has had no effect on the effectiveness of the leadership of the students.

Since ratings were used, it might be difficult to defend the idea that the measurement was interval in nature, but it is possible to order the students from most effective to least effective. Since all students can be ordered, it would be a simple task to determine the median of the group—i.e., identify the students whose rating is such that 50 percent of the group are rated as being more effective leaders and 50 percent rated as being less effective. Under the null hypothesis, we would expect that half of the trained leaders would be above the median and half below; similarly, half of the untrained leaders would be above the median of effectiveness and half below. Suppose the data came out as shown below.

	Untrained leaders	Trained leaders	
Above median effectiveness	3	17	19
Below median effectiveness	13	6	39
	16	23	

The data are of the frequency type, and it would be possible to compute a χ^2. Using formula 11–3, the sociologist obtains a χ^2 of 9.37. From the table of the chi-square distribution (table J of the Appendix), he finds that for $df = 1$, this value of χ^2 has a probability of occurrence of less than 1 in 100 ($p < .01$). He would reject the null hypothesis and conclude that the training did make a difference in the later leadership effectiveness. From this, the sociologist could not conclude that the training experience itself was necessarily the factor that produced the difference, but he could state that those who had had the special training became more effective leaders. (Having had the training might have given those 16 students more opportunity for leadership positions; hence, they may have more postworkshop experience than those not selected for training.)

We can refer back to the study concerning persistence of an old, no longer adaptive response in white rats which was presented in chapter

2 (p. 7). One group of 17 animals received no persistence training. A second group of ten animals received intermittent electric shock and were tested in a zero-second delay condition. These data are analyzed by use of the median test in example 11–5.

Example 11–5   Use of the median test with persistence data

	No-shock group	Zero-second group	
Median or above	6	8	14
Below median	11	2	13
	17	10	27

Using formula 11–4:

$$\chi^2 = \frac{N(|AD - BC| - .5N)^2}{(A + B)(C + D)(A + C)(B + D)}$$

$$\chi^2 = \frac{27[|(6)(2) - (8)(11)| - (.5)(27)]^2}{(14)(13)(17)(10)} = \frac{27(|12 - 88| - 13.5)^2}{30,940} = \frac{105,468.75}{30,940}$$

$$\chi^2 = 3.41; \, df = 1; \, p > .05$$

Since the number of subjects in the experiment shown in example 11–5 was an odd number, one score was right at the median. In this case we divide the groups into those at or above the median and those below the median. In the example given, the probability of occurrence of a χ^2 of this magnitude by chance alone is greater than .05. We would not reject the null hypothesis.

The Mann-Whitney U test (independent samples)

The median test with two independent samples uses only the information as to whether or not an individual is above or below the overall median. If it is possible to rank all individuals in the two groups, then it should be possible to utilize this information concerning order of the individuals. The Mann-Whitney U test does use all of the order information which is in the data.

We must first rank all observations from both groups into one common ranking. In this ranking we use the algebraic value of the numbers—that is, the largest negative number (if any) would have the lowest rank. The statistic U is computed by counting the number of cases in one sample which rank lower than each case in the other sample. This presents a minor problem. Depending upon which sample we use as the reference sample we will obtain two different values. The smaller of these is U and the larger is designated as U'. Tables L and M of the Appendix show the probability of the smaller value (U) under the null hypothesis. An easy check on the accuracy of the counting is given by

$$U = N_1N_2 - U' \qquad (11\text{–}8)$$

If the null hypothesis is correct, then we would expect the samples from the two groups to be throughly mixed. That is, a member of one group would precede or follow a member of the other group about the same number of times, or $U = U'$.

Suppose that a committee on natural resources of a state legislature consisted of nine members. On many issues involving ecology, the vote is split five to four. A political scientist believes that the vote reflects a city versus rural split on the committee. The voting record of each committee member is examined as to the percent of times a vote was cast "for" ecology. The investigator does not believe that the percentages indicate distance between legislators on this issue but rather indicate an order from "less concern" to "greater concern" for ecology. This hypothetical case is shown in example 11–6.

U can always be computed by counting the number of times members of one sample precede the members of the other sample. For large Ns in the two samples, this counting procedure becomes quite cumbersome. The values of U and U' can be obtained from the following

$$U = N_1N_2 + \frac{N_1(N_1 + 1)}{2} - R_1 \qquad (11\text{–}9)$$

or

$$U = N_1N_2 + \frac{N_2(N_2 + 1)}{2} - R_2 \qquad (11\text{–}10)$$

where

N_1 is the number of scores in group one
N_2 is the number of scores in group two
R_1 is the sum of the ranks of the scores in group one
R_2 is the sum of the ranks of the scores in group two

Example 11-6   Analysis of the voting record of rural (R) and city (C) legislators, using the Mann-Whitney U test.

The scores of the legislators are the percent of "favorable" votes on ecology issues. $N_1 = 4$, $N_2 = 5$, $\alpha = .05$, two-tailed.

Scores	.10	.15	.33	.52	.57	.77	.83	.91	.97
Rank	1	2	3	4	5	6	7	8	9
C or R	C	C	C	R	C	R	R	R	R
R precedes C	0	0	0	—	1	—	—	—	—
C precedes R	—	—	—	3	—	4	4	4	4

$$\Sigma \,(R \text{ precedes } C) = 1$$
$$\Sigma \,(C \text{ precedes } R) = 19$$
$$U = 1; \; U' = 19$$
$$\text{Check: } U = N_1 N_2 - U' = (5)(4) - 19 = 1$$

Entering table L with $N_1 = 4$, $N_2 = 5$, and $U = 1$: $p = .016$. Since table L gives one-tailed probabilities, the tabled value would be doubled for the two-tailed probability and $p = .032$. H_0 would be rejected and we would conclude that the rural-city difference does make a difference in voting behavior.

Using formulas 11-9 and 11-10 with the same problem

$$U = N_1 N_2 + \frac{N_1(N_1 + 1)}{2} - R_1 = (5)(4) + \frac{5(5 + 1)}{2} - 34 = 1$$
$$U' = N_1 N_2 + \frac{N_2(N_2 + 1)}{2} - R_2 = (5)(4) + \frac{4(4 + 1)}{2} - 11 = 19$$

Formulas 11–9 and 11–10 will not yield equal results. One will yield U and the other will yield U'. Both should be computed, and the *smaller* value is U. The use of formulas 11–9 and 11–10 is shown in the lower portion of example 11–6.

The use of the table of probability values for the U test requires three items of information: N_1, N_2, and U. Since it is extremely difficult to build a two-dimensional table that requires three items for entry to the table, six different tables have been grouped together for values up to $N_1 = N_2 = 8$. These are grouped together as table L in the Appendix. For values up to $N_1 = N_2 = 20$, four tables have been grouped together as table M of the Appendix. Table M does not give the exact probability level for each value of U but rather gives the critical values for rejection at different levels of α. The computation of U for values of N_1 or N_2 greater than 8 but less than 20 is shown in example 11–7. Note in exam-

Example 11-7   Use of the Mann-Whitney *U* test with persistence data

Score	Rank	Subject group	Number of times E precedes C	Number of times C precedes E
48	17	E	—	8
47	16	E	—	8
44	15	C	7	—
43	14	C	7	—
42	12.5	C	7	—
42	12.5	C	7	—
40	11	C	7	—
38	10	E	—	3
34	9	E	—	3
32	8	E	—	3
28	6.5	C	3	—
28	6.5	E	—	2
22	5	C	3	—
16	3.5	C	3	—
16	3.5	E	—	0
8	1.5	E	—	0
8	1.5	E	—	0
	$N_1 = 9$		$U' = \overline{44}$	$U = \overline{27}$
	$N_2 = 8$			

$\alpha = .05$ two tailed from Table K, $p > .05$; H_0 not rejected since a U of 15 or less is needed with $N_1 = 9$, $N_2 = 8$, and $\alpha = .05$.

Using formula 11-9 and 11-10 involving the sum of ranks

$$U' = N_1N_2 + \frac{N_1(N_1 + 1)}{2} - R_1 = (9)(8) + \frac{(9)(9 + 1)}{2} - 73.0 = 44$$

$$U = N_1N_2 + \frac{N_2(N_2 + 1)}{2} - R_2 = (9)(8) + \frac{(8)(8 + 1)}{2} - 80.0 = 28$$

Since 28 is larger than the tabled value of $U = 15$, H_0 would not be rejected.

ple 11-7, that several scores are tied in the rankings. This may introduce a minor problem. The assumptions underlying the Mann-Whitney *U* test include the notion that the characteristic being measured is a continuous variable and that no ties would occur except through the crudeness of the measuring instrument. If the ties occur among members of the same group there is no problem. If the ties occur among members of the different groups, then a minor inaccuracy is introduced into the computation of *U*, if formulas 11-9 and 11-10 are used. The value of *U* computed from formulas 11-9 or 11-10 will be larger if there are tied ranks between the groups, and, as a consequence, formula 11-8 may not hold as a check on the computation of *U* and *U'*. This situation is also shown in example 11-7. It should be noted that a *U* which is larger than it should be, means that the probability of a Type I error has been *decreased*. This is referred to as a conservative test, which also means that the probability of the Type II error has been *increased*.

Methods of dealing with an excessive number of ties can be found in advanced textbooks in statistics. The impact of a few ties is small enough that no correction is warranted.

If both N_1 and N_2 become large (more than 20), the sampling distribution of U approaches that of the normal curve and a Z score approach can be used.

$$Z = \frac{U - \dfrac{N_1 N_2}{2}}{\sqrt{\dfrac{(N_1)(N_2)(N_1 + N_2 + 1)}{12}}} \qquad (11\text{--}11)$$

where

    U is the smaller value from formula 11–9 or formula 11–10
    N_1 is the sample size of the first group
    N_2 is the sample size of the second group

The probability of the occurrence of this size Z score under the null hypothesis is determined from table B of the Appendix.

THE SIGN TEST (CORRELATED SAMPLES)

When we observe an individual under two different conditions, we are willing to state that the individual behaves differently under each of the conditions (he has changed). We might be willing to go further and state the *direction* of change, but we would be unwilling to state the magnitude of change. In this type of situation the McNemar test of change could be used. However, if we can specify that the change is of the "greater than" or "less than" type, the sign test is to be preferred.

All that is needed is to observe each individual under two conditions, and then make the decision that under the second condition the person changes in one direction ($+$) or in the other direction ($-$). We now count the number of plus signs and the number of minus signs. Under the null hypothesis, we would expect as many plus signs as minus signs. Since this is a two-category situation with $p = q = .50$ (under the null hypothesis), we could use table I of the Appendix and determine the probability of occurrence of the number of observed plus and minus signs.

Suppose that a church group which is entirely Caucasian in its membership decides to invite a number of children from a different racial background to attend a summer camp with the children from the church—the assumption being that the daily contact of living together in the camp situation will change the attitudes of the children. An attitude scale is used and a score obtained for each of the Caucasian children before the summer camp experience. After the Caucasian children have had contact with the minority group youngsters, the attitude scale is administered a second time. The research worker is not willing to make any assumptions about the change in scores indicating magnitude of attitude change, but he is willing to state that plus or minus changes have occurred. In some cases, no change would have occurred. The differences in attitude would be scored plus, minus, or zero. For the sign test, all zero changes would be disregarded, and the research worker would concentrate on the number of plus and minus signs. If the total number of children whose attitudes changed was 25 or less, then table I of the Appendix would be used to determine the probability of this many or more changes occurring by chance. Table I will yield a one-tailed probability in this situation. If a two-tailed probability is desired (the research hypothesis was that the attitudes might change to either more favorable or more unfavorable, but they would change), then the cell value in table I should be doubled for the two-tailed probability. Thus, if the total number of children whose attitude changes is 20 ($N = 20$) and the number of minus signs is five ($x = 5$), the entry in table I is .021. Since this is a one-tailed probability, we would double the figure and state that the probability that 15 or more attitude changes would be in the same direction will occur by chance only 4.2 times in 100, or $p = .042$. On the basis of this evidence, the research worker would reject the hypothesis that the experience produced no changes in attitude. Since most changes were in the plus direction, we would conclude that the experience produced a favorable attitude change in most of the children.

In using the sign test, we had to observe the same individual in two situations (we could also have observed two different individuals in the two situations if these individuals had been carefully matched prior to the experiment). We also were unwilling to make any statements as to the amount of change in attitude in the children—we were willing to make only a statement of direction of change. If we are able to *rank* the amount of change, we can utilize statistics that are more powerful than the sign test.

THE WILCOXON MATCHED-PAIRS SIGNED-RANKS TEST (CORRELATED GROUPS)

If we can state that the difference between a pair of matched scores indicates more or less of the quality being observed, the sign test can be used. If we can go one step further and state that the difference between one pair of scores is more or less than the difference between a second pair of scores we can utilize the Wilcoxon test to analyze the data. The Wilcoxon is much more powerful than the sign test since it can utilize all of the order information in the data. This requires two kinds of ordering: (1) order of the difference between each pair of matched scores and (2) order of the differences among the pairs. A large difference between a matched pair will receive more weight than a small difference. The sign test treated all differences as the same. The Wilcoxon utilizes differences among the differences.

To use the Wilcoxon, we must first compute the difference between each pair of scores. These differences are then ranked from the smallest absolute difference (ignore the plus or minus sign) to the largest absolute difference. After the differences have been ranked, the direction (plus or minus) of the difference is assigned to the rank. If the null hypothesis is true, we would expect that the sum of the negative ranks would be equal to the sum of the positive ranks. To test this null hypothesis we need only compute the smaller sum of the ranks and use table O of the Appendix. To enter table O we shall need the N, but in this case N is equal to the number of pairs that differed from each other. All zero differences are disregarded in the Wilcoxon just as they were in using the sign test. The computation of Wilcoxon's T is given in example 11–8.

Table O of the Appendix gives critical values of T up to $N = 25$. As the number of pairs becomes large, the sampling distribution of T approaches the normal curve. For N greater than 25, the Z score approach may be used and table B used to determine probabilities. The Z score for use with the Wilcoxon T is

$$Z = \frac{T - \mu_T}{\sigma_T} = \frac{T - \dfrac{N(N+1)}{4}}{\sqrt{\dfrac{N(N+1)(2N+1)}{24}}} \qquad (11\text{–}12)$$

where

T is the smaller sum of signed ranks
N is the number of paired differences

Example 11-8   The Wilcoxon matched-pair signed-ranks test

The data used in example 10-4 are used here to illustrate the Wilcoxon T

Pair	Delay period		Difference in scores	Rank of D	Rank with less frequent sign
	0 seconds	30 seconds			
1	31	8	23	8.5	
2	37	13	24	10	
3	45	22	23	8.5	
4	28	25	3	1	
5	50	29	21	7	
6	37	31	6	2.5	
7	49	35	14	5	
8	25	38	−13	−4	4
9	36	42	−6	−2.5	2.5
10	69	52	17	6	
					$T = 6.5$

With $N = 10$, using table O of the appendix, a T of 8 or less is required to reject H_0 at the .05 level. Since the obtained value of T was 6.5, H_0 would be rejected.

Under circumstances in which the t statistic and the Wilcoxon T can be directly compared, the Wilcoxon is almost as powerful (ability to reject a false null hypothesis) as the t. Thus, the Wilcoxon has become an extremely valuable tool for analyzing ordered data when the two groups are related—either the same subjects are used in both conditions of the experiment, or matched subjects are used.

Summary

The large number of specialized nonparametric statistics which have been developed in recent years makes it impossible to include more than a small number in an introductory text. Most statistics developed for use with nominal or ordinal data require two items: (1) some theoretical or to-be-expected data and (2) the observed data. The statistic is then a statement of the probability of the occurrence of the observed set of data, given the hypothesized expected data.

In any two-category situation, if $p = q = .50$, the binomial test may be used. Although it is possible to use the binomial expansion discussed in chapter 6 for the solution of these types of problems, the probabilities of various combinations have been tabled and are presented in table I of the Appendix.

The chi-square distribution is based on the discrepancies between hypothesized, expected frequencies and observed frequencies in the same categories. All that is needed to compute a χ^2 is the observed frequency in each category, the expected frequency in the same category, and the number of degrees of freedom in the table. The degrees of freedom are *not* based on the sample size but rather are based on the number of *independent* $O - E$ differences. To use the chi-square distribution to determine the probability of the occurrence of a χ^2, the following conditions must be met:

1.   The data must be of the count or frequency type.

2.   Each observation must be independent of every other observation.

3.   The sampling distribution of the observed frequencies around the expected frequency should be normal.

4.   The sum of the observed frequencies must always equal the sum of the expected frequencies.

5.   The categories used to classify the data should be established prior to the collection of data.

When a 2×2 table is used for chi-square $(df = 1)$, the Yates correction should be used. This reduces the absolute size of each $O - E$ difference by one-half unit. If the expected frequency in any cell of a 2×2 table is less than five, the χ^2 should not be computed.

The McNemar test of change uses the chi-square distribution in the special situation in which only the individuals who changed in a "before and after" type of situation are utilized.

The Fisher-Yates exact-probability test, while laborious to compute if the smallest cell frequency is of any magnitude, is extremely useful when the expected frequency in a 2×2 table is less than 5. It is a useful alternative to χ^2 for small samples.

The median test (as the name implies) requires that all individuals in two groups be ordered to determine one common median. Each group is then divided into those members who are above the overall median and those below the overall median. The χ^2 statistic is used to determine if group membership results in differential ordering with respect to the median.

The Mann-Whitney U test uses all the order information from two independent samples. U is the number of times members of one sample precede members of the second sample when all are ordered from lowest to highest. This is one of the most powerful of the nonparametric statistics and is used as a very acceptable alternative to the t test when the assumptions of the t cannot be met.

The sign test is used when the same individuals are observed in two different situations (or matched individuals are each observed in one of the two situations). If we can state a direction of difference between the two situations (plus or minus), the sign test can be utilized. The number of changes in one direction $(+)$ is compared to the number of changes in the other direction $(-)$ by using the table of the binomial test (table I of the Appendix).

The Wilcoxon matched-paired signed-ranks test, like the Mann-Whitney U test, utilizes all of the order information in ranks. The Wilcoxon T is used when the same subjects are used in both conditions of the experiment or the subjects have been matched. T is the smaller sum of the signed ranks. The Wilcoxon T requires that order is evident between the members of a pair and the differences among the pairs must also be ordered. If these conditions hold, the Wilcoxon is a very powerful alternative to the t test for correlated groups.

For convenience, the nine different nonparametric tests presented in this chapter have been classified as to the number of groups involved (one sample or two samples), if the samples are independent or correlated, and the scaling characteristic of the dependent variable (nominal or ordinal).

	One sample	Two samples Independent	Two samples Correlated
Nominal scale	Binomial test Chi-square one-way classification	Chi-square two-way classification Fisher-Yates exact probability test	McNemar test of change
Ordinal scale		Median test Mann-Whitney U test	Sign test Wilcoxon matched-pairs signed-ranks T test

References

GENERAL REFERENCES FOR NONPARAMETRIC STATISTICS

SIDNEY SIEGEL, *Nonparametric Statistics for the Behavioral Sciences* (New York: McGraw-Hill Book Co., Inc., 1956). This is probably the best single source on nonparametric statistics available to the student in the behavioral sciences.

W. G. COCHRAN, "Some Methods for Strengthening the Common χ^2 Tests," *Biometrics,* Vol. 10 (1954), pp. 417–51. This article furnishes more information on χ^2.

A. E. MAXWELL, *Analyzing Qualitative Data* (New York: John Wiley & Sons, Inc., 1961).

FREDERICK MOSTELLER and ROBERT E. K. ROURKE, *Sturdy Statistics Nonparametric and Order Statistics* (Reading, Massachusetts: Addison-Wesley Publishing Company, 1973). This is more advanced than the Siegel book.

12

Introduction to the analysis of variance and multi-group nonparametric statistics

MOST of the tests of significance of differences mentioned thus far have dealt with differences in means or observed frequencies in different groups. The t test has as one of its requirements that the standard deviations of the two different groups must be unbiased estimates of the same population standard deviation, or $\sigma_1 = \sigma_2$. How can we test to determine if two standard deviations are significantly different from each other? Is it possible to test for the differences in standard deviations? Such a test has been developed and is called the F *ratio*.

Testing for differences between variances

If we drew a large number of samples from the same population and computed the standard deviation of each sample (unbiased), we could develop a sampling distribution of these standard deviations. We could go a step further and draw samples from the same population and find the differences between pairs of standard deviations. A sampling distribution of these differences in pairs of standard deviations could also be generated, and we could test to see if the difference between any two standard deviations could readily occur by chance if these two standard deviations were random samples from the same population.

Sir Ronald Fisher developed the sampling distribution of the differences in standard deviations. To use Fisher's approach, it is necessary to convert

259

the standard deviations into logarithms and work with the differences in logarithms of the respective standard deviations. G. W. Snedecor of Iowa State University modified the Fisher approach by using the square of each standard deviation (the variance). The sampling distribution of the *ratio* of the two variances was developed, and by looking in the appropriate table, we ·can determine if the ratio of the two variances is sufficiently large to reject the idea that these could be random samples from the same population. Snedecor named this statistic the F ratio in honor of Sir Ronald Fisher.

If two variances are based on random samples from the same population, we would expect each variance to be of approximately the same magnitude. We would *not* expect them to be *exactly* the same, but they should be highly similar. Under the null hypothesis (no difference), we would expect that the ratio of the two variances should be very close to one. If the variance used in the numerator were smaller than the variance in the denominator, the ratio would be slightly less than one. If the two variances were reversed, the ratio would be slightly larger than one. If the two sample variances were exactly the same size, the ratio would be one. If many random samples were drawn from the same population and ratios of variances were made, we would expect the mean of all F ratios to be one.

The decision as to which sample variance should be in the numerator of the F ratio would be arbitrary. We could consistently place the variance of the first sample in the numerator and the variance of the second sample in the denominator of the ratio. In this case, some ratios would be less than one, and some would be larger than one. Suppose that two samples were drawn so that $s_1^2 = 6$ and $s_2^2 = 8$. We wish to determine whether the difference between these two variances is larger than we would expect by chance. We could compute two different F ratios: (1) $F = 6/8 = .75$ and (2) $F = 8/6 = 1.33$. These two F ratios would mean exactly the same and would have the same probability of occurrence under the null hypothesis. Because the F ratio has the same meaning when it is less than one and when it is greater than one, table F in the Appendix is only one half of the table of the F ratio. Only the values greater than one have been tabled. This means that to test for the significance of differences between two variances, the larger of the two variances must be placed in the numerator of the F ratio. In this respect, table F of the Appendix is like the table of the normal curve.

The use of the half table to determine the probability of occurrence of the ratio of two variances does present one problem. The half table will yield a one-tailed probability statement. We are only examining variation in one direction from the expected value of one. We can compensate for this by doubling the tabled value to obtain a two-tailed probability for the occurrence of the observed ratio.

To illustrate the use of table F of the Appendix, a portion of that table has been reproduced as table 12–1.

TABLE 12–1
Table of the F ratio for .05 and .01 levels of significance

Degrees of freedom associated with the denominator		Degrees of freedom associated with the numerator							
		1	2	3	4	5	6	7	8
1	.05	161	200	216	225	230	234	237	239
	.01	4,052	5,000	5,403	5,625	5,724	5,859	5,928	5,982
. . .	. . .	. . . .	. . . . .	. . . . .	. . . . .	. . . . .	. . . . .	. . . . .	. . . . .
6	.05	5.99	. . . . .	. . . . .	. . . . .	. . . . .	. . . . .	4.21	. . . . .
	.01	13.7	. . . . .	. . . . .	. . . . .	. . . . .	. . . . .	8.26	. . . . .
. . .	. . .	. . . . .	. . . . .	. . . . .	. . . . .	. . . . .	. . . . .	. . . . .	. . . . .
10	.05	4.96	. . . . .	. . . . .	. . . . .	3.33	. . . . .	. . . . .	. . . . .
	.01	10.0	. . . . .	. . . . .	. . . . .	5.64	. . . . .	. . . . .	. . . . .

The first column of table 12–1 is labeled *degrees of freedom associated with the denominator*. The rows of the table are labeled *degrees of freedom associated with the numerator*. Obviously, we must determine the degrees of freedom associated with the two variances if we are to use the table of the F ratio. The degrees of freedom are those discussed in chapter 10 in connection with the t test—namely, $N_1 - 1$ for the first sample and $N_2 - 1$ for the second sample. Thus, if the larger of the two variances is based on 20 observations ($N_1 = 20$), the degrees of freedom associated with the numerator will be 19 ($N_1 - 1 = 20 - 1$). A similar procedure would be followed for the variance used as the denominator of the F ratio. Let us suppose that the F ratio of two variances is 5.64 and that the degrees of freedom associated with the numerator are 5 and the df for the denominator are 10. Examination of table 12–1 indicates that this ratio has a probability of occurrence by chance of .01 if the null hypothesis is true. But this is a one-tailed probability. If we had a full table available to us, we could have put the smaller variance in the numerator, in which case the F ratio would have been less than one. We can determine the two-tailed probability by doubling the *table value*. Thus the probability of the occurrence of this F ratio by chance is not .01, but .02.

The use of the F ratio to test for the significance of the difference between two variances is illustrated in example 12–1.

Example 12–1   Testing for the significance of differences between two variances

Using the data from example 10–3 (p. 221):

Group I	Group II
$\Sigma x^2 = 1,420.00$	$\Sigma x^2 = 889.98$
$N = 8$	$N = 7$
$s^2 = 1,420/(8 - 1) = 202.86$	$s^2 = 889.98/(7 - 1) = 148.33$

$$F = \frac{s^2}{s^2} = \frac{202.86}{148.33} = 1.37$$

$$df \text{ for numerator} = 8 - 1 = 7$$
$$df \text{ for denominator} = 7 - 1 = 6$$

Using table F of the Appendix, an *F* of 1.37 would not reach significance at any of the tabled values. We would not be able to reject the hypothesis that these samples came from populations whose true variances were equal.

If the F ratio or F test could be used only for testing for the significance of differences between two sample variances, it would be a valuable addition to the field of statistics and data analysis. Fortunately, the F ratio has much wider application and is one of the most valuable analytic tools available in the field of statistics.

The F test for testing differences among means

In chapter 10, we examined the use of the *t* test to test for the significance of the difference between two sample means. Suppose that we had five different groups in a study. We could compute a *t* test between each pair of means, but this would require 10 different *t* tests. With 10 groups in the study, it would require 45 different *t* tests. This presents a rather serious problem. We compute statistics in order to help us make decisions about the data. If we obtain a large *t* test, we reject the null hypothesis and conclude that the data indicate that a true difference exists between the groups. The decision to reject the null hypothesis could be in error.

We have committed a Type I error by rejecting the null hypothesis when it was true (see chap. 10). The probability of committing a Type I error for a single t test is small (whatever we set α to be). If we draw many random samples from the same population and compute many t tests, we expect that about 5 percent of all the t's will be larger than the tabled value for $\alpha = .05$ and about 1 percent of the obtained t's will be larger than the tabled value for $\alpha = .01$. If we compute a large number of t's in one study, the tabled probabilities are not valid. The t tests are not independent of each other. We need a statistical technique which will allow us to examine the difference among all sample means at one time. Fortunately, the F ratio can be used for this purpose.

ANALYZING VARIANCE

It should be reemphasized that variance is the square of the standard deviation. We work with the variance rather than the standard deviation because it simplifies the work by removing the radical from the standard deviation. Before looking at the analysis of variance, let us examine the analysis of variability. By variability, we mean observed differences among individuals—they vary. If there is any one universal law in the behavioral sciences, it is that people are different. They do vary. The problem of the behavioral scientist is to determine what underlies these differences. Is the variability among people due to genetic factors? Some differences undoubtedly are genetic. Are the differences due to social factors? Certainly some are. Political factors? Economic factors? All of these produce variability among people. The task of the behavioral scientist is to analyze this variability to determine the relative importance of different factors.

A first-grade teacher is very much aware of differences in the rate at which children learn to read. Some learn quickly, some slowly. What are the factors which lead to this variability among first-grade children? The teacher may notice that the older children seem to do better in reading than the younger children. Thus, age may be one variable which is related to the differences in reading skill. Further examination would reveal that age per se is not as important as mental age in the progress of the child in learning to read. Some of the variability among the children could be reduced by requiring *all* children to be at least six years of age to begin the first grade. There would still be variability among the children. A vision test may indicate that some children do not see well, and this leads to poor reading. Some children have a speech or hearing problem, and this interferes with their learning to read. All these factors will produce variation among the children with respect to their learning to read. The isolation of each factor will allow us to account for the variability observed among the children. We are analyzing the variability.

Everyone is constantly analyzing variability. Usually, this is done in a nonstatistical way. We are noticing differences among people and then attempting to determine what is responsible for these differences. The analysis of variance is a statistical approach to allow us to analyze differences and test to see if the factors involved are producing statistically significant differences. The approach is quite straightforward. We need to find one source of variance which reflects the variation *within* samples and a second source of variation *between* the different groups.

If we view the mean of each group as a single score, it should be possible to find the deviation of each group mean about the total mean. Thus, $(\bar{X}_g - \bar{X}_t)$ would be a deviation score. The larger the deviation, the more variation there is among the means of the groups in the experiment. If we squared all deviations, summed them, and divided by the degrees of freedom, we would have a term similar to the usual notion of variance. But this would be the variance *between* the means of the groups in the experiment.

The variance *within* groups would be based on the squared deviation of each score in a group about the mean of that particular group $(X - \bar{X}_g)^2$. When the sum of these squared deviations is divided by the degrees of freedom, it would be the variance within the specific group.

The analysis of variance involves splitting up the total variance in a study into two component parts: (1) the variance *between the means* of the various groups in the study and (2) the variance *within* the individual groups in the study. The next five sections of this chapter show how the total variance is split into *between* and *within* sources of variance and how the F ratio is used to test for significance of differences between these two sources of variance.

DIVIDING MEASUREMENTS INTO ADDITIVE COMPONENTS. To determine the variance of any set of data, we must first determine the deviation of the scores about the mean $(X - \bar{X} = x)$. These deviations are squared (x^2) and then summed (Σx^2). When the sum of the squared deviations is divided by the degrees of freedom, we have computed an unbiased estimate of the population variance. An analysis of variance must then focus on that part of the variance which reflects differences between individuals and differences between groups. We must analyze the deviation. If we drew two samples from the same population, we could combine all measurements and compute a mean of the total group. We would have three means: (1) the mean of one sample, $\bar{X}_1$; (2) the mean of the second sample, $\bar{X}_2$; and (3) the mean of the total group, $\bar{X}_t$. We could determine the variance of the total group by determining the deviation of each score in the total group about the mean of the total group:

$$x = X - \bar{X}_t$$

But the individual score was obtained as part of a sample; thus, we can also obtain a deviation of that score from the mean of the sample in which it occurred:

$$x = X - \bar{X}_1$$

We can express the same score in terms of the deviation of that score about the mean of the sample and also the deviation of that score about the mean of the total. But we can go a step further. We can express the deviation of the score about the total mean in terms of two additive components—(1) the deviation of the score about the mean of the sample and (2) the deviation of the mean of the sample about the total mean:

$$x = (X - \bar{X}_t) = (X - \bar{X}_1) + (\bar{X}_1 - \bar{X}_t)$$

If the parentheses on the right-hand side of the second equal sign are removed, we see that this is an equality:

$$(X - \bar{X}_t) = X - \bar{X}_1 + \bar{X}_1 - \bar{X}_t$$

We can divide a deviation about the mean of the total into two additive components: the deviation about the mean of the sample, and the deviation of the mean of the sample about the mean of the total.

DEVELOPMENT OF THE SUM OF SQUARED DEVIATIONS FOR BETWEEN GROUPS AND WITHIN GROUPS. Analyzing variance required us to square the deviations of scores about the mean and sum these squared deviations. Since the deviation itself can be broken into two parts, we shall square and sum these parts and then express the total sum of squared deviations in terms of components. The deviation of any score in sample 1 may be expressed as:

$$(X - \bar{X}_t) = (X - \bar{X}_1) + (\bar{X}_1 - \bar{X}_t)$$

Squaring both sides of the equation:

$$(X - \bar{X}_t)^2 = (X - \bar{X}_1)^2 + 2(X - \bar{X}_1)(\bar{X}_1 - \bar{X}_t) + (\bar{X}_1 - \bar{X}_t)^2$$

A similar equation could be written for every score in sample 1. We would obtain N_1 different equations expressing the deviation of each score about the mean of the total:

$$(X_1 - \bar{X}_t)^2 = (X_1 - \bar{X}_1)^2 + 2(X_1 - \bar{X}_1)(\bar{X}_1 - \bar{X}_t) + (\bar{X}_1 - \bar{X}_t)^2$$
$$(X_2 - \bar{X}_t)^2 = (X_2 - \bar{X}_1)^2 + 2(X_2 - \bar{X}_1)(\bar{X}_1 - \bar{X}_t) + (\bar{X}_1 - \bar{X}_t)^2$$
$$\cdot \cdot \cdot \cdot \cdot \cdot \cdot \cdot \cdot \cdot \cdot \cdot \cdot \cdot \cdot \cdot \cdot \cdot \cdot \cdot \cdot \cdot \cdot \cdot \cdot \cdot \cdot$$
$$(X_i - \bar{X}_t)^2 = (X_i - \bar{X}_1)^2 + 2(X_i - \bar{X}_1)(\bar{X}_1 - \bar{X}_t) + (\bar{X}_1 - \bar{X}_t)^2$$

Summing all the squared deviations in sample 1:

$$\sum_{}^{N_1} (X - \bar{X}_t)^2 = \sum_{}^{N_1} (X - \bar{X}_1)^2 + 2(\bar{X}_1 - \bar{X}_t) \sum_{}^{N_1} (X - \bar{X}_1) + N_1(\bar{X}_1 - \bar{X}_t)^2$$

The expression $\sum_{}^{N_1} (X - \bar{X}_t)^2$ is the sum of the squared deviations (the deviations about the *mean of the total*) for the first sample. The expression

$$\sum_{}^{N_1} (X - \bar{X}_1)^2$$

is the sum of the squared deviations (*the deviations about the mean of the first sample*) for the first sample. The expression

$$2(\bar{X}_1 - \bar{X}_t) \sum_{}^{N_1} (X_1 - \bar{X}_1)$$

is equal to zero. In adding all of these cross product terms for the N_1 scores in sample 1, the number two is a constant, and the deviations of the mean of sample 1 about the mean of the total will also be a constant; thus, they are written to the left of the summation sign. The term $\sum_{}^{N_1} (X - \bar{X}_1)$ is the sum of the deviations of a set of scores about the mean of that set and is equal to zero. Thus the entire middle term goes to zero. The expression $N_1(\bar{X}_1 - \bar{X}_t)^2$ indicates the deviation of the mean of the first sample about the mean of the total; but since this is a constant for all N_1 scores in the first sample, this "sum" may be written as N_1 times the constant. The entire equation may be written as

$$\sum_{}^{N_1} (X - \bar{X}_t)^2 = \sum_{}^{N_1} (X - \bar{X}_1)^2 + N_1(\bar{X}_1 - \bar{X}_t)^2$$

A similar equation could be written for sample 2; and if a third sample had been drawn, this could be expressed in the same form. Thus

For sample 1: $\displaystyle\sum^{N_1} (X - \bar{X}_t)^2 = \sum^{N_1} (X - \bar{X}_1)^2 + N_1(\bar{X}_1 - \bar{X}_t)^2$

For sample 2: $\displaystyle\sum^{N_2} (X - \bar{X}_t)^2 = \sum^{N_2} (X - \bar{X}_2)^2 + N_2(\bar{X}_2 - \bar{X}_t)^2$

For sample 3: $\displaystyle\sum^{N_3} (X - \bar{X}_t)^2 = \sum^{N_3} (X - \bar{X}_3)^2 + N_3(\bar{X}_3 - \bar{X}_t)^2$

. . . .

For sample k: $\displaystyle\sum^{N_k} (X - \bar{X}_t)^2 = \sum^{N_k} (X - \bar{X}_k)^2 + N_k(\bar{X}_k - \bar{X}_t)^2$

Summing across all k samples:

$$\sum^{k} \sum^{N_g} (X - \bar{X}_t)^2 = \sum^{k} \sum^{N_g} (X - \bar{X}_g)^2 + \sum^{k} N_g(\bar{X}_g - \bar{X}_t)^2$$

Let us examine the first part of the expression given above.

$$\sum^{k} \sum^{N_g} (X - \bar{X}_t)^2 \qquad (12\text{--}1)$$

where

$X - \bar{X}_t$ is the deviation of a score about the mean of the total

$\displaystyle\sum^{N_g}$ indicates that these deviations are summed for all scores for each group

$\displaystyle\sum^{k}$ indicates that the sums for each group are added together for all k groups

Formula 12–1 is the total sum of squared deviations of all scores. When this is divided by the total number of degrees of freedom, we obtain an unbiased estimate of the population variance. It is this total sum of squared deviations that can be divided into two additive components:

$$\sum^{k} \sum^{N_g} (X - \bar{X}_g)^2 \quad \text{(sum of squared deviations \textit{within} the groups)} \qquad (12\text{--}2)$$

where

$X - \bar{X}_g$ is the deviation of a score about the mean of the group in which the score is located

$\displaystyle\sum^{N_g}$ indicates that these deviations are summed for all the scores within the particular group

$\displaystyle\sum^{k}$ indicates that the sum of the squared deviations within each group is added over all such groups and

$$\sum_{}^{k} N_g(\bar{X}_g - \bar{X}_t)^2 \text{ (sum of squared deviations } between \text{ the groups)} \quad (12\text{–}3)$$

where

$\bar{X}_g - \bar{X}_t$ is the deviation of a sample mean about the total mean

N_g is the number of observations in the specific group

$\sum^{k}$ indicates that the product of the sample size times the squared deviation of the sample mean about the total mean is summed for all samples

Expression 12–2 is usually referred to as the sum of squares *within* groups (SS_{wg}), and expression 12–3 is called the sum of squares *between* groups (SS_{bg}).

DEGREES OF FREEDOM FOR BETWEEN-GROUP AND WITHIN-GROUP SUMS OF SQUARES. The sum of squared deviations about the mean of the total is based on all observations, but the deviations are taken about the mean of the total. Thus the total degrees of freedom are the total number of observations minus one ($N_1 + N_2 + N_3 + \cdots + N_k - 1$). This may be expressed as $N_t - 1$, where N_t is the total number of scores. What are the degrees of freedom associated with the sums of squares *within* groups? Examination of the within-group sums of squares indicates that we have a total of N_t different scores, but within each group the deviation of these scores is about the mean of their own sample. Thus, we lose one degree of freedom for each sample mean. The degrees of freedom associated with the within-group sums of squares are $N_t - k$, where N_t is the total number of observations and k is the number of different samples (groups). Examination of the sums of squares *between* groups indicates that we have k different sample means, but we are determining the deviation of these sample means about the mean of the total. Thus, we have $k - 1$ degrees of freedom associated with the between-group sums of squares. If we add the sum of squares *between* groups and the sum of squares *within* groups, this will add to the total sums of squares. If we add the degrees of freedom *between* groups ($k - 1$) and the degrees of freedom *within* groups ($N_t - k$), this will add to the total degrees of freedom ($N_t - 1$). Not only can the total sum of squared deviations be broken into two additive components, but the degrees of freedom can also be divided into additive components associated with the sums of squares *within* groups and the sums of squares *between* groups.

MEAN SQUARES WITHIN GROUPS AND MEAN SQUARES BETWEEN GROUPS.   To this point, we have been able to divide the total sum of squared deviations into two additive components: (1) the sum of squared deviations within the groups and (2) the sum of squared deviations between the groups. We have also been able to divide the total degrees of freedom into two additive components: (1) *df within* groups and (2) *df between* groups. When we divide a sum of squared deviations by its degrees of freedom, we obtain a ratio called the *mean square*. Thus the mean square *within* groups is:

$$MS_{wg} = \frac{\sum\limits_{}^{k} \sum\limits_{}^{N_g} (X - \bar{X}_g)^2}{N_t - k} \tag{12-4}$$

where

$\sum\limits_{}^{k} \sum\limits_{}^{N_g} (X - \bar{X}_g)^2$ is the sum of squared deviations *within* groups

$N_t - k$ is the degrees of freedom *within* groups

The mean square *between* groups is

$$MS_{bg} = \frac{\sum\limits_{}^{k} N_g(\bar{X}_g - \bar{X}_t)^2}{k - 1} \tag{12-5}$$

where

$\sum\limits_{}^{k} N_g(\bar{X}_g - \bar{X}_t)^2$ is the sum of squared deviations *between* groups

$k - 1$ is the degrees of freedom *between* groups

THE F RATIO FOR TESTING DIFFERENCES AMONG MEANS

These mean squares are of particular interest because under certain conditions they can be considered to be unbiased estimates of the same population variance (σ^2). When these two mean squares can be considered as unbiased estimates of the same population variance, we can use the F ratio and the table of the F test to test for the differences among the several sample means.

Let us examine the mean squares *within* groups first. For simplicity, let us assume that the number of observations in each sample is the same ($N_1 = N_2 =$ etc.). For sample 1, we can compute the sum of squared deviations within that sample; and if this sum of squared deviations is divided by the number of degrees of freedom for sample 1, we shall have obtained an unbiased estimate of the population variance for the population from which sample 1 was obtained:

$$\frac{\sum\limits_{}^{N_1} (X - \bar{X}_1)^2}{N_1 - 1} = s_1^2 \text{ (unbiased estimate of } \sigma_1^2\text{)}$$

Similarly, for sample 2:

$$\frac{\sum\limits_{}^{N_2} (X - \bar{X}_2)^2}{N_2 - 1} = s_2^2 \text{ (unbiased estimate of } \sigma_2^2\text{)}$$

In the same fashion, the sum of squared deviations within each sample when divided by the degrees of freedom for that sample would be an unbiased estimate of the variance of the population from which that sample was obtained. If all samples are random samples from the *same population,* each of the unbiased estimates is an estimate of the *same population variance.* Thus the mean of all unbiased estimates would become the single best estimate of the population variance. The mean square *within* groups is the mean of all unbiased estimates obtained from each sample. Under the assumption that all samples were obtained from the same population or the assumption that the samples came from different populations whose variances were equal, the mean square *within* groups is the single best estimate of the common population variance.

If the null hypothesis is true (there are no differences in the means of the populations from which the samples were drawn—$\mu_1 = \mu_2 = \mu_3$, etc.), it can be shown that the mean square between groups is also an unbiased estimate of the population variance. If the null hypothesis is false ($\mu_1 \neq \mu_2 \neq \mu_3$, etc.), the mean squares *between* groups should be *greater than* the unbiased estimate of the population variance. This allows us to use the F ratio to determine if the null hypothesis is false:

$$F = \frac{MS_{bg}}{MS_{wg}} = \frac{\sum\limits_{}^{k} N_g (\bar{X}_g - \bar{X}_t)^2 / (k - 1)}{\sum\limits_{}^{k} \sum\limits_{}^{N_g} (X - \bar{X}_g)^2 / (N_t - k)} \qquad (12\text{--}6)$$

If the null hypothesis is true, we would expect the ratio of the two mean squares to be about one. If the null hypothesis is false, we would expect the F ratio to be greater than one. How much greater than one? What evidence shall we need to reject the null hypothesis? This will depend upon the number of degrees of freedom associated with the numerator of the F test and the number of degrees of freedom associated with the denominator term. In addition, it will depend upon how much risk we are willing to run in deciding to reject the null hypothesis when it is true (Type I error). Table F of the Appendix can be used for determining the probability of an F ratio occurring by chance when the null hypothesis is true. In testing the *differences between means* of groups, table F gives *two-tailed* probabilities.

ASSUMPTIONS FOR THE ANALYSIS OF VARIANCE. Like the *t* statistic and the chi-square distribution, use of the analysis of variance to test for the difference between means has several conditions or assumptions.

1. *The data should be reasonably close to interval or ratio.* Since the analysis of variance requires the use of deviation scores, this assumes that the operations of addition may be carried out on the numbers.

2. *The samples used are random samples from their respective populations.*

3. *The populations from which these samples were drawn are normally distributed.*

4. *The variances of the underlying populations are equal.* This condition is especially important for the mean squares *within* groups to be an unbiased estimate of the population variance.

Computation of the analysis of variance

To compute the analysis of variance, we need to determine the mean squares between groups and the mean squares within groups. This could be done by using formula 12–2 for the sums of squares within groups and formula 12–3 for the sums of squares between groups. However, special computing formulas have been developed which do not require working with each deviation score. By using the whole-score approach, we can eliminate both the labor and much of the possibility of arithmetic error. The sum of squared deviations for the total may be found by

$$SS_t = \sum^{N_t} X^2 - \frac{\left(\sum^{N_t} X\right)^2}{N_t} \tag{12–7}$$

where

$$\sum^{N_t} X^2 \quad \text{is the sum of the squared scores for all cases}$$

$$\left(\sum^{N_t} X\right)^2 \quad \text{is the square of the sum of the scores for all individuals}$$

N_t is the total number of scores in all groups combined

The sum of squared deviations *between* groups may be found by

$$SS_{bg} = \sum^{k} \frac{\left(\sum^{N_g} X_g\right)^2}{N_g} - \frac{\left(\sum^{N_t} X\right)^2}{N_t} \qquad (12\text{--}8)$$

where

$$\left(\sum^{N_g} X_g\right)^2 \quad \text{is the square of the sum of the scores in each specific group}$$

N_g is the number of cases in each specific group

$$\left(\sum^{N_t} X\right)^2 \quad \text{is the square of the sum of the scores for all cases in the study}$$

N_t is the total number of cases in the entire study
k is the number of groups in the study

The sum of squared deviation *within* groups may be found by

$$SS_{wg} = \sum^{k} \left[\sum^{N_g} X_g^2 - \frac{\left(\sum^{N_g} X_g\right)^2}{N_g} \right] \qquad (12\text{--}9)$$

where

$$\sum^{N_g} X_g^2 \quad \text{is the sum of the squared scores in each specific sample}$$

$$\left(\sum^{N_g} X_g\right)^2 \quad \text{is the square of the sum of scores in each specific sample}$$

N_g is the number of cases in each specific sample
k is the number of groups in the study

Note that the term subtracted from formula 12–7 is the same as that subtracted in formula 12–8. When the term $\left(\sum^{N_t} X\right)^2/N_t$ has been computed it can be used in both formula 12–7 and formula 12–8. A similar situation is true for formulas 12–8 and 12–9. The terms of the form $\left(\sum^{N_g} X_g\right)^2/N_g$ can be used in both formula 12–8 and 12–9.

Since the sum of squared deviations for the total is the sum of squared deviations for *between* groups plus the sum of squared deviations for *within* groups, a useful check on the computations can be made by

$$SS_t = SS_{bg} + SS_{wg} \qquad (12\text{--}10)$$

After the sum of squared deviations *between* groups and *within* groups has been determined, the mean squares can be obtained by dividing each sum of squares by the appropriate degrees of freedom. These two mean squares can then be placed in an F ratio to test for the significance of differences among the several sample means. The computation of an analysis of variance is shown in example 12–2.

Example 12–2   Computation of the analysis of variance

Group 1		Group 2		Group 3		Group 4	
X	X²	X	X²	X	X²	X	X²
52	2,704	72	5,184	90	8,100	68	4,624
45	2,025	63	3,969	78	6,084	63	3,969
35	1,225	52	2,704	69	4,761	58	3,364
34	1,156	51	2,601	63	3,969	54	2,916
26	676	47	2,209	62	3,844	50	2,500
14	196	46	2,116	58	3,364	44	1,936
12	144	42	1,764			32	1,024
		40	1,600			26	676
		36	1,296				
		21	441				
218	8,126	470	23,884	420	30,122	395	21,009

$$N_1 = 7 \qquad N_2 = 10 \qquad N_3 = 6 \qquad N_4 = 8 \qquad N_t = 31 \qquad \sum^{N_t} X = 1{,}503$$

$$\bar{X}_1 = 31.14 \quad \bar{X}_2 = 47.00 \quad \bar{X}_3 = 70.00 \quad \bar{X}_4 = 49.38 \quad \bar{X}_t = 48.48 \quad \sum^{N_t} X^2 = 83{,}141$$

Using formula 12–7

$$SS_t = \sum^{N_t} X^2 - \frac{\left(\sum^{N_t} x\right)^2}{N_t}$$

$$SS_t = 83{,}141 - (1{,}503)^2/31 = 83{,}141 - 72{,}870.97 = 10{,}270.03$$

Using formula 12–8

$$SS_{bg} = \sum^k \frac{\left(\sum^{N_g} x_g\right)^2}{N_g} - \frac{\left(\sum^{N_t} x\right)^2}{N_t}$$

$$SS_{bg} = [(218)^2/7] + [(470)^2/10] + [(420)^2/6] + [(395)^2/8] - [(1{,}503)^2/31]$$
$$= 6{,}789.14 + 22{,}090.00 + 29{,}400.00 + 19{,}503.13 - 72{,}870.97$$
$$= 77{,}782.27 - 72{,}870.97 = 4{,}911.30$$

Example 12-2 *Continued*

Using formula 12-9

$$SS_{wg} = \sum \left[\sum^{N} X_g{}^2 - \frac{\left(\sum^{N_g} X_g\right)^2}{N_g} \right]$$

$$SS_{wg} = [8,126 - (218)^2/7] + [23,884 - (470)^2/10] + [30,122 - (420)^2/6]$$
$$+ [21,009 - (395)^2/8]$$

$$= 1,336.86 + 1,794.00 + 722.00 + 1,505.87$$
$$= 5,358.73$$

Using formula 12-10 as a check

$$SS_t = SS_{bg} + SS_{wg} \qquad \begin{array}{l} SS_{bg} = 4,911.30 \\ SS_{wg} = 5,358.73 \\ \hline 10,270.03 = SS_t \end{array}$$

Dividing the SS_{bg} and the SS_{wg} by their appropriate *df* will yield the MS_{bg} and MS_{wg}:

$$MS_{bg} = \frac{SS_{bg}}{k-1} = \frac{4,911.30}{4-1} = 1,637.10$$

$$MS_{wg} = \frac{SS_{wg}}{N_t - k} = \frac{5,358.73}{31 - 4} = 198.47$$

The ratio between the MS_{bg} and MS_{wg} will form the F ratio:

$$F = \frac{MS_{bg}}{MS_{wg}} = \frac{1,637.10}{198.47} = 8.25; \quad df = 3, 27$$

Using table F in the Appendix with *df* for the numerator equal to 3 and *df* for the denominator equal to 27: $p < .01$.

The analysis of variance shown in example 12-2 yielded an F ratio of 8.25 and with 3 and 27 degrees of freedom. An *F* of this magnitude will occur less than 1 time in 100 if the null hypothesis is true. Based on this evidence, we would reject the null hypothesis and state that there is a significant difference among the four group means.

In presenting data to others, we do not usually show all the steps in the analysis of the data. We usually summarize the information, so that the reader can interpret the results quite easily. In a study of the

TABLE 12-2
Summary of the analysis of variance

Source of variance	df	Sum of squares	Mean squares	F
Between groups	3	4,911.30	1,637.10	8.25*
Within groups	27	5,358.73	198.47	
Total	30	10,270.03		

$*p < .01$

type illustrated in example 12–2, we would normally present the mean of each of the groups, the sample sizes (N_g), and the standard deviation of each group. The summary of the analysis of variance is usually presented in table form. This type of summary is shown in table 12–2.

TESTING FOR SIGNIFICANT DIFFERENCES BETWEEN PAIRS OF MEANS

After computing the F ratio and finding that the differences among the group means are statistically significant, we need to go one step further. The overall F test allows us to reject the null hypothesis that all groups have equal population means. The significant F ratio does not tell us which of the individual means are significantly different from the others. A quick glance at the means of example 12–2 would indicate that group means 2 and 4 are probably not different from each other, but means 1 and 3 are very likely from different populations.

After the F test has been computed and significant results obtained, it is then possible to examine each pair of group means. We can use the t test for this, provided we are not going to make too many different comparisons of pairs of means. We shall use the t test, but with two important differences: (1) the denominator of each t will be based on the MS_{wg} computed from the analysis of variance, and (2) the degrees of freedom for each t test will be the degrees of freedom *within* groups from the analysis of variance.

In chapter 10, it was pointed out that in order to use the t test, the assumption of homogeneity of variance has to hold true for the data. It should be noted that exactly the same assumption is true for the analysis of variance. If each within-group variance is an unbiased estimate of the same population variance, then the best estimate will be a combination of all within-group variance estimates. The mean square within groups is the average of all the individual unbiased estimates of the population variance. Since we have estimates available from each of the groups in the study, the composite estimate is a "better" estimate than that based on just the two groups on which the t test is based. Since we are going to use all of the within-group variances to determine the denominator of the t test, we should use all the degrees of freedom on which this term is based. It can be shown that the t test is a special case of the F test. If we have only two groups in a study and compute the F ratio with one degree of freedom for the numerator, the value of F obtained will be equal to t^2 computed from the same data, or the reverse $t = \sqrt{F}$.

This can be verified by using table F and table G of the Appendix. Referring to table F with $1df$ for the numerator and $10df$ for the denominator, an F of 4.96 is necessary to reject H_0 at $\alpha = .05$. From table G, with $10df$, a t of 2.228 is necessary to reject H_0 at $\alpha = .05$. The square root of 4.96 is 2.23; the square of 2.228 is 4.96. The equation for the t test after having found a significant F ratio is

$$t = \frac{\bar{X}_a - \bar{X}_b}{\sqrt{MS_{wg}\left(\frac{1}{N_a} + \frac{1}{N_b}\right)}} \tag{12-11}$$

where

$\bar{X}_a$ is the mean of one group
$\bar{X}_b$ is the mean of a second group
MS_{wg} is that obtained from the analysis of variance
N_a is the number of observations in the one group
N_b is the number of observations in the second group

The use of the t test for testing differences between the pairs of means after obtaining a significant F ratio is illustrated in example 12–3. With four different groups, there are six different intercomparisons between pairs of means. We shall compare the mean of group 1 with each of the other three means. The mean of group 2 can then be compared with the mean of group 3 and the mean of group 4; and finally, the mean of group 3 can be compared with the mean of group 4.

The use of the t statistic to test for differences between pairs of means has certain limitations. In general, the t test should be used only if the F ratio indicates that the group of means are statistically different from each other. If the F ratio is nonsignificant, the use of the t test on pairs of means may be quite misleading. If the F ratio is not significant, the t test should be used only if the research worker has hypothesized that specific means will be different, and then the t should be applied *only to those differences which were specified in advance of the collection of data.*

If the F ratio is significant and t tests have been made on the differences in pairs of means, the specific differences which are significant can be shown in concise table form. Table 12–3 shows a summary of the t test given in example 12–3.

The pattern in table 12–3 shows clearly that group 3 is significantly different from all three other groups. In addition, group 1 is different from all other groups. The difference between the means of groups 2 and 4 is not statistically significant.

Example 12-3 Use of the t test in comparing pairs of means after the analysis of variance (data from example 12-2)

$\bar{X}_1 = 31.14$	$\bar{X}_2 = 47.00$	$\bar{X}_3 = 70.00$	$\bar{X}_4 = 49.38$	$MS_{wg} = 198.47$
$N_1 = 7$	$N_2 = 10$	$N_3 = 6$	$N_4 = 8$	$df_{wg} = 31 - 4 = 27$

Using formula 12–11:

$$t = \frac{\bar{X}_a - \bar{X}_b}{\sqrt{MS_{wg}\left(\dfrac{1}{N_a} + \dfrac{1}{N_b}\right)}}$$

$$t_{1\text{ vs. }2} = \frac{31.14 - 47.00}{\sqrt{198.47\left(\dfrac{1}{7} + \dfrac{1}{10}\right)}} = \frac{-15.86}{\sqrt{198.47(.24)}} = \frac{15.86}{6.90} = 2.30*$$

$$t_{1\text{ vs. }3} = \frac{31.14 - 70.00}{\sqrt{198.47\left(\dfrac{1}{7} + \dfrac{1}{6}\right)}} = \frac{-38.86}{\sqrt{198.47(.31)}} = \frac{38.86}{7.84} = 4.96\dagger$$

$$t_{1\text{ vs. }4} = \frac{31.14 - 49.38}{\sqrt{198.47\left(\dfrac{1}{7} + \dfrac{1}{8}\right)}} = \frac{-18.24}{\sqrt{198.47(.27)}} = \frac{18.24}{7.32} = 2.49*$$

$$t_{2\text{ vs. }3} = \frac{47.00 - 70.00}{\sqrt{198.47\left(\dfrac{1}{10} + \dfrac{1}{6}\right)}} = \frac{-23.00}{\sqrt{198.47(.27)}} = \frac{23.00}{7.32} = 3.14\dagger$$

$$t_{2\text{ vs. }4} = \frac{47.00 - 49.38}{\sqrt{198.47\left(\dfrac{1}{10} + \dfrac{1}{8}\right)}} = \frac{-2.38}{\sqrt{198.47(.22)}} = \frac{2.38}{6.61} = .36$$

$$t_{3\text{ vs. }4} = \frac{70.00 - 49.38}{\sqrt{198.47\left(\dfrac{1}{6} + \dfrac{1}{8}\right)}} = \frac{20.62}{\sqrt{198.47(.29)}} = \frac{20.62}{7.59} = 2.72*$$

$*p < .05$
$\dagger p < .01$

Each of the six t tests would involve 27 degrees of freedom. Entering table G of the Appendix with $df = 27$, we find that the difference between groups 1 and 2, 1 and 4, and 3 and 4 reach significance at the .05 level. The differences between the means of groups 1 and 3 and the differences between those of groups 2 and 3 reach significance beyond the .01 level.

The analysis of variance allows us to make one overall test of the significance of the differences among a set of means. Wise use of the t test after the analysis of variance allows us to pinpoint where differences are occurring.

If a large number of internal comparisons are made after the F test, than the use of the t test can lead to a large Type I error. Part of

TABLE 12–3
Summary of t tests after the analysis of variance

		Groups		
		2	3	4
Groups	1	*	†	*
	2		†	
	3			*

* Indicates that the difference in means was significant at $p < .05$.
† Indicates that the difference in means was significant at $p < .01$.

this increase in the probability of a Type I error is due to the fact that the different comparisons are not independent of each other. The t test between groups 1 and 3 and the t test between groups 1 and 2 both use the mean of group 1. The sampling errors which will occur with group 1 will influence both t tests; thus, the t tests are not independent, and the tabled values for the significance of each t will not be accurate.

The other part of this problem is related to the number of t tests being computed. If the null hypothesis is true, and we compute 100 *independent* t tests, we would expect 5 of these to exceed an α level of .05 and 1 of them to exceed an α level of .01. As we compute more and more *independent* t tests, the probability that at least one of these tests will exceed the decision point for rejection of the null hypothesis becomes very high. The probability of the Type I error becomes very large. When the different t tests are *not* independent, the probability of a Type I error increases even more rapidly.

Multiple, internal comparison is a topic of rather wide dimensions. There are several different techniques available which permit multiple, internal comparisons in a set of data without doing violence to the probability of the Type I error. Comparison techniques developed by Duncan, Newman-Kuels, Schefé, and Tukey are all useful, but each is designed for a slightly different purpose. There is no one multiple internal-comparison technique which is preferred under all circumstances. These tests will not be discussed in this book, but the references at the end of the chapter provide a thorough discussion of the advantages and limitations of each of these approaches.

The t test can be used to make multiple internal comparisons between means provided that: (1) the overall F test allows the rejection of the null hypothesis and (2) not too many such t tests are to be made. If either of these two conditions are not met, the reader is urged to consult the references listed for mulitple internal comparisons.

COMPLEX ANALYSIS OF VARIANCE

The analysis of variance presented in this chapter is often referred to as a *one-way classification* or a *simple analysis* of variance. As the name implies, the one-way classification involves setting up groups on only one dimension. The categories of manic-depression, schizophrenic, and character disorders might be used and a drug administered to individuals within each of these categories. The efficacy of the drug could then be determined for each of these categories and the one-way analysis of variance used to determine if the drug had differential effects in the groups. The same approach can be extended to classification systems with two or more dimensions. Using the three diagnostic categories mentioned above, it would be possible to use several different drugs. In this case, each diagnostic group would be subdivided and a different drug administered to each of the subgroups. This would represent a two-way classification. From this two-way classification, it would be possible to determine (1) which of the diagnostic categories seemed to be most amenable to drug therapy, (2) which of the drugs produced the greatest changes in the patients, and (3) if certain combinations of diagnostic category and drugs were markedly superior (or inferior) to other combinations of diagnostic category and drug. This is referred to as an *interaction*.

The more advanced uses of the analysis of variance are proving to be of great value to the behavioral scientist in his attempts to analyze and understand complex behavior, especially interactions among variables. It is unfortunate that it is not possible to explore these advanced uses in an introductory book; however, the references at the end of this chapter present most of the advanced techniques which have proved useful in the behavioral sciences.

Comparisons of several groups with nominal or ordinal scaling

The analysis of variance just discussed is often referred to as a one-way classification or a simple analysis of variance. As the name implies, the one-way classification involves setting up groups on only one dimension. If the measurements do not meet the requirements of interval scaling then some nonparametric approach must be used to analyze the data.

This section will present five nonparametric statistics which may be used with more than two groups in the study. The five statistics are (1) the χ^2, (2) the median test, (3) the Kruskal-Wallis one-way analysis of variance for ranks, (4) the Cochran Q test, and (5) the Friedman analysis of variance for ranks. All five of these statistics use the sampling distribution of chi-square. Two of these are extensions of statistical tests presented in chapter 11.

TESTING FOR SIGNIFICANCE OF DIFFERENCES AMONG k INDEPENDENT GROUPS WITH NOMINAL DATA
THE CHI-SQUARE TEST

In chapter 11, an example was used comparing the library usage of biology majors and English majors. This same type of approach can be used with more than two different categories of majors. Suppose the study had been larger and included mathematics majors and sociology majors. Using the two categories of use and nonuse of the library will yield a 2×4 table or eight different cells. We shall have to determine the two items of information: (1) how many individuals were observed in each cell and (2) how many individuals were expected in each cell. With this information a χ^2 can be computed using formula 11–1 which is repeated below

$$\chi^2 = \sum \frac{(O - E)^2}{E}$$

where

O is the observed frequency in each cell
E is the expected frequency in each cell

The observed frequency, of course, comes directly from the cells. The expected frequency is computed in the same manner as before: the row frequency and column frequency are multiplied together and divided by the total frequency to yield the expected frequency for the specific cell. The same procedure is repeated for each cell. Example 12–4 shows the computation of the χ^2 for this fictitious data. Note that the expected frequency for the biology majors who used the library would be $(188)(105)/353 = 19,740/353 = 55.9$. Similarly the expected frequency for the mathematics majors who did not use the library would be $(165)(53)/353 = 8745/353 = 24.8$.

Example 12-4   Use of the χ^2 with more than two groups.

MAJOR FIELD

	Biology	English	Mathematics	Sociology	
Used library	47 (55.9)	62 (53.8)	20 (28.2)	59 (50.1)	188
Did not use library	58 (49.1)	39 (47.2)	33 (24.8)	35 (43.9)	165
	105	101	53	94	353

$$\chi^2 = \sum \frac{(O-E)^2}{E} = \frac{(47-55.9)^2}{55.9} + \frac{(62-53.8)^2}{53.8} + \frac{(20-28.2)^2}{28.2}$$

$$+ \frac{(59-50.1)^2}{50.1} + \frac{(58-49.1)^2}{49.1} + \frac{(39-47.2)^2}{47.2} + \frac{(33-24.8)^2}{24.8}$$

$$+ \frac{(35-43.9)^2}{43.9} = \frac{79.21}{55.9} + \frac{67.24}{53.8} + \frac{67.24}{28.2} + \frac{79.21}{50.1} + \frac{79.21}{49.1}$$

$$+ \frac{67.24}{47.2} + \frac{67.24}{24.8} + \frac{79.21}{43.9} = 1.42 + 1.25 + 2.38 + 1.58$$

$$+ 1.61 + 1.42 + 2.71 + 1.80 = 14.17$$

Using formula 11-2 for the degrees of freedom

$$df = (r-1)(c-1) = (2-1)(4-1) = 3;$$

and table J of the Appendix, a χ^2 of 14.17 with 3df has $p < .01$

Example 12-4 was for a 2×4 table. The χ^2 is not limited by the number of categories in either the rows or columns. Thus if we had been interested in possible differences in major areas as it might be related to year in college, we might have a 4×4 table with the four majors on one dimension and the four categories—freshman, sophomore, junior, and senior—on the other dimension.

With a large number of degrees of freedom in a χ^2 analysis, it is sometimes desirable to analyze subsections of the table. Techniques for partitioning the df in a χ^2 analysis are beyond the scope of this book but may be found in the references listed at the end of this chapter.

TESTING FOR SIGNIFICANCE OF DIFFERENCES AMONG k INDEPENDENT GROUPS WITH ORDINAL DATA
THE MEDIAN TEST FOR MORE THAN TWO GROUPS

In the previous section it was shown that the χ^2 test is not limited to only two groups. This should make it immediately apparent that the median test can also be used in a multi-group situation. The median test requires that all members of the different groups must be ranked in one, common, overall ranking. From this overall ranking a common median can be determined. The median is then used to form two subgroups within each of the k different major groups—a subgroup below the median and a subgroup above the median. Thus we will have a $2 \times k$ table, and the regular procedures of the χ^2 are used to determine if the k groups differ as to the proportions above and below the common median.

Suppose the coach of a football team were interested in the effectiveness ratings of the players on the team. Each week the assistant coaches rated each member of the team as to his degree of effectiveness on a scale from 1 to 100. The coach was interested in seeing if there was any difference in effectiveness as a function of the player's time on the squad. The four categories of freshman, sophomore, junior, and senior were used. The overall median of the 80 players was established, and the results are given in example 12–5.

Example 12–5   Use of the median test with four groups

CLASS YEAR

	Freshman	Sophomore	Junior	Senior	
Median and above in effectiveness	5 (11.5)	6 (9)	10 (8)	19 (11.5)	40
Below-median effectiveness	18 (11.5)	12 (9)	6 (8)	4 (11.5)	40
	23	18	16	23	80

$$\chi^2 = \frac{(5-11.5)^2}{11.5} + \frac{(6-9)^2}{9} + \frac{(6-8)^2}{8} + \frac{(19-11.5)^2}{11.5}$$
$$+ \frac{(18-11.5)^2}{11.5} + \frac{(12-9)^2}{9} + \frac{(6-8)^2}{8} + \frac{(4-11.5)^2}{11.5}$$

$\chi^2 = 3.67 + 1.00 + .50 + 4.80 + 3.67 + 1.00 + .50 + 4.80$
$\chi^2 = 19.94;\ df = (2-1)(4-1) = 3;$ from table J: $p < .001$

The Kruskal-Wallis One-way Analysis of Variance for Ranks

The median test only uses the information relative to the median of the groups. If rankings have been made of all the members, then the information contained in those ranks (other than above or below the common median) is not used. What is needed is a procedure which will utilize all of the ranking information. The Kruskal-Wallis does just that.

To understand the Kruskal-Wallis, we must go back to the basic idea of just what is *expected* if there are no differences among the groups—the null hypothesis is true. If all the groups are random samples from the same population, we would expect variation in the *ranks* assigned to the different observations, but we would also expect that the *average* (mean) *rank* in each group would be about the same size as the average rank in every other group. Further we would expect that the overall average rank would be about the same as the average rank for each group. We can compare the expected, average rank with the observed, average rank. Kruskal and Wallis have shown that if H_0 is true, the statistic H follows the chi-square distribution with $k - 1$ degrees of freedom. The statistic H can be found by use of formula 12–12

$$H = \frac{12}{N(N + 1)} \sum_{j}^{k} \frac{R_j^2}{N_j} - 3(N + 1) \qquad (12\text{–}12)$$

where

N is the total number of cases in all groups combined
R_j is the *sum of ranks* in the jth group
N_j is the number of cases in the jth group
k is the number of groups

Formula 12–12 will hold if the number of cases in each of the groups is not less than five. Siegel (listed at the end of the chapter) presents tables for use when the number of cases in any group is less than five and there are three groups in the study.

In the previous section a hypothetical example was given involving the rating of effectiveness of football players and the relationship of these ratings to the class year of the players. The median test was used to analyze those data. These same kinds of rating data are presented for fraternity members in example 12–6. Included in the example are the participation ratings (on a scale from 0–100). The Kruskal-Wallis, as mentioned above, utilizes all of the order information in the data.

Example 12-6   The Kruskal-Wallis one-way analysis of variance for ranks

The fictitious data presented are the participation ratings of fraternity members arranged by the individual's year in school.

FRESHMAN		SOPHOMORE		JUNIOR		SENIOR	
Rating	*Rank*	*Rating*	*Rank*	*Rating*	*Rank*	*Rating*	*Rank*
81	27	88	39	84	33.5	95	41
79	24	87	37.5	83	32	89	40
78	21.5	78	21.5	82	29.5	87	37.5
77	20	74	18	82	29.5	85	35.5
73	16.5	72	14.5	80	26	85	35.5
73	16.5	68	10.5	68	10.5	84	33.5
71	13	56	3.5	62	7	82	29.5
70	12	48	1.5	56	3.5	82	29.5
66	9	48	1.5			79	24
64	8					79	24
59	6					76	19
58	5					72	14.5
$R_1 = 178.5$		$R_2 = 147.5$		$R_3 = 171.5$		$R_4 = 363.5$	
$N_1 = 12$		$N_2 = 9$		$N_3 = 8$		$N_4 = 12$	

Using formula 12-12:

$$H = \frac{12}{N(N+1)} \sum_{}^{k} \frac{R_j^2}{N_j} - 3(N+1)$$

$$H = \frac{12}{41(41+1)} \left[\frac{(178.5)^2}{12} + \frac{(147.5)^2}{9} + \frac{(171.5)^2}{8} + \frac{(363.5)^2}{12} \right] - 3(41+1)$$

$$= (.00697)(2655.17 + 2417.36 + 3676.53 + 11{,}011.02) - 126$$
$$= (.00697)(19{,}760.08) - 126 = 137.73 - 126$$
$$H = 11.73; \ df = k - 1 = 3; \text{ using table J of Appendix: } p < .01$$

Correction for ties using formulas 12-13 and 12-14: $T = t^3 - t$; scores 48, 56, 68, 72, 73, 78, 84, 85, and 87 are each tied with one other score. For each, $T = 2^3 - 2 = 6$. Score 82 has four scores tied; thus $T = 4^3 - 4 = 60$ and $\Sigma T = (9)(6) + 60 = 54 + 60 = 114$

Using formula 12-13:

$$1 - \frac{\Sigma T}{N^3 - N} = 1 - \frac{114}{(41)^3 - 41} = 1 - \frac{114}{68{,}921 - 41} = 1 - .0017 = .9983$$
$$H_c = 11.73/.9983 = 11.75$$

CORRECTION FOR TIES.   When two or more scores are tied for the same rank, the effect is to *decrease* the value of H. An H computed by formula 12–12, and not corrected for ties will be a conservative test of the null hypothesis. The probability of a Type I error (α) will be less than the investigator has selected; however, this means that the probability of the Type II error (β) will be a little higher. A correction for ties may be made by *dividing* the results of formula 12–12 by the correctional term which is computed by

$$1 - \frac{\Sigma T}{N^3 - N} \qquad (12\text{--}13)$$

where

N is the total number of cases in all groups combined
T is found by

$$T = t^3 - t \qquad (12\text{--}14)$$

where

t is the number of scores in a tied group of scores

The Kruskal-Wallis H, corrected for ties, is given in the lower part of example 12–6. Note that there are two scores of 48 and they are tied for the ranks 1 and 2, hence each received the rank 1.5. For this tie, $t = 2$ and $T = 2^3 - 2 = 8 - 2 = 6$. Both scores happen to be in the same group (sophomore). There are two scores of 56 (one in the sophomore group and one in the junior group). Each received the rank of 3.5. Again $t = 2$ and $T = 6$. There are four scores of 82 (two in the junior group and two in the senior group). The ranks of 28, 29, 30, and 31 go with these four scores, thus each is given the average of the four ranks or 29.5. For this tie, $t = 4$ and $T = 4^3 - 4 = 64 - 4 = 60$.

There is some question if the correction for ties has any practical consequences. Notice that in example 12–6, 22 of the 41 observations were tied, yet the change in the value of H was only 2/100. If the computed value of H exceeds the decision point established by the investigator, then the correction will be of no consequence since the value of H will become larger when corrected. If the value of H is quite small relative to that needed to reach statistical significance at the decision point, a correction will rarely increase H enough for rejection of the null hypothesis. The important determiner of T is not the total number of ties but rather the number of scores involved in each tie. Suppose that all 22 scores that were involved in ties were tied with each other; then, $T = 22^3 - 22 = 10,626$. The correction factor would be $1 - (10,626)/(68,880) = 1 - .1542 = .8457$, and the "corrected" H would be $11.37/.8457 = 13.87$. This yields an 18 percent increase in the value of H as compared to a 0.17 percent increase in example 12–6.

Clearly, it is the length of ties in each set of ties that is important, rather than the total number of observations that are involved in ties.

The Kruskal-Wallis is one of the most powerful alternatives to the simple analysis of variance and is the preferred statistic to use if ranking information is available concerning three or more groups. Since it uses all the order information, it is clearly a more powerful tool than the extension of the median test which only uses the information above or below the group median.

Earlier, it was pointed out that after the analysis of variance had been computed and if the null hypothesis had been rejected for all the means considered at one time, it was sometimes valuable to compare specific mean differences to determine which groups differed. The same situation is true after the Kruskal-Wallis H has been computed. Obviously the t test would be inappropriate. If the data do not meet the requirements for the analysis of variance (F test), then they will not meet the requirements for the t test. The Mann-Whitney U test presented in chapter 11 would be an appropriate tool to test for the differences between the groups, taken two at a time. This should be done only if the Kruskal-Wallis H is statistically significant, or if the investigator specified tests between specific groups *before the data were collected.*

TESTING FOR SIGNIFICANCE OF DIFFERENCES AMONG k CORRELATED GROUPS WITH NOMINAL DATA THE COCHRAN Q TEST

The observation of the same subject under several different conditions offers many advantages to the research worker. The problems of possible extraneous variables which might influence the study are minimized. The relative advantages and disadvantages of using subjects under more than one condition in a study (or using matched subjects in the different conditions) is discussed more fully in chapter 13. To this point we have discussed only the McNemar test of change which is suitable for use with nominal scales when the same subjects are observed twice. The Cochran Q test can be used with three or more conditions in the study, provided the observations can be placed in a dichotomy.

Most test items are scored on a pass–fail basis (one point for a correct answer, zero points for an incorrect answer) and since all individuals respond to all items, the Cochran Q test can be used to test for differences in the difficulty of the items. Voting can be considered a dichotomized situation. The voter either votes for a candidate or does not vote for the candidate.

Individuals running for political office are concerned about their "message getting across" to the voters. To test the impact of different appeals to the voters, it would be possible to select a sample of voters and ask each one if he intended to vote for candidate A. These voters could be given information in a specific form about candidate A and again asked about their preference. This same procedure could be repeated several times using different informational approaches. Since the response of the voters could be recorded as "yes" or "no" and the same individuals respond each time, the Cochran Q test could be used to determine the differential effects of the different informational approaches. The same thing could be done using matched individuals rather than the same person questioned several times.

A table could be arranged with the different conditions of the study as columns and the different individuals (or matched groups) as rows. The obtained data would be entered as zeros and ones. Cochran has shown that if the zeros and ones are randomly distributed in the rows and columns of the table (the null hypothesis is true and the type of condition under which the individual was observed does not make a difference), the statistic Q is distributed as chi-square with $k - 1$ degrees of freedom. Q can be determined by means of formula 12–15.

$$Q = \frac{(k-1)\left[k \sum_{j}^{k} G_j{}^2 - \left(\sum_{j}^{k} G_j\right)^2\right]}{k \sum_{i}^{N} L_i - \sum_{i}^{N} L_i{}^2} \tag{12–15}$$

where

k is the number of conditions in the study
N is the number of individuals (or matched groups)
G_j is the sum of the entries in the jth column
L_i is the sum of the entries in the ith row.

Suppose that a political organization selects a group of voters to test the appeal of four different campaign approaches. The voters are carefully matched in groups of four. Each group of four individuals is matched as to political party, sex, age, and social-economic level. Sixteen such groups (64 individuals) are established. Each member of each group is given a different informational approach and then asked if he would or would not vote for candidate *A*. Fictitious data from such a situation are presented in example 12–7 to illustrate the computation of the Cochran *Q* test.

Note the similarity in the approach between the Cochran *Q* test and the analysis of variance. In the analysis of variance, if the null hypothesis is true, we would expect the means of the different treatment conditions

---

Example 12-7   Computation of the Cochran Q test

Sixteen groups of four matched individuals in each group are presented information in different forms concerning candidate *A*. Each individual is asked if he will vote for the candidate. A *yes* vote is recorded as a "1" and a *no* vote as a "0." This fictitious data is arranged as follows.

Voter group	Form A	Form B	Form C	Form D	L_i	L_i^2
1	0	1	0	1	2	4
2	1	0	0	1	2	4
3	1	1	0	1	3	9
4	1	1	0	1	3	9
5	0	1	0	1	2	4
6	0	0	1	1	2	4
7	1	1	0	1	3	9
8	0	0	1	1	2	4
9	0	0	1	1	2	4
10	1	1	0	1	3	9
11	0	1	0	1	2	4
12	1	1	0	1	3	9
13	1	1	1	1	4	16
14	0	1	0	1	2	4
15	0	1	0	0	1	1
16	0	1	0	1	2	4
Sum	$G_1 = 7$	$G_2 = 12$	$G_3 = 4$	$G_4 = 15$	$\sum\limits^{N} L_i = 38$	$\sum\limits^{k} L_i^2 = 98$
G_j^2	49	144	16	225		$\sum\limits^{k} G_j^2 = 434$

Note that $\sum\limits^{N} L_i = \sum\limits^{k} G_j$

Using formula 12–15:

$$Q = \frac{(k-1)\left[k\sum\limits^{k} G_j^2 - \left(\sum\limits^{k} G_j\right)^2\right]}{k\sum\limits^{N} L_i - \sum\limits^{N} L_i^2} = \frac{(4-1)[(4)(434) - (38)^2]}{(4)(38) - 98}$$

$$Q = \frac{876}{54} = 16.22; \; df = (k-1) = (4-1) = 3; \text{ from table J, } p < .01$$

The four information presentations have differing effects.

---

to be of about the same value. In the Cochran Q test, if the null hypothesis is true, we would expect the column sums to be of about the same value.

The Cochran Q test is an extremely useful statistical test for correlated groups (or the same individuals observed more than once) when the observations are in the form of a dichotomy. If the observations are in the form of an interval or ratio scale, then the Cochran is very wasteful of information. The analysis of variance for correlated groups should be used. If the observations can be ranked, a test based on ordinal data should be used.

TESTING FOR SIGNIFICANCE OF DIFFERENCES AMONG k CORRELATED GROUPS WITH ORDINAL DATA
THE FRIEDMAN ANALYSIS OF VARIANCE FOR RANKS

The Friedman test is similar to the Cochran Q test in that it was developed for use with correlated groups (or the same subject observed under several different conditions). The Friedman, however, requires that the different observations on the one individual (or matched group) must be capable of being ordered (ranked). To use the Friedman, we must be able to state that an individual did the "best" under one condition of the study, and "second best" under a different condition. Thus we must be able to rank each individual (or matched group) from 1 to k, where k is the number of conditions in the study under which each individual (or group) was observed. If we could measure the differences on an interval or ratio scale, then the Friedman would not be the appropriate statistic to use—we should use the analysis of variance for correlated samples. If the observations can be ranked, the Friedman is a very powerful analytic tool.

To use the Friedman, the data must be arranged in a two-dimensional table. The conditions of the study should be set up as the columns in the two-way table, and the different individuals (or matched groups) as the rows. The observations on each individual (or matched group) are ranked from 1 to k. Friedman has shown that if the null hypothesis is true, we would expect the sum of the ranks of each column (treatment condition) to be of about the same size and the statistic χ_r^2 is distributed as chi-square with $df = k - 1$. The statistic χ_r^2 can be computed by use of formula 12–16 as follows:

$$\chi_r^2 = \frac{12}{Nk(k+1)} \sum_{}^{k} R_j^2 - 3N(k+1) \qquad (12\text{–}16)$$

where

N is the number of rows (individuals or matched groups)
k is the number of columns (treatment conditions)
R_j is the sum of the ranks in the *j*th column (treatment conditions)

The computation of the Friedman analysis of variance for ranks is shown in example 12–8.

Example 12-8   Computation of the Friedman Test

In many beauty contests, the young ladies are judged in four different events: (1) appearance in bathing suits, (2) appearance in formal dresses, (3) talent, and (4) poise in an interview situation. Since each individual appears in each of the four situations it would be possible to utilize the Friedman test to judge the relative importance of these four situations. Ficticious data for ten contestants is presented below. Each contestant is ranked from 1 to 4 as to her "best" category.

Different situation for judgement

Contestants	Bathing suits	Formal dress	Talent	Poise in interview
1	1	2	4	3
2	1	2	3	4
3	1	2	4	3
4	1	2	4	3
5	2.5	2.5	4	1
6	1	2	3	4
7	1	3	4	2
8	1	2	4	3
9	1	2	4	3
10	4	2	1	3
R_j	14.5	21.5	35	29

Using formula 12–16:

$$\chi_r^2 = \frac{12}{Nk(k+1)} \sum^k R_j^2 - 3N(k+1)$$

$$\chi_r^2 = \frac{12}{(10)(4)(4+1)} [(14.5)^2 + (21.5)^2 + (35)^2 + (29)^2] - (3)(10)(4+1)$$

$$\chi_r^2 = (.06)(2738.5) - 150 = 14.31; \ df = (k-1) = (4-1) = 3$$

From table J of the Appendix, $p < .001$. We would reject the hypothesis that these four attributes are ranked equally across the ten contestants.

In the event an individual is judged to be equal in two or more categories, the average of the ranks is assigned. Friedman reports that ties do not affect the sampling distribution of χ_r^2. When there are only three categories and N is less than 10 or there are four categories but N is less than 4, the sampling distibution of χ_r^2 does not follow the chi-square distribution and table J of the Appendix should not be used. Table P in the Appendix was developed for these small-sample situations.

Summary

The significance of the difference between two standard deviations can be determined by use of the F ratio. The variance of each sample is computed, and the ratio of these two variances is formed by placing the larger of the two variances in the numerator of the F ratio and the smaller variance in the denominator. When used in this manner, the table of the F ratio is a half table, and the probability of the occurrence of the observed ratio is one-tailed. The tabled probability should be doubled to determine the two-tailed probability of the occurrence of the observed ratio.

The basic idea behind the analysis of variance is to divide the observed difference into components and relate these components to the factors which are producing these differences. The deviation of an individual score from the mean of all scores can be broken into additive components: (1) the variation of the score about the mean of the subgroup in which it occurs and (2) the variation of the mean of the subgroup about the mean of all scores. Squaring these deviations and summing them for all individuals divides the total sum of squared deviations into two additive parts: (1) the sum of squared deviations *within* the subgroups and (2) the sum of squared deviations *between* the subgroups.

The total degrees of freedom can also be divided into two additive parts: (1) the *df within* groups and (2) the *df between* groups. If each sum of squared deviations is divided by its degrees of freedom, the resulting *mean squares* can be placed in the F ratio. The mean square *between* groups is always placed in the numerator of the F ratio, and the mean square *within* groups is placed in the denominator. If the null hypothesis is true (no true difference between the means), we would expect the value of the F ratio to be about one. If the null hypothesis is false (there is a true difference among the means), we would expect the F ratio to be larger than one. The table of the F statistic, when used for testing *differences among means,* is a two-tailed table.

Use of the table of the F statistic (like any other statistical test) requires that the data obtained meet the assumptions of the mathematical model. For the F statistic, these assumptions are:

1.   The data should be interval or ratio.

2.   The samples are random samples from their respective populations.

3.   The populations from which these samples were drawn are normally distributed.

4.   The variances of the underlying populations are equal.

After an F ratio has been computed to determine if sample means are significantly different from each other, it is next necessary to determine the specific differences among the means. If the total number of pairs of means to be compared is not too large, the t test can be used for making these pairs of comparisons. When the t is used in this manner, the mean squares *within* groups should be used in the denominator of the t. Since the mean square *within* groups is an estimate of the common population variance, it is a better estimate than the within-group variance of only the two groups being compared. The degrees of freedom for these t tests should be the degrees of freedom *within* groups.

The χ^2 test can be extended for any number of groups. This could be a one-way classification in which case the expected frequencies must be obtained from a source outside of the data. The two-way classification can be extended indefinitely in both directions. In this situation, the expected frequencies in each cell are obtained by multiplying the row and column totals for that cell and dividing the product by the total frequency. This extended χ^2 is interpreted as any χ^2.

Since the median test is a special case of the two-way classification for χ^2, it may be extended for use with any number of groups. The median test does not use all of the order information which is in complete ranking.

The Kruskal-Wallis one-way analysis of variance for ranks does use all of the order information in ranks. The scores in all k independent groups are put in one common ranking. The sum of the ranks within each group is determined. If the null hypothesis were true then we would expect the average rank for each group to be about the same as the average rank for any other group. The statistic H follows the chi-square sampling distribution with $k - 1$ degrees of freedom. The Kruskal-Wallis is a powerful test for use with ranking data. Ties in the ranking tend to make the Kruskal-Wallis a conservative test. A correction for tied ranks is available, but the use of the adjustment has little practical impact unless a large proportion of the ranks are tied at the same value.

Two tests were presented for use with correlated observations when the measurements do not meet the assumptions of interval or ratio scaling. The Cochran Q test is appropriate when the information can be placed in two categories. One category is scored "one" and the other category is scored "zero." The sampling distribution of the Q follows the chi-square distribution with $k - 1$ degrees of freedom, hence table J of the Appendix may be used to determine the probability of the Q of a given size or larger occurring by chance.

The Friedman analysis of variance for ranks also requires that all k observations are on the same individual or on matched individuals. The observations for each individual (or matched group) are ranked from one to k. If the null hypothesis is true, we would expect about the same sum of ranks in each of the different conditions under which the individuals were observed. The statistic χ_r^2 is distributed as chi-square with $k - 1$ degrees of freedom, hence table J of the Appendix may be used with larger samples. For small samples table P of the Appendix must be used.

References

INTERNAL ANALYSIS OF DATA AFTER AN F TEST

W. S. RAY, *Introduction to Experimental Design* (New York: Macmillan Co., 1960), chap. xv.

B. J. WINER, *Statistical Principles in Experimental Design* (2nd ed.; New York: McGraw-Hill Book Co., Inc., 1971), pp. 170–204.

A. L. EDWARDS, *Experimental Design in Psychological Research* (3d ed.; New York: Holt, Rinehart & Winston, Inc., 1968), chap. viii.

MORE COMPLEX USES OF THE ANALYSIS OF VARIANCE

All three of the references cited above are good sources on the use of the analysis of variance. Ray is perhaps the easiest book to read. Winer is fairly advanced but does not require mathematics beyond algebra. The book by Edwards is especially useful in setting up computing formulas for more complex uses of the analysis of variance.

NONPARAMETRIC STATISTICS FOR USE WITH k GROUPS

W. G. COCHRAN, "Some Methods for Strengthening the Common χ^2 Tests," *Biometrics,* Vol. 10 (1954), pp. 417–51. This article includes a discussion of the situation in which the expected values in some cells is less than five.

A. E. MAXWELL, *Analyzing Qualitative Data* (New York: John Wiley & Sons, Inc., 1961). This book includes information about splitting out individual degrees of freedom in χ^2 for making multiple internal analysis of a contingency table.

FREDERICK MOSTELLER and ROBERT E. K. ROURKE, *Sturdy Statistics Nonparametric and Order Statistics* (Reading, Massachusetts: Addison-Wesley Publishing Company, 1973).

SIDNEY SIEGEL, *Nonparametric Statistics for the Behavioral Sciences* (New York: McGraw-Hill Book Co., Inc., 1956). Chapter viii deals with k independent samples; chapter vii deals with k correlated samples.

13

Planning and interpreting experiments

THROUGHOUT this book the emphasis has been upon the acquisition of various statistical techniques and their application to problems in the behavioral sciences. Chapter 9 presented some specific applications of statistics to the area of tests and measurements. The present chapter will introduce some of the applications of statistics to the planning of experiments.

Because this book has emphasized statistics, it is easy for the student to believe that research and more specifically experimental techniques in the behavioral sciences have developed along with statistics. This is not the case either for the behavioral sciences or for other areas which today depend upon statistics for both the planning and the interpretation of experiments.

Characteristics of experiments

INDEPENDENT AND DEPENDENT VARIABLES

Whenever an experiment is conducted, the experimenter is attempting to determine if a change in one (or more) variable(s) will produce some change in a second variable. Thus an experiment on the rate of learning with distributed practice would involve some change in the degree of distribution of practice (length of rest period between trials) and a measure of the number of trials necessary to reach some criterion of learning. In even the simplest experiment, at least two variables are present: the *independent* variable and the *dependent* variable. The independent variable is the one the expeimenter changes intentionally (in the illustration the degree of distribution is the independent variable). The dependent variable is the variable that changes as a result of the experimenter's manipulation of the independent variable (in the illustration the rate of learning—number of trials to learn the material—is the dependent variable). Every experiment will include these two variables, and if the experiment is well designed, the research worker will be able to state the relationship between distributed practice and rate of learning in the type of situation used.

Suppose, in the experiment to investigate the rate of learning as a function of distribution of practice, we decide to use two degrees of distribution—no rest between trials and a 30-second rest between trials. The task to be used is the learning of a list of nonsense syllables. If we use one list of nonsense material, we certainly can't have the subjects learn the list under one condition (zero seconds rest) and then learn the list again under conditions of a 30-second rest between trials. We must use either two different lists of nonsense syllables or two different groups of subjects. If we use two different lists, we must somehow make sure that the lists are comparable. Similarly, if two groups of subjects are used, we must assure comparability of the two groups. These problems of assuring equality for the different experimental conditions on all variables except the independent variable are problems in experimental design and may in many cases be quite independent of statistics per se.

The marriage of experimental design and statistics has been a relatively recent development; but as with all good unions, its strength has increased with time. Today, in many areas, especially in agriculture and medicine, there is a marked degree of consultation between statisticians and research workers during the planning stages of experimentation to insure (1) that the experiment will be as efficient as possible in terms of time, money, facilities, etc.; and (2) that when the experiment is completed, the results will not only be capable of analysis but will answer the question the research worker wanted to ask. Far too often, a research worker will complete an experiment and then find that existing statistics cannot be utilized in the analysis of the data. Every hour spent in planning an experiment with statistical analysis as one factor in planning may save countless hours later in trying to salvage something meaningful from the data.

EXTRANEOUS VARIABLES

But let us return to the problem of rate of learning as a function of the degree of distribution of practice. After the experiment is completed, the analysis will involve a comparison of the mean number of trials for the zero-second-rest group to learn the list with the mean number of trials for the 30-second-rest group to reach the criterion of learning. Any differences between these two means (other than differences to be anticipated through random sampling) we shall attribute to the differences in length of rest period. This conclusion is warranted only if we can be sure that no other variables are affecting the two groups in different ways.

Suppose that we obtained all the data from the zero-second group in the morning and all the data from the 30-second group in the afternoon. Let us further suppose that during the morning the building was noisy, with many students going to and from classes, banging doors, etc., but that in the afternoon the building was quiet. Differences between the two groups could be a result of the length of the rest period, but these differences could also be the result of differences in distraction, or a combination of distraction and rest period. We would not be able to say that the length of rest was necessarily the factor producing the differences in learning scores. The observed difference might have been due to one or more *extraneous* variables.

Every experiment will include a *dependent* variable, an *independent* variable, and one or more potential *extraneous* variables. The validity of the results of the experiment will depend in large part upon the experimenter's handling of these extraneous variables.

There are two aspects to the validity of the results of an experiment. The first of these is *bias*. Earlier, biased and unbiased estimators were discussed. An unbiased estimator is one which does not *systematically* either underestimate or overestimate the true value in which we are interested. Merchants are required by law to have an inspection sticker on their scales to indicate that the scale is accurate, i.e. unbiased. If the scale were set to read zero when there were two lbs. of weight on the scale, then any measurement would be two pounds too light and the merchant would undercharge for the goods purchased this way. On the other hand, if the scale read two pounds when there was no weight on the scale, the customers would be systematically overcharged by the value of two pounds. Experimenters, too, must be concerned about possible bias in their measurements.

The second aspect of validity is the *precision* of the experiment. The concept of precision is a little more difficult to understand than bias. Precision deals with variability. Suppose that we are attempting to determine the true weight of some object. We have a scale which has been calibrated and is known to be unbiased. We weigh the object not once but many times. If we have a precise measuring instrument, there will be very little variation from one measurement to the next. If we find a great deal of variation from one measurement to the next, we have a problem. The average of all the measurements will yield the true value (the measures are unbiased), but we cannot trust any one measurement because of the variability. The concept of precision implies the reduction of unintended variability.

Experimental design involves two basic ideas:

1.   Elimination of bias so that differences observed between the groups represent differences created by the independent variable.

2.   Increase in precision.

MINIMIZING THE EFFECTS OF EXTRANEOUS VARIABLES.   The most straightforward technique is to *eliminate* these extraneous variables. The distracting noise in a crowded building could be eliminated by sound-proofing the experimental room. In the use of electronic recording equipment, it is sometimes necessary to shield the entire area in wire mesh to eliminate various electrical signals that might influence the equipment. Elimination is generally the preferred technique, if feasible. Elimination of an extraneous variable has the obvious effect of removing a possible source of bias. Less obvious, but just as important, elimination of extraneous variables will probably increase the precision. A noise which is distracting will seldom be uniform. At one instant the noise may effect the performance of a subject. A few moments later the noise may have no effect. This will tend to increase the variability and hence decrease the precision.

In some cases, if elimination is too expensive or perhaps too difficult to achieve, it might be possible to *equalize* the effects of the extraneous variables. In the study of rate of learning as a function of degree of distribution of practice, all subjects could be tested in the morning or all in the afternoon, so that the disturbing effects would be equalized for both groups. Another procedure, if both morning and afternoon sessions are necessary, is to be sure that half of each group is tested during each period of the day. This form of equalizing is called *counterbalancing*. Another technique to equalize differences is to use the *principle of randomization*.

The use of the principle of randomization for obtaining comparable groups of subjects is relatively recent in the history of experimentation. If we wish to select two groups of subjects—a zero-second rest and a 30-second rest—we can assign subjects to these groups by utilizing this principle of randomization. Random assignment of subjects will not insure that each group will be exactly equivalent to the other group, but it does insure that the variation between the groups will be within limits which can be predicted by sampling theory.

Equalizing the effects of extraneous variables, either through counter-balancing of conditions or through randomization, will remove bias but will not increase precision.

MATCHED SAMPLES. As was mentioned earlier, these techniques for minimizing the effects of extraneous variables have been known and used for a long time, but only recently has it been possible to assess and statistically eliminate many of these effects from the analysis of the results of experiments. Matching of subjects is one way to equalize the effects of extraneous variables which may be a characteristic of the subject. If intelligence differences is an extraneous variable in a study, then using the same subject in the different conditions of the experiment will equalize intelligence across conditions. Matching subjects will accomplish the same thing. The use of correlated samples will remove any bias effects associated with the matching variable(s) but it will also increase precision. Perhaps the easiest way to demonstrate this is to use the t statistic for illustration. Formula 10–9 for the t statistic is given below for convenience. The

$$ t = \frac{D_{\bar{X}} - \mu}{s_{D_{\bar{X}}}} $$

impact of using correlated subjects is in the denominator of the t statistic. The standard error of the difference in means for *independent* groups (formula 10–3) is given below

$$ s_{D_{\bar{X}}} = \sqrt{\frac{s^2}{N_1} + \frac{s^2}{N_2}} $$

and the standard error of the difference in means for *correlated* groups is (formula 10–10)

$$ s_{D_{\bar{X}}} = \sqrt{\frac{s^2}{N} + \frac{s^2}{N} - 2r\frac{s^2}{N}} $$

The difference in these two expressions is the correlational term

$$ -2r\frac{s^2}{N} $$

The impact of successful matching (or using the same subjects in the different conditions of the experiment) is to reduce the size of the standard error of the difference in means as well as remove bias. Matching is a powerful technique.

The use of matched samples has been discussed, using the *t* statistic for illustration, but the same principles apply to other statistical tests. The F test presented in chapter 12 should be used only when the groups are independent of each other. A more elaborate procedure is required to use the F test when the subjects have been matched, but the principles involved are the same.

In summary, we can say that experimental design involves two basic ideas:

1. *Elimination of bias so that the differences observed between the groups represent differences created by the independent variable.* Any difference due to extraneous variables is called *bias* and seriously changes the conclusion that can be obtained from an experiment. The elimination of bias is most often accomplished by eliminating or equalizing the extraneous variables.

2. *Increase in precision.* This is accomplished by matching subjects on one or more variables which are highly correlated with the dependent variable. The loss of degrees of freedom through the matching procedure is more than offset if the correlation is high.

NONCOMPARABLE GROUPS

Often, in behavioral science research, it is impossible to select subjects independent of each other. If a research worker wished to perform a study within a given grade school system, he might discover that he could utilize an entire grade or classroom but that he would have to work with all students in the particular class. The selection of individual subjects would not be independent. This is called *sampling by intact groups*—either all subjects in a given group are used, or none of them are used. When this type of sampling is used, it is generally done with many groups; the type of statistical analysis that is used may be found in advanced statistic textbooks.

Occasionally, it is necessary to use intact groups in research, and too few of these groups are available to permit the assignment of several intact groups to each treatment condition of the experiment. In this case it is sometimes possible to determine how the groups differ with respect to extraneous variables. If we can compute the correlation between the extraneous variable(s) and the dependent variable, it may be possible statistically to equate the groups. This technique—a combination of correlation and analysis of variance—is known as *covariance adjustment*. The procedures to be followed in making a covariance adjustment may be found in advanced statistics textbooks.

Use of statistics in planning experiments

Most experiments are performed because the experimenter thinks that one variable has an effect upon a second. Seldom does the experimenter start with an "I wonder what would happen if. . . ." type of hypothesis. The research hypothesis is far more likely to be: "I think the mean of group 1 will be larger than the mean of group 2." In fact, if the previous information is sufficient, the experimenter may even be attempting to verify the amount of difference between groups.

Obviously, an experiment with only one subject in each group and one observation of that subject would be of little value. We know that differences among individuals are too great to permit us to say with any confidence that the observed differences are due to the experimental differences and not due to differences which exist between subjects independent of any experimental manipulations. Knowledge of statistics can help us to plan the experiment, so that if a true difference does exist, we shall be able to detect this difference from an experiment.

ESTIMATING SAMPLE SIZE

INDEPENDENT SAMPLES.   An easy way to understand the use of statistics in planning an experiment is to use an illustration. We shall use the equation for *t* for unmatched groups to determine the size of the sample that would be necessary to establish significant differences between two groups if a true difference does exist in the populations from which these samples were selected:

$$t = \frac{\bar{X}_1 - \bar{X}_2}{\sqrt{\dfrac{s^2}{N_1} + \dfrac{s^2}{N_2}}}$$

Since we are going to use the knowledge of statistics to plan an experiment, we can make some simplifying assumptions for convenience. Let

N_1 be equal to N_2; with equal N's in the samples, the denominator of the t ratio becomes

$$s_{D_{\bar{x}}} = \sqrt{2\left(\frac{s^2}{N}\right)}$$

Squaring both sides of the equation for the t statistic and solving for N

$$N = \frac{2s^2t^2}{D^2_{\bar{x}}} \tag{13-1}$$

The use of formula 13–1 is shown in example 13–1.

Example 13–1   Estimating the sample size for use with the t statistic (independent samples)

We have prior information that $D_{\bar{x}}$ will be approximately equal to 5, and that s will be approximately equal to 10. If it is decided that $p = .05$ will be sufficient for rejection of the null hypothesis, we can use a t value of 2 as an approximation. Substituting these values in formula 13–1

$$N = \frac{2s^2t^2}{D^2_{\bar{x}}} = \frac{2(10)^2(2)^2}{(5)^2} = \frac{800}{25}$$

$$N = 32$$

Since there are two groups in the study, a minimum of 64 subjects would be required.

From example 13–1 we would need at least 32 subjects in each group or a total of at least 64 subjects to establish that a significant difference exists. If would be a waste of time to run an experiment with only ten subjects if we had evidence that the difference in means would be about five and the standard deviation ten.

There is a more precise technique available for arriving at an estimate of the minimum N necessary, but the above technique will work reasonably well by simply increasing the actual sample size. In the above illustration we would probably use 40 to 45 subjects in each group, thus assuring that the sample size is large enough to detect a true difference if it exists. If the cost per subject is high, then the more precise technique should be used; but if the cost per subject is reasonably low, the above technique is accurate enough for most purposes.

At this point the student may well ask: "But where do you get the estimate of $\bar{X}_1 - \bar{X}_2$ and s?" The answer is: "From previous research." There is no substitute for sound scholarship in research and the determination of the previous work that has been done in the area. The library is the backbone of any planned research. There are many occasions, however, when no such data are available. In the latter case, pilot studies to attempt to estimate some of the various conditions can be of tremendous aid. In any case the researcher should avail himself of all information available from different sources to plan the experiment. The time and energy expended prior to the actual experimentation will more than pay for itself. From this type of planning can come decisions as to whether it is more economical of research resources to eliminate extraneous variables or equalize their effects, or increase the number of subjects to achieve the same precision. Intelligent planning is essential to experimentation, and intelligent planning can be facilitated by an understanding of the statistics to be used in the analysis of the research results.

MATCHED SAMPLES. To illustrate further this type of planning, we shall use the t test for matched groups. Again, let us assume prior information which would lead us to expect that there is a difference in means of five and that the standard deviation of each group will be approximately ten. Let us further assume that the correlation between the matching variable and the dependent variable is .8. Again, we shall accept $t = 2$ for rejection of the null hypothesis. Using a little algebra and rearranging the equation for the t statistic for correlated groups (formula 10–10) we can solve for N

$$N = \frac{2s^2t^2(1 - r)}{D^2_{\bar{x}}} \tag{13–2}$$

Example 13–2 demonstrates the solution for estimating N with matched groups.

Example 13–2 Estimating the sample size for use with the t statistic (correlated groups)

$$D_{\bar{x}} = 5; \quad s = 10; \quad t = 2; \quad r = .80$$

Using formula 13–2

$$N = \frac{2s^2t^2(1 - r)}{D^2_{\bar{x}}} = \frac{2(10)^2(2)^2(1 - .80)}{(5)^2} = \frac{(800)(.2)}{25}$$

$$N = 6.4 \text{ rounded to 7 pairs of subjects}$$

If we are using matched subjects we would need a minimum of 14 (7 in each group). If we are using the same subject in each of the two conditions of the study, we would need 7 subjects (observed twice).

From example 13–2 we have determined that we would need 7 subjects per group, or a total of 14 subjects, to establish significance at the 5 percent level by using matched groups if $r = .80$. Thus, by matching, we could achieve the same precision in the study with only 22 percent as many subjects. A decision to match subjects or to use a larger N would be based on the overall economy involved. In some cases, matching might be a difficult or expensive operation, but the cost per unmatched subject might be small. In that case it might be wiser to use two randomly selected groups with more subjects per group. In other cases the difficulty or cost factors may be reversed.

USING SUBJECTS AS THEIR OWN CONTROLS

The number of potential extraneous variables in any research is always large. People will differ in intelligence, amount of prior education, socio-economic status, and other characteristics which may or may not be related to the research. Matching on many variables becomes a difficult job and may require an extremely large pool of available individuals to achieve the desired matching. One solution to this problem is to match each subject in the study with himself! Who else has the same intelligence, same prior education, same socioeconomic level? The use of the same subjects in two or more conditions of a study is a special case of matching subjects, and the statistics for matched groups should be used in analyzing the data.

Using subjects as their own controls is especially useful in "before and after" studies. The subjects are measured (dependent variable scores obtained) before some condition or treatment is introduced and then after the treatment condition. This yields correlated groups under the two conditions. One caution is necessary: Although using subjects as their own control would be an excellent way to increase the precision of the study, often this is not possible because exposure to one treatment condition in the study will destroy the usefulness of that subject for the other treatment condition.

THE EXPERIMENT AS A "FAIR" TEST OF THE HYPOTHESIS

As has been stressed repeatedly, the essence of experimentation is to design a test situation which will yield answers we can trust. The test must be fair. The measurements must be unbiased and this requires proper control of extraneous variables. The study must have sufficient precision to detect differences, if they exist, and yet not be wasteful of precious resources. When we use the null hypothesis and the logic of decision theory, we wish to fail to reject H_0 if the null hypothesis is true. This we do by careful consideration of the level at which α is set. Equally important, we wish to reject H_0 if it is false. Rejection of a false null hypothesis, requires us to make the experiment as powerful as possible (without wasting resources). The experimenter, before collecting data, must be concerned about the probability of the Type II error (β) and take the necessary planning steps to reduce β to an acceptable level.

In chapter 10 page 214 the factors affecting the probability of a Type II error (β) were summarized. That summary, slightly expanded, is repeated here. The probability of a Type II error (β) is affected by:

1. *The α level set by the research worker for rejecting the null hypothesis.* If the research worker sets a very small value for α and thus makes it more difficult to reject the null hypothesis, this will increase the probability of the occurrence of a Type II error. The decision as to the decision point for α is an important one which the research worker should consider *before the research is begun.*

2. *The size of the true difference between population means* ($\mu_1 - \mu_2$). Since a Type II error can occur only if the null hypothesis is false (a true difference does exist), the probability of the occurrence of a Type II error will depend in part on the magnitude of the true difference. Careful selection of treatment conditions will make this difference large, if a true difference does exist. Small differences are difficult to detect; large differences are relatively easy to detect. The research should be planned to maximize any possible difference.

3. *The amount of variability in the research material.* The denominator of most statistics is based on some estimate of the variability of the research material. Any technique which will reduce variability will reduce the probability of a Type II error and thus increase the power of the study. The techniques for reducing variability were discussed in this chapter:

   a) Elimination of extraneous variables, which will not only reduce variability but help guard against bias.

   b) Matching of subjects, which will also reduce variability and guard against bias. However, matching of subjects will reduce the number of degrees of freedom on which the statistic is based; thus the research worker should be sure that the matching variable is highly correlated with the dependent variable in the study.

4. *The sample size.* The larger the sample on which the study is based, the lower will be the probability of a Type II error. However, research consumes resources and the larger the sample size the greater will be the resource expenditure. The investigator should attempt to determine the optimum sample size prior to doing the experiment.

5. *The statistic used to analyze the results of the study.* The various statistical techniques presented utilize differing amounts of information contained in the numbers. The more information a statistical technique can utilize, the more powerful the technique is. However, an important factor is the scaling characteristics of the numbers being analyzed. Those statistics which require interval or ratio scaling will be able to detect a true difference more readily than will statistics based on ordinal or nominal scaling, but the investigator must be concerned that the statistic is appropriate to the scaling level of the data.

Characteristics of the dependent variable

The outcome of any research is a series of numbers obtained by observing the subjects of the research. These numbers are the dependent variable. The entire purpose for doing the research is to obtain these numbers and from them draw conclusions as to the impact of the independent variable upon the dependent variable. Obviously much care and thought should be expended on the selection of a dependent variable. Far too often measurements which are easy to obtain or are "traditional" are used without questioning if these measures reflect the behavior that is of interest to the investigator. Concepts such as aggression, cooperation, mental health, empathy are important in the behavioral sciences, yet adequate measures are not yet in existance for these phenomena. It has been suggested that perhaps the most pressing task of the behavioral scientist is to concentrate on the development of measures of these concepts. Before we can measure the different lengths of objects, we need a good ruler.

RELIABILITY OF THE DEPENDENT VARIABLE

In chapter 9, the reliability of tests was discussed. It was pointed out that the basic idea of reliability is the notion of consistency. A measurement to be useful must be consistent. An elastic ruler may be an interesting device to amuse friends, but it is less than useless in making important measurements of length. The behavioral sciences may never achieve the degree of consistency in their measures that has been achieved in the physical sciences, but we must keep trying. At this stage we must at least be aware of reliability as a problem. We cannot afford to assume that the measurements in the behavioral sciences are reliable. We must present evidence as to the reliability of our dependent variables.

On page 195 of chapter 9, formula 9–3 showed the relationship among the reliability of two measures, the correlation between the true scores of these measures, and the obtained correlation. This relationship is repeated below

$$r_{xy} = r_{tt} \sqrt{r_{xx}} \sqrt{r_{yy}}$$

When we manipulate an independent variable and observe the effect of these changes on the dependent variable, we are looking for a correlation—not a Pearson product-moment correlation necessarily, but correlation in the broad sense, that as the independent variable changes, the dependent variable also changes in a predictable way. This is correlation. In any experiment, the investigator wishes to uncover the relationship between the independent variable and the dependent variable. Examination of formula 9 3 shows that if the reliability of the dependent variable (r_{yy}) is low, then regardless of the relationship between the true scores (r_{tt}), the relationship obtained in the study *cannot be high*. If there were a perfect relationship between the true scores of both variables ($r_{tt} = 1.00$) and the reliability of the independent variable were perfect ($r_{xx} = 1.00$), then the obtained relationship between the independent variable and the dependent variable cannot exceed $\sqrt{r_{yy}}$! Before investing resources in an experiment, the investigator has the responsibility of determining that the reliability of the dependent variable is acceptable.

SCALE CHARACTERISTICS OF THE DEPENDENT VARIABLE

By now the reader is well aware that numbers contain differing amounts of information. It would be nice if all dependent variables in the behavioral sciences reflected interval or ratio scales, but they do not; so, we must utilize whatever level of measurement we have achieved. In planning an experiment, the investigator should, before collecting data, consider the level of measurement achieved in the dependent variable that will be used in the experiment. To use a statistic which requires interval or ratio information in the numbers can be very misleading, yet using a statistic for nominal information may be extremely wasteful of information and will lead to an increase in the probability of a Type

II error. The variety of different statistical tests presented in this book can be confusing unless the reader fits them into a framework. This framework has only three dimensions: (1) what is the scalar characteristic of the dependent variable, (2) how many groups are in the study (number of categories on the independent variable), and (3) are the groups independent of each other or correlated? Table 13–1 was designed to aid

TABLE 13–1
Statistical tests of H_0

Level of measurement	One-sample case	Two-sample case		k-sample case	
		Independent samples	Correlated samples	Independent samples	Correlated samples
Nominal	Binomial test p. 223 χ^2 one-way classification p. 235	χ^2 test for two-way classification p. 237 Fisher-Yates test p. 243	McNemar test of change p. 241	χ^2 test p. 280	Cochran Q test p. 286
Ordinal	See references Chapter 11	Median test p. 246 Mann-Whitney U test p. 248	Sign test p. 252 Wilcoxon matched-pairs signed-ranks test p. 254	Extension of median test p. 282 Kruskal-Wallis one-way analysis of variance for ranks p. 283	Friedman analysis of variance by ranks p. 289
Interval or ratio	Student's t test p. 216	Student's t test p. 216	Student's t test p. 221 Sandler's A p. 226	Analysis of variance p. 262	See references Chapter 12

the reader in fitting these different tools for testing hypotheses into one framework.

The various statistics available for hypothesis testing are more numerous than presented in this book and hence in table 13–1. Two cells in the table do not have any statistical tests, but instead the reader is directed to appropriate references. The tests available for the one-sample case with ordinal scaling, should present no problem for the reader who has mastered the material in this book. The k-sample case with correlated groups with interval or ratio scaling is an extension of the analysis of variance into at least two dimensions and generally three. This topic of the complex analysis of variance is too lengthy for adequate treatment in a first course in applied statistics.

Interpreting research results

Published reports of research contain an introduction to the particular problem, a description of what the investigator did, the results of the study and finally the conclusion that the investigator drew from the work. The description of the procedures and the results obtained should be straightforward. The interpretation of the results becomes a very debatable point. Is it reasonable for the investigator to conclude what he or she did from this data? Do the procedures leave some questions about the results as well as the conclusion? The most important factor in interpreting the results of research is a sound background in the area being investigated. With a sound background the reader can "fit" these findings into the general area of knowledge. A knowledge of statistics can be helpful, independent of the subject matter of the specific study. This section will deal with some principles of statistics and probability theory which can be used in interpreting research results.

RANDOM SAMPLING

Early in the book a distinction was made between descriptive statistics and inferential statistics. Statistics are used to describe populations. The mean of the population (μ), the standard deviation of the population (σ), and other summary statements about populations are important. Inferential statistics always involves the use of one or more samples drawn from a population(s). From the sample(s) inferences are made concerning the characteristics of the population. Whenever a sample is used to make inferences about the population, we need to be concerned about the manner in which the sample was drawn—is it representative of the population? Will the conclusions made on the basis of the samples (and the probability statements) hold true for the population? Part of the trust that we may have in the results of research is related to the manner in which samples are drawn if inferences are to be made to a larger population.

In survey work, such as political polls, the necessity for proper sampling procedures is readily apparent. If some type of randomization procedure is not used, we have no way of knowing if the results of the sampling reflect the characteristics of the population. It is the responsibility of the investigator to specify, in detail, the procedures followed in obtaining the sample. This allows the consumer of the research results the opportunity to evaluate what was done and to decide if the results of the investigation can be trusted. Can we generalize back to the population of interest?

A somewhat similar problem is related to generalization of experimental findings to the population from which the sample was drawn. Generally, experimenters do not select a random sample from all possible subjects. If the subjects are people, the sample is not selected from all people in the world, or all the people in the United States. Often the subjects available are enrolled in undergraduate courses. (It has been said that research in psychology is almost all based on the white rat and the college sophomore.) Certainly college students do not constitute a random sample of all people, or even all people of that age range.

If we have 40 college students (or white rats) and we use a table of random numbers to assign 20 subjects to each of 2 groups, we have used randomization. Does this mean that we can use inferential statistics and generalize the results to a much larger population such as *all* college students (or *all* white rats)? The answer is both *yes* and *no*. We can use inferential statistics but we cannot generalize the results beyond the 40 subjects used in the study.

With 40 subjects and 2 different groups in the experiment there are 780 ways in which we could assign the 40 subjects to these 2 groups. This is found by using the expression for N things taken x at a time $\binom{N}{x}$. In this case

$$\binom{40}{2} = \frac{40!}{2!(40-2)!} = 780.$$

With all conditions exactly the same and with these 40 subjects there are 780 different experiments that could have been run and we have drawn a random sample from the *population of possible experiments*. Suppose that the weight of the subjects was an important extraneous variable. One possible arrangement of these 40 subjects would be the 20 heaviest subjects in one group and the 20 lightest subjects in the other. By chance, this can occur once in 780 times. A very low probability, but it could occur. Random assignment of subjects will allow us to use statistics based on probability theory, but it *will not insure* that the results of the randomization will always yield equality between the two groups. This problem can be solved in two ways. The subjects can be matched as to weight, and one subject from each pair would be randomly assigned to each group. This will insure the equality of the extraneous variable of weight in the two groups; and since random assignment was used, the particular arrangement is a random sample of the 190 different arrangements that were possible, based on $\binom{20}{2}$. The other approach is that after subjects have been randomly assigned to the different conditions of the experiment, check to see if the particular random assignment did lead to differences on the extraneous variable which are statistically significant. If not, the experiment can be run with these groups. If the difference in the extraneous variable is significant, the subjects should again be randomly assigned and the same checks made.

Random assignment of subjects is necessary if we wish to use inferential statistics and be able to trust the probability statements obtained, but random assignment does not in any way guarantee that the results of the randomization will produce equality between the groups.

The random assignment of a given group of subjects to different treatment conditions will permit us to utilize statistics which depend upon sampling theory, but we should be very cautious in generalizing the research findings beyond this immediate population of possible replications of the same experiment. Any faith in the universality of research findings will have to be based on logic other than that of sampling theory.

Even if we could draw a random sample of all humans on the earth, the results of that research could only be generalized to the population that existed at the instant the samples were drawn. New individuals are being born; others are dying. Most populations in which behavioral scientists have an interest, are constantly changing. Is there no hope that we can generalize to populations of interest? The relative universality of research findings is to be found in replications of the research in other settings. If reinforcement works with laboratory rats in California, England, New Zealand, and Japan, reinforcement will probably also work with laboratory rats in Illinois and Sweden. This is based, not on theoretical probability of sampling theory, but on the usual human use of inductive logic. The principle of reinforcement does work with a great variety of different lower animals. It also works with a great variety of different humans under a great variety of different conditions. Inductive logic would lead us to believe that the principle of reinforcement would also work in a new situation which was somewhat similar to the ones that had been studied.

We should be very cautious in generalizing research findings beyond the immediate research situation. But if the investigator did not use random assignment of the subjects to the different conditions of the experiment, we cannot be sure of the results even in that research setting! It is the obligation of the investigator to specify in sufficient detail, just what procedures were followed in assigning subjects to different conditions in the experiment. This will permit the reader of the research to decide how much faith can be placed in the conclusions drawn from the study.

SAMPLING WITHOUT REPLACEMENT. In chaper 6, the problems of sampling without replacement from small populations was introduced. As each member of the sample is drawn, the probability of the selection of the remaining members of the population changes. If the population is large, the amount of error introduced into a statistic is negligible. If the population is small, the error can be serious.

In assigning a small number of subjects to two or more conditions in an experiment, (i.e., 40 subjects to 2 conditions), it would appear that we should correct for sampling without replacement. The question is: What is the population from which we are sampling? Is it the small number of subjects available to us? No. We are drawing one random sample from the population of different arrangements that could have been obtained with this small number of subjects. If we were to work with 40 subjects to be placed into two groups, the population of possible arrangements would be $\binom{40}{2} = 780$. We would be drawing one sample from a population of 780. If we had only 20 subjects and they were assigned randomly to two groups, we would be working with one sample from a population of 190 different arrangements. With 20 subjects assigned to four groups, the population of different possible arrangements would be 4,845.

In assigning a small number of available subjects to different groups in an experiment, it is not necessary to use the correction for sampling from a small finite population *provided* the subjects were assigned randomly. We can generalize from the results of the one experiment back to the population of different experimental arrangements that could have been made with these subjects. Generalization beyond this population does not depend on sampling theory. This issue was discussed in the previous section.

Proportion of variance accounted for

It is unfortunate that the term *significance* is used in rejecting the null hypothesis. The word tends to carry the connotation of importance. Unfortunately this is not true. Let us review the meaning of the statement *statistically significant difference*. Under the null hypothesis we have stated that there is no difference in the underlying populations. We obtain samples from the specific populations and then, based on the probability of occurrence of the particular statistic, we may reject H_0. What have we done? We have concluded that the population values do differ. So what? Is this an important difference? Is this a trivial difference? The term *significant difference* means we believe that there *is* a difference. It may be trivial. It may be important. All the statistical test can tell us is that the population values differ (within the possibility of a Type I error).

The importance of a statistically significant difference, like most interpretations, must be based on the background of the reader. If the finding fits into a large amount of prior information in the area, the reader is in a better position to judge the importance of the findings. A knowledge of statistics, while no substitute for a solid knowledge of the content area, can be helpful.

CORRELATION COEFFICIENTS. To state that a correlation is statistically significant may be one of the least important conclusions one can draw. To state that variables X and Y are related and to stop there, is saying little. Of paramount interest is the question: "What does our knowledge of the X variable tell us about the Y variable?" The correlation coefficient itself does not tell us much, but the square of the Pearson product-moment correlation coefficient (r^2) may tell us a great deal. The Pearson r^2 tells us the proportion of the total variance in Y which can be predicted from our knowledge of X. Suppose an investigator reports that the correlation between family income and success in school is significant for grade school children. He reports this significant correlation as $r = .20$. This means that family income and success in school are related—the correlation is not zero. What proportion of the total variance in school performance can be predicted from our knowledge of family income? Four percent. Of the variation in school performance 96 percent cannot be predicted from our knowledge of family income. Suppose that there were a cause and effect relationship between family income and school success (remember that one of the problems with the correlational approach is that we cannot draw cause and effect relationships from correlations). With a cause and effect relationship, if all family incomes were made the same we would observe only a 4 percent reduction in the variance in school success. This discussion is, of course, limited to the *linear* relationship between these two variables. If the relationship is curvilinear, the proportion of variance may be much higher. But if the relationship is curvilinear, the r should not have been reported. The statistic eta (η) is a measure of curvilinear relationship and η^2 can be interpreted as r^2, but for curvilinear relationships.

Unfortunately, the behavioral sciences are full of statistically significant trivial correlations. The writer once read a study in which the investigator reported a significant correlation of .10 and interpreted this as important. If we square .10 and convert this to a percentage, we have accounted for 1 percent of the total variance. The square of the Pearson r can be most useful in interpreting the importance of a reported finding.

THE ANALYSIS OF VARIANCE.   Knowledge of the proportion of variance accounted for by use of the square of the Pearson r has been around for a long time and has been widely used. A similar statement of proportion of variance can be made with the analysis of variance (and Student's t), but for some reason this has not been widely used in the behavioral sciences. The ratio *of the sum of squares for between groups* to the *sum of squares for total* can be interpreted in a manner similar to r^2 (actually η^2). A significant F ratio tells us that the group of means do not come from a single population. The true means of the different groups are different. It is sometimes interesting to ask: "Of all of the variations in the total study, what proportion can be attributed to the differences in means?" The ratio of SS_b to SS_t will tell us this.

In chapter 12, an analysis of variance was computed to demonstrate the procedures involved. Table 12–2 gave a summary of the analysis of variance. From table 12–2 we can obtain $SS_b = 4{,}911.30$ and $SS_t = 10{,}270.03$. The ratio between these two sums of squares is .4782 or 48 percent of the total variance can be attributed to the differences in means.

This ratio of SS of treatment to SS of total should be used far more often than it is in interpreting the meaning (importance) of studies using the analysis of variance or t test. However, the implications of a particular finding must be viewed in the context of the content area. Small differences may be extremely important in some areas of investigation. Large differences may be trivial in other cases. This judgment should be made on the basis of the knowledge of the area, and an understanding of statistics.

BEHAVIORAL SCIENCES AND HUMAN WELFARE

Knowledge of the physical sciences and, hence, the ability to predict and even control physical variables have revolutionized the human condition. The understanding and harnessing of physical energy has freed humans from back-breaking drudgery. Knowledge of disease conditions has removed much pain and suffering and greatly extended the human life span.

Knowledge in the behavioral sciences has not developed to the same point as the physical and biological sciences. But the demands for changes in the areas of human endeavor covered by the behavioral sciences are great. There are demands that the school systems must change but no good evidence that suggested changes will, in fact, produce desired ends. Methods of child rearing are proposed with only vague hopes for the impact of the methods. Before expensive changes are made on a mass basis, we need good information as to the effect of these variables. This means more research, but research that is carefully planned to yield the maximum amount of information with the minimum expenditure of resources.

Statistics can be used not only in the analysis of the experimental results but also in planning the experiment so as to make optimum use of the research resources at hand. This demands both a firm knowledge of the content area in which one wishes to do research and a degree of sophistication in the use of modern statistics.

Summary

Research workers are generally concerned with determining the relationship between one or more *independent variables* and a *dependent variable*. The independent variable is either deliberately changed by the researcher or allowed to change, and the effect of these changes is noted by observing changes in the dependent variable. The validity of the statements relating independent and dependent variables will be directly related to the control of *extraneous variables* in the study.

Two techniques are used to remove the effects of extraneous variables: (1) eliminating the extraneous variables or (2) counterbalancing the effects of these variables across the different conditions of the research.

One way to remove the effects of extraneous variables is through matching subjects for the different conditions of the study. Matching will increase the precision of the study if the matching variable has a high correlation with the dependent variable. If the correlation is low, the precision is reduced because the number of degrees of freedom is reduced by one half through the matching procedure.

Increasing the sample size is one way to increase the precision of the study, but too large a sample is wasteful of resources. Too small a sample size may lead to a Type II error. The minimum sample size to detect a true difference—if it exists—can be estimated if it is possible to estimate the difference between means to be anticipated in the study and the amount of variability to be anticipated. These can usually be determined by studying the previous research in similar situations or by using small pilot studies.

Using the same subjects in different portions of the study is an excellent way of increasing precision by matching. The statistic used to analyze the data must be one which takes into account the fact that matching has occurred. Using subjects as their own control is especially useful in "before and after" studies. The same subjects should not be used in more than one condition of the study if in so doing their usefulness is destroyed for the other conditions.

Knowledge of statistics can be used to plan studies before they are carried out. Most investigators are interested in finding true differences between groups if these differences exist. This requires that Type II errors (failure to reject the null hypothesis when it is false) be kept as small as feasible.

The probability of the occurrence of a Type II error can be kept small if the investigator uses the information that β is a function of:

1. The α level set for rejecting the null hypothesis
2. The size of the true difference between population means
3. The amount of variability in the research material
4. The sample size
5. The statistic used to analyze the data.

Two characteristics of the dependent variable should be examined prior to the beginning of data collection. The reliability of the dependent variable must be reasonably high if there is to be any possibility of showing that a relationship exists between the dependent variable and the independent variable. The scaling characteristics of the dependent variable should be explored and appropriate statistical tests identified. Table 13–1 was developed for use in locating the particular statistical test appropriate for a given study.

The interpretation of the results of any research requires first a good background of prior knowledge in the area; and understanding of statistics is also important in interpreting research results. The reader of research reports should judge if proper *randomization* occurred when generalizations are made to larger populations. If correlation coefficients are reported, the *square of the Pearson r* will indicate the proportion of variance in the dependent variable that can be predicted from the knowledge of the independent variable. The *ratio of the SS$_b$ to the SS$_t$* can be used in the analysis of variance to determine the proportion of total variance that is accounted for by the treatment differences in the independent variable.

Much additional research is needed in the behavioral sciences, but it must be research based on a solid knowledge of the content area and use knowledge of statistics for intelligent planning and interpretation of the research.

References

DESIGN OF EXPERIMENTS (NONSTATISTICAL)

B. J. UNDERWOOD, *Psychological Research* (New York: Appleton-Century-Crofts, Inc., 1957).

L. FESTINGER and D. KATZ (eds.), *Research Methods in the Behavioral Sciences* (New York: Dryden Press, 1953).

COVARIANT ADJUSTMENT

QUINN MCNEMAR, *Psychological Statistics* (4th ed.; New York: John Wiley & Sons, Inc., 1969), chap. xviii.

W. S. RAY, *Introduction to Experimental Design* (New York: Macmillan Co., 1960), chap. x.

S. H. EVANS and E. J. ANASTASIO, "Misuse of Analysis of Covariance when Treatment Effect and Covariate Are Confounded," *Psychological Bulletin,* Vol. 69 (1968), pp. 225–34.

CONCEPT OF PRECISION IN RESEARCH

D. R. COX, *Planning of Experiments* (New York: John Wiley & Sons, Inc., 1958), chap. i.

appendix: part A

Tables

Table of squares and square roots of the numbers from 1 to 1,000

Number	Square	Square root	Number	Square	Square root
1	1	1.000	31	9 61	5.568
2	4	1.414	32	10 24	5.657
3	9	1.732	33	10 89	5.745
4	16	2.000	34	11 56	5.831
5	25	2.236	35	12 25	5.916
6	36	2.449	36	12 96	6.000
7	49	2.646	37	13 69	6.083
8	64	2.828	38	14 44	6.164
9	81	3.000	39	15 21	6.245
10	1 00	3.162	40	16 00	6.325
11	1 21	3.317	41	16 81	6.403
12	1 44	3.464	42	17 64	6.481
13	1 69	3.606	43	18 49	6.557
14	1 96	3.742	44	19 36	6.633
15	2 25	3.873	45	20 25	6.708
16	2 56	4.000	46	21 16	6.782
17	2 89	4.123	47	22 09	6.856
18	3 24	4.243	48	23 04	6.928
19	3 61	4.359	49	24 01	7.000
20	4 00	4.472	50	25 00	7.071
21	4 41	4.583	51	26 01	7.141
22	4 84	4.690	52	27 04	7.211
23	5 29	4.796	53	28 09	7.280
24	5 76	4.899	54	29 16	7.348
25	6 25	5.000	55	30 25	7.416
26	6 76	5.099	56	31 36	7.483
27	7 29	5.196	57	32 49	7.550
28	7 84	5.292	58	33 64	7.616
29	8 41	5.385	59	34 81	7.681
30	9 00	5.477	60	36 00	7.746

TABLE A—*Continued*

Number	Square	Square root	Number	Square	Square root
61	37 21	7.810	99	98 01	9.950
62	38 44	7.874	100	100 00	10.000
63	39 69	7.937			
64	40 96	8.000	101	1 02 01	10.050
65	42 25	8.062	102	1 04 04	10.100
			103	1 06 09	10.149
66	43 56	8.124	104	1 08 16	10.198
67	44 89	8.185	105	1 10 25	10.247
68	46 24	8.246			
69	47 61	8.307	106	1 12 36	10.296
70	49 00	8.367	107	1 14 49	10.344
			108	1 16 64	10.392
71	50 41	8.426	109	1 18 81	10.440
72	51 84	8.485	110	1 21 00	10.488
73	53 29	8.544			
74	54 76	8.602	111	1 23 21	10.536
75	56 25	8.660	112	1 25 44	10.583
			113	1 27 69	10.630
76	57 76	8.718	114	1 29 96	10.677
77	59 29	8.775	115	1 32 25	10.724
78	60 84	8.832			
79	62 41	8.888	116	1 34 56	10.770
80	64 00	8.944	117	1 36 89	10.817
			118	1 39 24	10.863
81	65 61	9.000	119	1 41 61	10.909
82	67 24	9.055	120	1 44 00	10.954
83	68 89	9.110			
84	70 56	9.165	121	1 46 41	11.000
85	72 25	9.220	122	1 48 84	11.045
			123	1 51 29	11.091
86	73 96	9.274	124	1 53 76	11.136
87	75 69	9.327	125	1 56 25	11.180
88	77 44	9.381			
89	79 21	9.434	126	1 58 76	11.225
90	81 00	9.487	127	1 61 29	11.269
			128	1 63 84	11.314
91	82 81	9.539	129	1 66 41	11.358
92	84 64	9.592	130	1 69 00	11.402
93	86 49	9.644			
94	88 36	9.695	131	1 71 61	11.446
95	90 25		132	1 74 24	11.489
			133	1 76 89	11.533
96	92 16	9.798	134	1 79 56	11.576
97	94 09	9.849	135	1 82 25	11.619
98	96 04	9.899			

TABLE A—*Continued*

Number	Square	Square root	Number	Square	Square root
136	1 84 96	11.662	174	3 02 76	13.191
137	1 87 69	11.705	175	3 06 25	13.229
138	1 90 44	11.747			
139	1 93 21	11.790	176	3 09 76	13.266
140	1 96 00	11.832	177	3 13 29	13.304
			178	3 16 84	13.342
141	1 98 81	11.874	179	3 20 41	13.379
142	2 01 64	11.916	180	3 24 00	13.416
143	2 04 49	11.958			
144	2 07 36	12.000	181	3 27 61	13.454
145	2 10 25	12.042	182	3 31 25	13.491
			183	3 34 89	13.528
146	2 13 16	12.083	184	3 38 56	13.565
147	2 16 09	12.124	185	3 42 25	13.601
148	2 19 04	12.166			
149	2 22 01	12.207	186	3 45 96	13.638
150	2 25 00	12.247	187	3 49 69	13.675
			188	3 53 44	13.711
151	2 28 01	12.288	189	3 57 21	13.748
152	2 31 04	12.329	190	3 61 00	13.784
153	2 34 09	12.369			
154	2 37 16	12.410	191	3 64 81	13.820
155	2 40 25	12.450	192	3 68 64	13.856
			193	3 72 49	13.892
156	2 43 36	12.490	194	3 76 36	13.928
157	2 46 49	12.530	195	3 80 25	13.964
158	2 49 64	12.570			
159	2 52 81	12.610	196	3 84 16	14.000
160	2 56 00	12.649	197	3 88 09	14.036
			198	3 92 04	14.071
161	2 59 21	12.689	199	3 96 01	14.107
162	2 62 44	12.728	200	4 00 00	14.142
163	2 65 69	12.767			
164	2 68 96	12.806	201	4 04 01	14.177
165	2 72 25	12.845	202	4 08 04	14.213
			203	4 12 09	14.248
166	2 75 56	12.884	204	4 16 16	14.283
167	2 78 89	12.923	205	4 20 25	14.318
168	2 82 24	12.961			
169	2 85 61	13.000	206	4 24 36	14.353
170	2 89 00	13.038	207	4 28 49	14.387
			208	4 32 64	14.422
171	2 92 41	13.077	209	4 36 81	14.457
172	2 95 84	13.115	210	4 41 00	14.491
173	2 99 29	13.153			

TABLE A—*Continued*

Number	Square	Square root	Number	Square	Square root
211	4 45 21	14.526	249	6 20 01	15.780
212	4 49 44	14.560	250	6 25 00	15.811
213	4 53 60	14.595			
214	4 57 96	14.629	251	6 30 01	15.843
215	4 62 25	14.663	252	6 35 04	15.875
			253	6 40 09	15.906
216	4 66 56	14.697	254	6 45 16	15.937
217	4 70 89	14.731	255	6 50 25	15.969
218	4 75 24	14.765			
219	4 79 61	14.799	256	6 55 36	16.000
220	4 84 00	14.832	257	6 60 49	16.031
			258	6 65 64	16.062
221	4 88 41	14.866	259	6 70 81	16.093
222	4 92 84	14.900	260	6 76 00	16.125
223	4 97 29	14.933			
224	5 01 76	14.967	261	6 81 21	16.155
225	5 06 25	15.000	262	6 86 44	16.186
			263	6 91 69	16.217
226	5 10 76	15.033	264	6 96 96	16.248
227	5 15 29	15.067	265	7 02 25	16.279
228	5 19 84	15.100			
229	5 24 41	15.133	266	7 07 56	16.310
230	5 29 00	15.166	267	7 12 89	16.340
			268	7 18 24	16.371
231	5 33 61	15.199	269	7 23 61	16.401
232	5 38 24	15.232	270	7 29 00	16.432
233	5 42 89	15.264			
234	5 47 56	15.297	271	7 34 41	16.462
235	5 52 25	15.330	272	7 39 84	16.492
			273	7 45 29	16.523
236	5 56 96	15.362	274	7 50 76	16.553
237	5 61 69	15.395	275	7 56 25	16.583
238	5 66 44	15.427			
239	5 71 21	15.460	276	7 61 76	16.613
240	5 76 00	15.492	277	7 67 29	16.643
			278	7 72 84	16.673
241	5 80 81	15.524	279	7 78 41	16.703
242	5 85 64	15.556	280	7 84 00	16.733
243	5 90 49	15.588			
244	5 95 36	15.620	281	7 89 61	16.763
245	6 00 25	15.652	282	7 95 24	16.793
			283	8 00 89	16.823
246	6 05 16	15.684	284	8 06 56	16.852
247	6 10 09	15.716	285	8 12 25	16.882
248	6 15 04	15.748			

TABLE A—*Continued*

Number	Square	Square root	Number	Square	Square root
286	8 17 96	16.912	324	10 49 76	18.000
287	8 23 69	16.941	325	10 56 25	18.028
288	8 29 44	16.971			
289	8 35 21	17.000	326	10 62 76	18.055
290	8 41 00	17.029	327	10 69 29	18.083
			328	10 75 84	18.111
291	8 46 81	17.059	329	10 82 41	18.138
292	8 52 64	17.088	330	10 89 00	18.166
293	8 58 49	17.117			
294	8 64 36	17.146	331	10 95 61	18.193
295	8 70 25	17.176	332	11 01 24	18.221
			333	11 08 89	18.248
296	8 76 16	17.205	334	11 15 56	18.276
297	8 82 09	17.234	335	11 22 25	18.303
298	8 88 04	17.263			
299	8 94 01	17.292	336	11 28 96	18.330
300	9 00 00	17.321	337	11 35 69	18.358
			338	11 42 44	18.385
301	9 06 01	17.349	339	11 49 21	18.412
302	9 12 04	17.378	340	11 56 00	18.439
303	9 18 09	17.407			
304	9 24 16	17.436	341	11 62 81	18.466
305	9 30 25	17.464	342	11 69 64	18.493
			343	11 76 49	18.520
306	9 36 36	17.493	344	11 83 36	18.547
307	9 42 49	17.521	345	11 90 25	18.574
308	9 48 64	17.550			
309	9 54 81	17.578	346	11 97 16	18.601
310	9 61 00	17.607	347	12 04 09	18.628
			348	12 11 04	18.655
311	9 67 21	17.635	349	12 18 01	18.682
312	9 73 44	17.664	350	12 25 00	18.708
313	9 79 69	17.692			
314	9 85 96	17.720	351	12 32 01	18.735
315	9 92 25	17.748	352	12 39 04	18.762
			353	12 46 09	18.788
316	9 98 56	17.776	354	12 53 16	18.815
317	10 04 89	17.804	355	12 60 25	18.841
318	10 11 24	17.833			
319	10 17 61	17.861	356	12 67 36	18.868
320	10 24 00	17.889	357	12 74 49	18.894
			358	12 81 64	18.921
321	10 30 41	17.916	359	12 88 81	18.947
322	10 36 84	17.944	360	12 96 00	18.974
323	10 43 29	17.972			

TABLE A—*Continued*

Number	Square	Square root	Number	Square	Square root
361	13 03 21	19.000	399	15 92 01	19.975
362	13 10 44	19.026	400	16 00 00	20.000
363	13 17 69	19.053			
364	13 24 96	19.079	401	16 08 01	20.025
365	13 32 25	19.105	402	16 16 04	20.050
			403	16 24 09	20.075
366	13 39 56	19.131	404	16 32 16	20.100
367	13 46 89	19.157	405	16 40 25	20.125
368	13 54 24	19.183			
369	13 61 61	19.209	406	16 48 36	20.149
370	13 69 00	19.235	407	16 56 49	20.174
			408	16 64 64	20.199
371	13 76 41	19.261	409	16 72 81	20.224
372	13 83 84	19.287	410	16 81 00	20.248
373	13 91 29	19.313			
374	13 98 76	19.339	411	16 89 21	20.273
375	14 06 25	19.363	412	16 97 44	20.298
			413	17 05 69	20.322
376	14 13 76	19.391	414	17 13 96	20.347
377	14 21 29	19.416	415	17 22 25	20.372
378	14 28 84	19.442			
379	14 36 41	19.468	416	17 30 56	20.396
380	14 44 00	19.494	417	17 38 89	20.421
			418	17 47 24	20.445
381	14 51 61	19.519	419	17 55 61	20.469
382	14 59 24	19.545	420	17 64 00	20.494
383	14 66 89	19.570			
384	14 74 56	19.596	421	17 72 41	20.518
385	14 82 25	19.621	422	17 80 84	20.543
			423	17 89 29	20.567
386	14 89 96	19.647	424	17 97 76	20.591
387	14 97 69	19.672	425	18 06 25	20.616
388	15 05 44	19.698			
389	15 13 21	19.723	426	18 14 76	20.640
390	15 21 00	19.748	427	18 23 29	20.664
			428	18 31 84	20.688
391	15 28 81	19.774	429	18 40 41	20.712
392	15 36 64	19.799	430	18 49 00	20.736
393	15 44 49	19.824			
394	15 52 36	19.849	431	18 57 61	20.761
395	15 60 25	19.875	432	18 66 24	20.785
			433	18 74 89	20.809
396	15 68 16	19.900	434	18 83 56	20.833
397	15 76 09	19.925	435	18 92 25	20.857
398	15 84 04	19.950			

TABLE A—Continued

Number	Square	Square root	Number	Square	Square root
436	19 00 96	20.881	474	22 46 76	21.772
437	19 09 69	20.905	475	22 56 25	21.794
438	19 18 44	20.928			
439	19 27 21	20.952	476	22 65 76	21.817
440	19 36 00	20.976	477	22 75 29	21.840
			478	22 84 84	21.863
441	19 44 81	21.000	479	22 94 41	21.886
442	19 53 64	21.024	480	23 04 00	21.909
443	19 62 49	21.048			
444	19 71 36	21.071	481	23 13 61	21.932
445	19 80 25	21.095	482	23 23 24	21.954
			483	23 32 89	21.977
446	19 89 16	21.119	484	23 42 56	22.000
447	19 98 09	21.142	485	23 52 25	22.023
448	20 07 04	21.166			
449	20 16 01	21.190	486	23 61 96	22.045
450	20 25 00	21.213	487	23 71 69	22.068
			488	23 81 44	22.091
451	20 34 01	21.237	489	23 91 21	22.113
452	20 43 04	21.260	490	24 01 00	22.136
453	20 52 09	21.284			
454	20 61 16	21.307	491	24 10 81	22.159
455	20 70 25	21.331	492	24 20 64	22.181
			493	24 30 49	22.204
456	20 79 36	21.354	494	24 40 36	22.226
457	20 88 49	21.378	495	24 50 25	22.249
458	20 97 64	21.401			
459	21 06 81	21.424	496	24 60 16	22.271
460	21 16 00	21.448	497	24 70 09	22.293
			498	24 80 04	22.316
461	21 25 21	21.471	499	24 90 01	22.338
462	21 34 44	21.494	500	25 00 00	22.361
463	21 43 69	21.517			
464	21 52 96	21.541	501	25 10 01	22.383
465	21 62 25	21.564	502	25 20 04	22.405
			503	25 30 09	22.428
466	21 71 56	21.587	504	25 40 16	22.450
467	21 80 89	21.610	505	25 50 25	22.472
468	21 90 24	21.633			
469	21 99 61	21.656	506	25 60 36	22.494
470	22 09 00	21.679	507	25 70 49	22.517
			508	25 80 64	22.539
471	22 18 41	21.703	509	25 90 81	22.561
472	22 27 84	21.726	510	26 01 00	22.583
473	22 37 29	21.749			

TABLE A—*Continued*

Number	Square	Square root	Number	Square	Square root
511	26 11 21	22.605	549	30 14 01	23.431
512	26 21 44	22.627	550	30 25 00	23.452
513	26 31 69	22.650			
514	26 41 96	22.672	551	30 36 01	23.473
515	26 52 25	22.694	552	30 47 04	23.495
			553	30 58 09	23.516
516	26 62 56	22.716	554	30 69 16	23.537
517	26 72 89	22.738	555	30 80 25	23.558
518	26 83 24	22.760			
519	26 93 61	22.782	556	30 91 36	23.580
520	27 04 00	22.804	557	31 02 49	23.601
			558	31 13 64	23.622
521	27 14 41	22.825	559	31 24 81	23.643
522	27 24 84	22.847	560	31 36 00	23.664
523	27 35 29	22.869			
524	27 45 76	22.891	561	31 47 21	23.685
525	27 56 25	22.913	562	31 58 44	23.707
			563	31 69 69	23.728
526	27 66 76	22.935	564	31 80 96	23.749
527	27 77 29	22.956	565	31 92 25	23.770
528	27 87 84	22.978			
529	27 98 41	23.000	566	32 03 56	23.791
530	28 09 00	23.022	567	32 14 89	23.812
			568	32 26 24	23.833
531	28 19 61	23.043	569	32 37 61	23.854
532	28 30 24	23.065	570	32 49 00	23.875
533	28 40 89	23.087			
534	28 51 56	23.108	571	32 60 41	23.896
535	28 62 25	23.130	572	32 71 84	23.917
			573	32 83 29	23.937
536	28 72 96	23.152	574	32 94 76	23.958
537	28 83 69	23.173	575	33 06 25	23.979
538	28 94 44	23.195			
539	29 05 21	23.216	576	33 17 76	24.000
540	29 16 00	23.238	577	33 29 29	24.021
			578	33 40 84	24.042
541	29 26 81	23.259	579	33 52 41	24.062
542	29 37 64	23.281	580	33 64 00	24.083
543	29 48 49	23.302			
544	29 59 36	23.324	581	33 75 61	24.104
545	29 70 25	23.345	582	33 87 24	24.125
			583	33 98 89	24.145
546	29 81 16	23.367	584	34 10 56	24.166
547	29 92 09	23.388	585	34 22 25	24.187
548	30 03 04	23.409			

TABLE A—Continued

Number	Square	Square root	Number	Square	Square root
586	34 33 96	24.207	624	38 93 76	24.980
587	34 45 69	24.228	625	39 06 25	25.000
588	34 57 44	24.249			
589	34 69 21	24.269	626	39 18 76	25.020
590	34 81 00	24.290	627	39 31 29	25.040
			628	39 43 84	25.060
591	34 92 81	24.310	629	39 56 41	25.080
592	35 04 64	24.331	630	39 69 00	25.100
593	35 16 49	24.352			
594	35 28 36	24.372	631	39 81 61	25.120
595	35 40 25	24.393	632	39 94 24	25.140
			633	40 06 89	25.159
596	35 52 16	24.413	634	40 19 56	25.179
597	35 64 09	24.434	635	40 32 25	25.199
598	35 76 04	24.454			
599	35 88 01	24.474	636	40 44 96	25.219
600	36 00 00	24.495	637	40 57 69	25.239
			638	40 70 44	25.259
601	36 12 01	24.515	639	40 83 21	25.278
602	36 24 04	24.536	640	40 96 00	25.298
603	36 36 09	24.556			
604	36 48 16	24.576	641	41 08 81	25.318
605	36 60 25	24.597	642	41 21 64	25.338
			643	41 34 49	25.357
606	36 72 36	24.617	644	41 47 36	25.377
607	36 84 49	24.637	645	41 60 25	25.397
608	36 96 64	24.658			
609	37 08 81	24.678	646	41 73 16	25.417
610	37 21 00	24.698	647	41 86 09	25.436
			648	41 99 04	25.456
611	37 33 21	24.718	649	42 12 01	25.475
612	37 45 44	24.739	650	42 25 00	25.495
613	37 57 69	24.759			
614	37 69 96	24.779	651	42 38 01	25.515
615	37 82 25	24.799	652	42 51 04	25.534
			653	42 64 09	25.554
616	37 94 56	24.819	654	42 77 16	25.573
617	38 06 89	24.839	655	42 90 25	25.593
618	38 19 24	24.860			
619	38 31 61	24.880	656	43 03 36	25.612
620	38 44 00	24.900	657	43 16 49	25.632
			658	43 29 64	25.652
621	38 56 41	24.920	659	43 42 81	25.671
622	38 68 84	24.940	660	43 56 00	25.690
623	38 81 29	24.960			

TABLE A—*Continued*

Number	Square	Square root	Number	Square	Square root
661	43 69 21	25.710	699	48 86 01	26.439
662	43 82 44	25.729	700	49 00 00	26.458
663	43 95 69	25.749			
664	44 08 96	25.768	701	49 14 01	26.476
665	44 22 25	25.788	702	49 28 04	26.495
			703	49 42 09	26.514
666	44 35 56	25.807	704	49 56 16	26.533
667	44 48 89	25.826	705	49 70 25	26.552
668	44 62 24	25.846			
669	44 75 61	25.865	706	49 84 36	26.571
670	44 89 00	25.884	707	49 98 49	26.589
			708	50 12 64	26.608
671	45 02 41	25.904	709	50 26 81	26.627
672	45 15 84	25.923	710	50 41 00	26.646
673	45 29 29	25.942			
674	45 42 76	25.962	711	50 55 21	26.665
675	45 56 25	25.981	712	50 69 44	26.683
			713	50 83 69	26.702
676	45 69 76	26.000	714	50 97 96	26.721
677	45 83 29	26.019	715	51 12 25	26.739
678	45 96 84	26.038			
679	46 10 41	26.058	716	51 26 56	26.758
680	46 24 00	26.077	717	51 40 89	26.777
			718	51 55 24	26.796
681	46 37 61	26.096	719	51 69 61	26.814
682	46 51 24	26.115	720	51 84 00	26.833
683	46 64 89	26.134			
684	46 78 56	26.153	721	51 98 41	26.851
685	46 92 25	26.173	722	52 12 84	26.870
			723	52 27 29	26.889
686	47 05 96	26.192	724	52 41 76	26.907
687	47 19 69	26.211	725	52 56 25	26.926
688	47 33 44	26.230			
689	47 47 21	26.249	726	52 70 76	26.944
690	47 61 00	26.268	727	52 85 29	26.963
			728	52 99 84	26.981
691	47 74 81	26.287	729	53 14 41	27.000
692	47 88 64	26.306	730	53 29 00	27.019
693	48 02 49	26.325			
694	48 16 36	26.344	731	53 43 61	27.037
695	48 30 25	26.363	732	53 58 24	27.055
			733	53 72 89	27.074
696	48 44 16	26.382	734	53 87 56	27.092
697	48 58 09	26.401	735	54 02 25	27.111
698	48 72 04	26.420			

TABLE A—*Continued*

Number	Square	Square root	Number	Square	Square root
736	54 16 96	27.129	774	59 90 76	27.821
737	54 31 69	27.148	775	60 06 25	27.839
738	54 46 44	27.166			
739	54 61 21	27.185	776	60 21 76	27.857
740	54 76 00	27.203	777	60 37 29	27.875
			778	60 52 84	27.893
741	54 90 81	27.221	779	60 68 41	27.911
742	55 05 64	27.240	780	60 84 00	27.928
743	55 20 49	27.258			
744	55 35 36	27.276	781	60 99 61	27.946
745	55 50 25	27.295	782	61 15 24	27.964
			783	61 30 89	27.982
746	55 65 16	27.313	784	61 46 56	28.000
747	55 80 09	27.331	785	61 62 25	28.018
748	55 95 04	27.350			
749	56 10 01	27.368	786	61 77 96	28.036
750	56 25 00	27.386	787	61 93 69	28.054
			788	62 09 44	28.071
751	56 40 01	27.404	789	62 25 21	28.089
752	56 55 04	27.423	790	62 41 00	28.107
753	56 70 09	27.441			
754	56 85 16	27.459	791	62 56 81	28.125
755	57 00 25	27.477	792	62 72 64	28.142
			793	62 88 49	28.160
756	57 15 36	27.495	794	63 04 36	28.178
757	57 30 49	27.514	795	63 20 25	28.196
758	57 45 64	27.532			
759	57 60 81	27.550	796	63 36 16	28.213
760	57 76 00	27.568	797	63 52 09	28.231
			798	63 68 04	28.249
761	57 91 21	27.586	799	63 84 01	28.267
762	58 06 44	27.604	800	64 00 00	28.284
763	58 21 69	27.622			
764	58 36 96	27.641	801	64 16 01	28.302
765	58 52 25	27.659	802	64 32 04	28.320
			803	64 48 09	28.337
766	58 67 56	27.677	804	64 64 16	28.355
767	58 82 89	27.695	805	64 80 25	28.373
768	58 98 24	27.713			
769	59 13 61	27.731	806	64 96 36	28.390
770	59 29 00	27.749	807	65 12 49	28.408
			808	65 28 64	28.425
771	59 44 41	27.767	809	65 44 81	28.443
772	59 59 84	27.785	810	65 61 00	28.460
773	59 75 29	27.803			

TABLE A—*Continued*

Number	Square	Square root	Number	Square	Square root
811	65 77 21	28.478	849	72 08 01	29.138
812	65 93 44	28.496	850	72 25 00	29.155
813	66 09 69	28.513			
814	66 25 96	28.531	851	72 42 01	29.172
815	66 42 25	28.548	852	72 59 04	29.189
			853	72 76 09	29.206
816	66 58 56	28.566	854	72 93 16	29.223
817	66 74 89	28.583	855	73 10 25	29.240
818	66 91 24	28.601			
819	67 07 61	28.618	856	73 27 36	29.257
820	67 24 00	28.636	857	73 44 49	29.275
			858	73 61 64	29.292
821	67 40 41	28.653	859	73 78 81	29.309
822	67 56 84	28.671	860	73 96 00	29.326
823	67 73 29	28.688			
824	67 89 76	28.705	861	74 13 21	29.343
825	68 06 25	28.723	862	74 30 44	29.360
			863	74 47 69	29.377
826	68 22 76	28.740	864	74 64 96	29.394
827	68 39 29	28.758	865	74 82 25	29.411
828	68 55 84	28.775			
829	68 72 41	28.792	866	74 99 56	29.428
830	68 89 00	28.810	867	75 16 89	29.445
			868	75 34 24	29.462
831	69 05 61	28.827	869	75 51 61	29.479
832	69 22 24	28.844	870	75 69 00	29.496
833	69 38 89	28.862			
834	69 55 56	28.879	871	75 86 41	29.513
835	69 72 25	28.896	872	76 03 84	29.530
			873	76 21 29	29.547
836	69 88 96	28.914	874	76 38 76	29.563
837	70 05 69	28.931	875	76 56 25	29.580
838	70 22 44	28.948			
839	70 39 21	28.965	876	76 73 76	29.597
840	70 56 00	28.983	877	76 91 29	29.614
			878	77 08 84	29.631
841	70 72 81	29.000	879	77 26 41	29.648
842	70 89 64	29.017	880	77 44 00	29.665
843	71 06 49	29.034			
844	71 23 36	29.052	881	77 61 61	29.682
845	71 40 25	29.069	882	77 79 24	29.698
			883	77 96 89	29.715
846	71 57 16	29.086	884	78 14 56	29.732
847	71 74 09	29.103	885	78 32 25	29.749
848	71 91 04	29.120			

TABLE A—*Continued*

Number	Square	Square root	Number	Square	Square root
886	78 49 96	29.766	924	85 37 76	30.397
887	78 67 69	29.783	925	85 56 25	30.414
888	78 85 44	29.799			
889	79 03 21	29.816	926	85 74 76	30.430
890	79 21 00	29.833	927	85 93 29	30.447
			928	86 11 84	30.463
891	79 38 81	29.850	929	86 30 41	30.480
892	79 56 64	29.866	930	86 49 00	30.496
893	79 74 49	29.833			
894	79 92 36	29.900	931	86 67 61	30.512
895	80 10 25	29.916	932	86 86 24	30.529
			933	87 04 89	30.545
896	80 28 16	29.933	934	87 23 56	30.561
897	80 46 09	29.950	935	87 42 25	30.578
898	80 64 04	29.967			
899	80 82 01	29.983	936	87 60 96	30.594
900	81 00 00	30.000	937	87 79 69	30.610
			938	87 98 44	30.627
901	81 18 01	30.017	939	88 17 21	30.643
902	81 36 04	30.033	940	88 36 00	30.659
903	81 54 09	30.050			
904	81 72 16	30.067	941	88 54 81	30.676
905	81 90 25	30.083	942	88 73 64	30.692
			943	88 92 49	30.708
906	82 08 36	30.100	944	89 11 36	30.725
907	82 26 49	30.116	945	89 30 25	30.741
908	82 44 64	30.133			
909	82 62 81	30.150	946	89 49 16	30.757
910	82 81 00	30.166	947	89 68 09	30.773
			948	89 87 04	30.790
911	82 99 21	30.183	949	90 06 01	30.806
912	83 17 44	30.199	950	90 25 00	30.822
913	83 35 69	30.216			
914	83 53 96	30.232	951	90 44 01	30.838
915	83 72 25	30.249	952	90 63 04	30.854
			953	90 82 09	30.871
916	83 90 56	30.265	954	91 01 16	30.887
917	84 08 89	30.282	955	91 20 25	30.903
918	84 27 24	30.299			
919	84 45 61	30.315	956	91 39 36	30.919
920	84 64 00	30.332	957	91 58 49	30.935
			958	91 77 64	30.952
921	84 82 41	30.348	959	91 96 81	30.968
922	85 00 84	30.364	960	92 16 00	30.984
923	85 19 29	30.381			

TABLE A—*Concluded*

Number	Square	Square root	Number	Square	Square root
961	92 35 21	31.000	981	96 23 61	31.321
962	92 54 44	31.016	982	96 43 24	31.337
963	92 73 69	31.032	983	96 62 89	31.353
964	92 92 96	31.048	984	96 82 56	31.369
965	93 12 25	31.064	985	97 02 25	31.385
966	93 31 56	31.081	986	97 21 96	31.401
967	93 50 89	31.097	987	97 41 69	31.417
968	93 70 24	31.113	988	97 61 44	31.432
969	93 89 61	31.129	989	97 81 21	31.448
970	94 09 00	31.145	990	98 01 00	31.464
971	94 28 41	31.161	991	98 20 81	31.480
972	94 47 84	31.177	992	98 40 64	31.496
973	94 67 29	31.193	993	98 60 49	31.512
974	94 86 76	31.209	994	98 80 36	31.528
975	95 06 25	31.225	995	99 00 25	31.544
976	95 25 76	31.241	996	99 20 16	31.559
977	95 45 29	31.257	997	99 40 09	31.575
978	95 64 84	31.273	998	99 60 04	31.591
979	95 84 41	31.289	999	99 80 01	31.607
980	96 04 00	31.305	1000	100 00 00	31.623

TABLE B
Percent of total area under the normal curve between mean ordinate and ordinate at any given Z score distance from the mean

Z	.00	.01	.02	.03	.04	.05	.06	.07	.08	.09
0.0	00.00	00.40	00.80	01.20	01.60	01.99	02.39	02.70	03.19	03.59
0.1	03.98	04.38	04.78	05.17	05.57	05.96	06.36	06.75	07.14	07.53
0.2	07.93	08.32	08.71	09.10	09.48	09.87	10.26	10.64	11.03	11.41
0.3	11.79	12.17	12.55	12.93	13.31	13.68	14.06	14.43	14.80	15.17
0.4	15.54	15.91	16.28	16.64	17.00	17.36	17.72	18.08	18.44	18.79
0.5	19.15	19.50	19.85	20.19	20.54	20.88	21.23	21.57	21.90	22.24
0.6	22.57	22.91	23.24	23.57	23.89	24.22	24.54	24.86	25.17	25.49
0.7	25.80	26.11	26.42	26.73	27.04	27.34	27.64	27.94	28.23	28.52
0.8	28.81	29.10	29.39	29.67	29.95	30.23	30.51	30.78	31.06	31.33
0.9	31.59	31.86	32.12	32.38	32.64	32.90	33.15	33.40	33.65	33.89
1.0	34.13	34.38	34.61	34.85	35.08	35.31	35.54	35.77	35.90	36.21
1.1	36.43	36.65	36.86	37.08	37.29	37.49	37.70	37.90	38.10	38.30
1.2	38.49	38.69	38.88	39.07	39.25	39.44	39.62	39.80	39.97	40.15
1.3	40.32	40.49	40.66	40.82	40.99	41.15	41.31	41.47	41.62	41.77
1.4	41.92	42.07	42.22	42.36	42.51	42.65	42.79	42.92	43.06	43.19
1.5	43.32	43.45	43.57	43.70	43.83	43.94	44.06	44.18	44.29	44.41
1.6	44.52	44.63	44.74	44.84	44.95	45.05	45.15	45.25	45.35	45.45
1.7	45.54	45.64	45.73	45.82	45.91	45.99	46.08	46.16	46.25	46.33
1.8	46.41	46.49	46.56	46.64	46.71	46.78	46.86	46.93	46.99	47.06
1.9	47.13	47.19	47.26	47.32	47.38	47.44	47.50	47.56	47.61	47.67
2.0	47.72	47.78	47.83	47.88	47.93	47.98	48.03	48.08	48.12	48.17
2.1	48.21	48.26	48.30	48.34	48.38	48.42	48.46	48.50	48.54	48.57
2.2	48.61	48.64	48.68	48.71	48.75	48.78	48.81	48.84	48.87	48.90
2.3	48.93	48.96	48.98	49.01	49.04	49.06	49.09	49.11	49.13	49.16
2.4	49.18	49.20	49.22	49.25	49.27	49.29	49.31	49.32	49.34	49.36
2.5	49.38	49.40	49.41	49.43	49.45	49.46	49.48	49.49	49.51	49.52
2.6	49.53	49.55	49.56	49.57	49.59	49.60	49.61	49.62	49.63	49.64
2.7	49.65	49.66	49.67	49.68	49.69	49.70	49.71	49.72	49.73	49.74
2.8	49.74	49.75	49.76	49.77	49.77	49.78	49.79	49.79	49.80	49.81
2.9	49.81	49.82	49.82	49.83	49.84	49.84	49.85	49.85	49.86	49.86
3.0	49.87									
3.5	49.98									
4.0	49.997									
5.0	49.99997									

TABLE C
Table of binomial coefficients

N	$\binom{N}{0}$	$\binom{N}{1}$	$\binom{N}{2}$	$\binom{N}{3}$	$\binom{N}{4}$	$\binom{N}{5}$	$\binom{N}{6}$	$\binom{N}{7}$	$\binom{N}{8}$	$\binom{N}{9}$	$\binom{N}{10}$
0	1										
1	1	1									
2	1	2	1								
3	1	3	3	1							
4	1	4	6	4	1						
5	1	5	10	10	5	1					
6	1	6	15	20	15	6	1				
7	1	7	21	35	35	21	7	1			
8	1	8	28	56	70	56	28	8	1		
9	1	9	36	84	126	126	84	36	9	1	
10	1	10	45	120	210	252	210	120	45	10	1
11	1	11	55	165	330	462	462	330	165	55	11
12	1	12	66	220	495	792	924	792	495	220	66
13	1	13	78	286	715	1,287	1,716	1,716	1,287	715	286
14	1	14	91	364	1,001	2,002	3,003	3,432	3,003	2,002	1,001
15	1	15	105	455	1,365	3,003	5,005	6,435	6,435	5,005	3,003
16	1	16	120	560	1,820	4,368	8,008	11,440	12,870	11,440	8,008
17	1	17	136	680	2,380	6,188	12,376	19,448	24,310	24,310	19,448
18	1	18	153	816	3,060	8,568	18,564	31,824	43,758	48,620	43,758
19	1	19	171	969	3,876	11,628	27,132	50,388	75,582	92,378	92,378
20	1	20	190	1,140	4,845	15,504	38,760	77,520	125,970	167,960	184,756

TABLE D
Values of *r* at the .05 and .01 levels of significance

Degrees of freedom (df)	5%	1%	Degrees of freedom (df)	5%	1%
1	.997	1.000	24	.388	.496
2	.950	.990	25	.381	.487
3	.878	.959	26	.374	.478
4	.811	.917	27	.367	.470
5	.754	.874	28	.361	.463
6	.707	.834	29	.355	.456
7	.666	.798	30	.349	.449
8	.632	.765	35	.325	.418
9	.602	.735	40	.304	.393
10	.576	.708	45	.288	.372
11	.553	.684	50	.273	.354
12	.532	.661	60	.250	.325
13	.514	.641	70	.232	.302
14	.497	.623	80	.217	.283
15	.482	.606	90	.205	.267
16	.468	.590	100	.195	.254
17	.456	.575	125	.174	.228
18	.444	.561	150	.159	.208
19	.433	.549	200	.138	.181
20	.423	.537	300	.113	.148
21	.413	.526	400	.098	.128
22	.404	.515	500	.088	.115
23	.396	.505	1000	.062	.081

Source: A portion of Table D is taken from Table VI of Fisher and Yates: *Statistical Tables for Biological, Agricultural and Medical Research*, published by Oliver and Boyd, Edinburgh, and by permission of the authors and publishers. Also reprinted by permission from *Statistical Methods*, by Snedecor and Cochran, 6th ed., © 1967, Iowa State University Press, Ames, Iowa.

TABLE E

Values of r_s (rank-order correlation coefficient) at the .05 and .01 levels of significance

N	5%	1%
5	1.000	.....
6	.886	1.000
7	.786	.929
8	.738	.881
9	.683	.833
10	.648	.794
12	.591	.777
14	.544	.715
16	.506	.665
18	.475	.625
20	.450	.591
22	.428	.562
24	.409	.537
26	.392	.515
28	.377	.496
30	.364	.478

Source: E. G. Olds, "Distribution of the Sum of Squares of Rank Differences for Small Numbers of Individuals," *Annals of Mathematical Statistics* 9 (1938): 133–48; and E. G. Olds, "The 5% Significance Levels for Sums of Squares of Rank Differences and a Correction," *Annals of Mathematical Statistics* 20 (1949): 117–18; by permission of the author and the Institute of Mathematical Statistics.

TABLE F
Values of F at the .05 and .01 significance levels

Degrees of freedom associated with the denominator		Degrees of freedom associated with the numerator								
		1	2	3	4	5	6	7	8	9
1	.05	161	200	216	225	230	234	237	239	241
	.01	4,052	5,000	5,403	5,625	5,764	5,859	5,928	5,982	6,022
2	.05	18.5	19.0	19.2	19.2	19.3	19.3	19.4	19.4	19.4
	.01	98.5	99.0	99.2	99.2	99.3	99.3	99.4	99.4	99.4
3	.05	10.1	9.55	9.28	9.12	9.01	8.94	8.89	8.85	8.81
	.01	34.1	30.8	29.5	28.7	28.2	27.9	27.7	27.5	27.3
4	.05	7.71	6.94	6.59	6.39	6.26	6.16	6.09	6.04	6.00
	.01	21.2	18.0	16.7	16.0	15.5	15.2	15.0	14.8	14.7
5	.05	6.61	5.79	5.41	5.19	5.05	4.95	4.88	4.82	4.77
	.01	16.3	13.3	12.1	11.4	11.0	10.7	10.5	10.3	10.2
6	.05	5.99	5.14	4.76	4.53	4.39	4.28	4.21	4.15	4.10
	.01	13.7	10.9	9.78	9.15	8.75	8.47	8.26	8.10	7.98
7	.05	5.59	4.74	4.35	4.12	3.97	3.87	3.79	3.73	3.68
	.01	12.2	9.55	8.45	7.85	7.46	7.19	6.99	6.84	6.72
8	.05	5.32	4.46	4.07	3.84	3.69	3.58	3.50	3.44	3.39
	.01	11.3	8.65	7.59	7.01	6.63	6.37	6.18	6.03	5.91
9	.05	5.12	4.26	3.86	3.63	3.48	3.37	3.29	3.23	3.18
	.01	10.6	8.02	6.99	6.42	6.06	5.80	5.61	5.47	5.35
10	.05	4.96	4.10	3.71	3.48	3.33	3.22	3.14	3.07	3.02
	.01	10.0	7.56	6.55	5.99	5.64	5.39	3.20	5.06	4.94
11	.05	4.84	3.98	3.59	3.36	3.20	3.09	3.01	2.95	2.90
	.01	9.65	7.21	6.22	5.67	5.32	5.07	4.89	4.74	4.63
12	.05	4.75	3.89	3.49	3.26	3.11	3.00	2.91	2.85	2.80
	.01	9.33	6.93	5.95	5.41	5.06	4.82	4.64	4.50	4.39
13	.05	4.67	3.81	3.41	3.18	3.03	2.92	2.83	2.77	2.71
	.01	9.07	6.70	5.74	5.21	4.86	4.62	4.44	4.30	4.19
14	.05	4.60	3.74	3.34	3.11	2.96	2.85	2.76	2.70	2.65
	.01	8.86	6.51	5.56	5.04	4.70	4.46	4.28	4.14	4.03
15	.05	4.54	3.68	3.29	3.06	2.99	2.79	2.71	2.64	2.59
	.01	8.68	6.36	5.42	4.89	4.56	4.32	4.14	4.00	3.89
16	.05	4.49	3.63	3.24	3.01	2.85	2.74	2.66	2.59	2.54
	.01	8.53	6.23	5.29	4.77	4.44	4.20	4.03	3.89	3.78
17	.05	4.45	3.59	3.20	2.96	2.81	2.70	2.61	2.55	2.49
	.01	8.40	6.11	5.18	4.67	4.34	4.10	3.93	3.79	3.68
18	.05	4.41	3.55	3.16	2.93	2.77	2.66	2.58	2.51	2.46
	.01	8.29	6.01	5.09	4.58	4.25	4.01	3.84	3.71	3.60
19	.05	4.38	3.52	3.13	2.90	2.74	2.63	2.54	2.48	2.42
	.01	8.18	5.93	5.01	4.50	4.17	3.94	3.77	3.63	3.52
20	.05	4.35	3.49	3.10	2.87	2.71	2.60	2.51	2.45	2.39
	.01	8.10	5.85	4.94	4.43	4.10	3.87	3.70	3.56	3.46

TABLE F—*Continued*

Degrees of freedom associated with the denominator		Degrees of freedom associated with the numerator								
		1	2	3	4	5	6	7	8	9
21	.05	4.32	3.47	3.07	3.84	2.68	2.57	2.49	2.42	2.37
	.01	8.02	5.78	4.87	4.37	4.04	3.81	3.64	3.51	3.40
22	.05	4.30	3.44	3.05	2.82	2.66	2.55	2.46	2.40	2.34
	.01	7.95	5.72	4.82	4.31	3.99	3.76	3.59	3.45	3.35
23	.05	4.28	3.42	3.03	2.80	2.64	2.53	2.44	2.37	2.32
	.01	7.88	5.66	4.76	4.26	3.94	3.71	3.54	3.41	3.30
24	.05	4.26	3.40	3.01	2.78	2.62	2.51	2.42	2.36	2.30
	.01	7.82	5.61	4.72	4.22	3.90	3.67	3.50	3.36	3.26
25	.05	4.24	3.39	2.99	2.76	2.60	2.49	2.40	2.34	2.28
	.01	7.77	5.57	4.68	4.18	3.86	3.63	3.46	3.32	3.22
26	.05	4.23	3.37	2.98	2.74	2.59	2.47	2.39	2.32	2.27
	.01	7.72	5.53	4.64	4.14	3.82	3.59	3.42	3.29	3.18
27	.05	4.21	3.35	2.96	2.73	2.57	2.46	2.37	2.31	2.25
	.01	7.68	5.49	4.60	4.11	3.78	3.56	3.39	3.26	3.15
28	.05	4.20	3.34	2.95	2.71	2.56	2.45	2.36	2.29	2.24
	.01	7.64	5.45	4.57	4.07	3.75	3.53	3.36	3.23	3.12
29	.05	4.18	3.33	2.93	2.70	2.55	2.43	2.35	2.28	2.22
	.01	7.60	5.42	4.54	4.04	3.73	3.50	3.33	3.20	2.09
30	.05	4.17	3.32	2.92	2.69	2.53	2.42	2.33	2.27	2.21
	.01	7.56	5.39	4.51	4.02	3.70	3.47	3.30	3.17	3.07
40	.05	4.08	3.23	2.84	2.61	2.45	2.34	2.25	2.18	2.12
	.01	7.31	5.18	4.31	3.83	3.51	3.29	3.12	2.99	2.89
60	.05	4.00	3.15	2.76	2.53	2.37	2.25	2.17	2.10	2.04
	.01	7.08	4.98	4.13	3.65	3.34	3.12	2.95	2.82	2.72
120	.05	3.92	3.07	2.68	2.45	2.29	2.18	2.09	2.02	1.96
	.01	6.85	4.79	3.95	3.48	3.17	2.96	2.79	2.66	2.56

Source: M. Merrington and C. M. Thompson, "Tables of Percentage Points of the Inverted Beta (F) Distribution," *Biometrika* 33 (1943): 73–88; by permission of the authors and publishers.

TABLE G
Distribution of t

df	$p = .10$	.05	.01	.001
1	6.314	12.706	63.657	636.619
2	2.920	4.303	9.925	31.598
3	2.353	3.182	5.841	12.941
4	2.132	2.776	4.604	8.610
5	2.015	2.571	4.032	6.859
6	1.943	2.447	3.707	5.959
7	1.895	2.365	3.499	5.405
8	1.860	2.306	3.355	5.041
9	1.833	2.262	3.250	4.781
10	1.812	2.228	3.169	4.587
11	1.796	2.201	3.106	4.437
12	1.782	2.179	3.055	4.318
13	1.771	2.160	3.012	4.221
14	1.761	2.145	2.977	4.140
15	1.753	2.131	2.947	4.073
16	1.746	2.120	2.921	4.015
17	1.740	2.110	2.898	3.965
18	1.734	2.101	2.878	3.922
19	1.729	2.093	2.861	3.883
20	1.725	2.086	2.845	3.850
21	1.721	2.080	2.831	3.819
22	1.717	2.074	2.819	3.792
23	1.714	2.069	2.807	3.767
24	1.711	2.064	2.797	3.745
25	1.708	2.060	2.787	3.725
26	1.706	2.056	2.779	3.707
27	1.703	2.052	2.771	3.690
28	1.701	2.048	2.763	3.674
29	1.699	2.045	2.756	3.659
30	1.697	2.042	2.750	3.646
40	1.684	2.021	2.704	3.551
60	1.671	2.000	2.660	3.460
120	1.658	1.980	2.617	3.373
∞	1.645	1.960	2.576	3.291

Source: Table G is taken from Table III of Fisher and Yates: *Statistical Tables for Biological, Agricultural, and Medical Research*, published by Oliver and Boyd, Edinburgh, and by permission of the authors and publishers.

Note: The values of p are given for a two-tailed test.

TABLE H
Critical values of Sandler's A

df	Level of significance for one-tailed test					N-1
	.05	.025	.01	.005	.0005	
	Level of significance for two-tailed test					
N-1	.10	.05	.02	.01	.001	N-1
1	0.5125	0.5031	0.50049	0.50012	0.5000012	1
2	0.412	0.369	0.347	0.340	0.334	2
3	0.385	0.324	0.286	0.272	0.254	3
4	0.376	0.304	0.257	0.238	0.211	4
5	0.372	0.293	0.240	0.218	0.184	5
6	0.370	0.286	0.230	0.205	0.167	6
7	0.369	0.281	0.222	0.196	0.155	7
8	0.368	0.278	0.217	0.190	0.146	8
9	0.368	0.276	0.213	0.185	0.139	9
10	0.368	0.274	0.210	0.181	0.134	10
11	0.368	0.273	0.207	0.178	0.130	11
12	0.368	0.271	0.205	0.176	0.126	12
13	0.368	0.270	0.204	0.174	0.124	13
14	0.368	0.270	0.202	0.172	0.121	14
15	0.368	0.269	0.201	0.170	0.119	15
16	0.368	0.268	0.200	0.169	0.117	16
17	0.368	0.268	0.199	0.168	0.116	17
18	0.368	0.267	0.198	0.167	0.114	18
19	0.368	0.267	0.197	0.166	0.113	19
20	0.368	0.266	0.197	0.165	0.112	20
21	0.368	0.266	0.196	0.165	0.111	21
22	0.368	0.266	0.196	0.164	0.110	22
23	0.368	0.266	0.195	0.163	0.109	23
24	0.368	0.265	0.195	0.163	0.108	24
25	0.368	0.265	0.194	0.162	0.108	25
26	0.368	0.265	0.194	0.162	0.107	26
27	0.368	0.265	0.193	0.161	0.107	27
28	0.368	0.265	0.193	0.161	0.106	28
29	0.368	0.264	0.193	0.161	0.106	29
30	0.368	0.264	0.193	0.160	0.105	30
40	0.368	0.263	0.191	0.158	0.102	40
60	0.369	0.262	0.189	0.155	0.099	60
120	0.369	0.261	0.187	0.153	0.095	120
∞	0.370	0.260	0.185	0.151	0.092	∞

Source: J. Sandler, "A test of the Significance of the Difference between the Means of Correlated Measures. Based on a Simplification of Student's *t*," *British Journal of Psychology* 46 (1955): 225-26; by permission of the author and publisher.
Note: For any given value of N − 1, the table shows the values of *A* corresponding to various levels of probability. *A* is significant at a given level if it is equal to or *less than* the value shown in the table.

TABLE I
Table of probabilities associated with values as small as observed values of x in binomial test

N \ x	0	1	2	3	4	5	6	7	8	9	10	11	12	13	14	15
5	031	188	500	812	969	‡										
6	016	109	344	656	891	984	‡									
7	008	062	227	500	773	938	992	‡								
8	004	035	145	363	637	855	965	996	‡							
9	002	020	090	254	500	746	910	980	998	‡						
10	001	011	055	172	377	623	828	945	989	999	‡					
11		006	033	113	274	500	726	887	967	994	‡	‡				
12		003	019	073	194	387	613	806	927	981	997	‡	‡			
13		002	011	046	133	291	500	709	867	954	989	998	‡	‡		
14		001	006	029	090	212	395	605	788	910	971	994	999	‡	‡	
15			004	018	059	151	304	500	696	849	941	982	996	‡	‡	‡
16			002	011	038	105	227	402	598	773	895	962	989	998	‡	‡
17			001	006	025	072	166	315	500	685	834	928	975	994	999	‡
18			001	004	015	048	119	240	407	593	760	881	952	985	996	999
19				002	010	032	084	180	324	500	676	820	916	968	990	998
20				001	006	021	058	132	252	412	588	748	868	942	979	994
21				001	004	013	039	095	192	332	500	668	808	905	961	987
22					002	008	026	067	143	262	416	584	738	857	933	974
23					001	005	017	047	105	202	339	500	661	798	895	953
24					001	003	011	032	076	154	271	419	581	729	846	924
25						002	007	022	054	115	212	345	500	655	788	885

Source: Helen M. Walker and Joseph Lev, *Statistical Inference* (New York: Holt, Rinehart and Winston, 1953), table IVB, p. 458 of the Appendix. Reprinted by permission of the publisher.

Note: Given in the body of this table are one-tailed probabilities under H_0 for the binomial test when $P = Q = \frac{1}{2}$. To save space, decimal points are omitted in the p's.
‡ 1.0 or approximately 1.0.

TABLE J
Critical values of chi-square

p	.50	.30	.20	.10	.05	.02	.01	.001
df								
1	.46	1.07	1.64	2.71	3.84	5.41	6.64	10.83
2	1.39	2.41	3.22	4.60	5.99	7.82	9.21	13.82
3	2.37	3.66	4.64	6.25	7.82	9.84	11.34	16.27
4	3.36	4.88	5.99	7.78	9.49	11.67	13.28	18.46
5	4.35	6.06	7.29	9.24	11.07	13.39	15.09	20.52
6	5.35	7.23	8.56	10.64	12.59	15.03	16.81	22.46
7	6.35	8.38	9.80	12.02	14.07	16.62	18.48	24.32
8	7.34	9.52	11.03	13.36	15.51	18.17	20.09	26.12
9	8.34	10.66	12.24	14.68	16.92	19.68	21.67	27.88
10	9.34	11.78	13.44	15.99	18.31	21.16	23.21	29.59
11	10.34	12.90	14.63	17.28	19.68	22.62	24.72	31.26
12	11.34	14.01	15.81	18.55	21.03	24.05	26.22	32.91
13	12.34	15.12	16.98	19.81	22.36	25.47	27.69	34.53
14	13.34	16.22	18.15	21.06	23.68	26.87	29.14	36.12
15	14.34	17.32	19.31	22.31	25.00	28.26	30.58	37.70
16	15.34	18.42	20.46	23.54	26.30	29.63	32.00	39.29
17	16.34	19.51	21.62	24.77	27.59	31.00	33.41	40.75
18	17.34	20.60	22.76	25.99	28.87	32.35	34.80	42.31
19	18.34	21.69	23.90	27.20	30.14	33.69	36.19	43.82
20	19.34	22.78	25.04	28.41	31.41	35.02	37.57	45.32
21	20.34	23.86	26.17	29.62	32.67	36.34	38.93	46.80
22	21.24	24.94	27.30	30.81	33.92	37.66	40.29	48.27
23	22.34	26.02	28.43	32.01	35.17	38.97	41.64	49.73
24	23.34	27.10	29.55	33.20	36.42	40.27	42.98	51.18
25	24.34	28.17	30.68	34.38	37.65	41.57	44.31	52.62
26	25.34	29.25	31.80	35.56	38.88	42.86	45.64	54.05
27	26.34	30.32	32.91	36.74	40.11	44.14	46.96	55.48
28	27.34	31.39	34.03	37.92	41.34	45.42	48.28	56.89
29	28.34	32.46	35.14	39.09	42.56	46.69	49.59	58.30
30	29.34	33.53	36.25	40.26	43.77	47.96	50.89	59.70

Source: Table J is taken from Table IV of Fisher and Yates: *Statistical Tables for Biological, Agricultural and Medical Research*, published by Oliver and Boyd, Edinburgh, and by permission of the authors and publishers.

TABLE K

Critical values of D (or C) in the Fisher-Yates test

Totals in right margin		B (or A)†	Level of significance			
			.05	.025	.01	.005
$A + B = 3$	$C + D = 3$	3	0	—	—	—
$A + B = 4$	$C + D = 4$	4	0	0	—	—
	$C + D = 3$	4	0	—	—	—
$A + B = 5$	$C + D = 5$	5	1	1	0	0
		4	0	0	—	—
	$C + D = 4$	5	1	0	0	—
		4	0	—	—	—
	$C + D = 3$	5	0	0	—	—
	$C + D = 2$	5	0	—	—	—
$A + B = 6$	$C + D = 6$	6	2	1	1	0
		5	1	0	0	—
		4	0	—	—	—
	$C + D = 5$	6	1	0	0	0
		5	0	0	—	—
		4	0	—	—	—
	$C + D = 4$	6	1	0	0	0
		5	0	0	—	—
	$C + D = 3$	6	0	0	—	—
		5	0	—	—	—
	$C + D = 2$	6	0	—	—	—
$A + B = 7$	$C + D = 7$	7	3	2	1	1
		6	1	1	0	0
		5	0	0	—	—
		4	0	—	—	—
	$C + D = 6$	7	2	2	1	1
		6	1	0	0	0
		5	0	0	—	—
		4	0	—	—	—
	$C + D = 5$	7	2	1	0	0
		6	1	0	0	—
		5	0	—	—	—
	$C + D = 4$	7	1	1	0	0
		6	0	0	—	—
		5	0	—	—	—
	$C + D = 3$	7	0	0	0	—
		6	0	—	—	—
	$C + D = 2$	7	0	—	—	—

Source: D. J. Finney, "The Fisher-Yates Test of Significance in 2 × 2 Contingency Tables," *Biometrika* 35: 149–54; with permission of the author and publisher.
    † When B is entered in the middle column, the significance levels are for D. When A is used in place of B, the significance levels are for C.

TABLE K—*Continued*

Totals in right margin		B (or A)†	Level of significance			
			.05	.025	.01	.005
$A + B = 8$	$C + D = 8$	8	4	3	2	2
		7	2	2	1	0
		6	1	1	0	0
		5	0	0	—	—
		4	0	—	—	—
	$C + D = 7$	8	3	2	2	1
		7	2	1	1	0
		6	1	0	0	—
		5	0	0	—	—
	$C + D = 6$	8	2	2	1	1
		7	1	1	0	0
		6	0	0	0	—
		5	0	—	—	—
	$C + D = 5$	8	2	1	1	0
		7	1	0	0	0
		6	0	0	—	—
		5	0	—	—	—
	$C + D = 4$	8	1	1	0	0
		7	0	0	—	—
		6	0	—	—	—
	$C + D = 3$	8	0	0	0	—
		7	0	0	—	—
	$C + D = 2$	8	0	0	—	—
$A + B = 9$	$C + D = 9$	9	5	4	3	3
		8	3	3	2	1
		7	2	1	1	0
		6	1	1	0	0
		5	0	0	—	—
		4	0	—	—	—
	$C + D = 8$	9	4	3	3	2
		8	3	2	1	1
		7	2	1	0	0
		6	1	0	0	—
		5	0	0	—	—
	$C + D = 7$	9	3	3	2	2
		8	2	2	1	0
		7	1	1	0	0
		6	0	0	—	—
		5	0	—	—	—

TABLE K—*Continued*

Totals in right margin		B (or A)†	Level of significance			
			.05	.025	.01	.005
$A + B = 9$	$C + D = 6$	9	3	2	1	1
		8	2	1	0	0
		7	1	0	0	—
		6	0	0	—	—
		5	0	—	—	—
	$C + D = 5$	9	2	1	1	1
		8	1	1	0	0
		7	0	0	—	—
		6	0	—	—	—
	$C + D = 4$	9	1	1	0	0
		8	0	0	0	—
		7	0	0	—	—
		6	0	—	—	—
	$C + D = 3$	9	1	0	0	0
		8	0	0	—	—
		7	0	—	—	—
	$C + D = 2$	9	0	0	—	—
$A + B = 10$	$C + D = 10$	10	6	5	4	3
		9	4	3	3	2
		8	3	2	1	1
		7	2	1	1	0
		6	1	0	0	—
		5	0	0	—	—
		4	0	—	—	—
	$C + D = 9$	10	5	4	3	3
		9	4	3	2	2
		8	2	2	1	1
		7	1	1	0	0
		6	1	0	0	—
		5	0	0	—	—
	$C + D = 8$	10	4	4	3	2
		9	3	2	2	1
		8	2	1	1	0
		7	1	1	0	0
		6	0	0	—	—
		5	0	—	—	—
	$C + D = 7$	10	3	3	2	2
		9	2	2	1	1
		8	1	1	0	0
		7	1	0	0	—
		6	0	0	—	—
		5	0	—	—	—

TABLE K—*Continued*

Totals in right margin		B (or A)†	Level of significance			
			.05	.025	.01	.005
$A + B = 10$	$C + D = 6$	10	3	2	2	1
		9	2	1	1	0
		8	1	1	0	0
		7	0	0	—	—
		6	0	—	—	—
	$C + D = 5$	10	2	2	1	1
		9	1	1	0	0
		8	1	0	0	—
		7	0	0	—	—
		6	0	—	—	—
	$C + D = 4$	10	1	1	0	0
		9	1	0	0	0
		8	0	0	—	—
		7	0	—	—	—
	$C + D = 3$	10	1	0	0	0
		9	0	0	—	—
		8	0	—	—	—
	$C + D = 2$	10	0	0	—	—
		9	0	—	—	—
$A + B = 11$	$C + D = 11$	11	7	6	5	4
		10	5	4	3	3
		9	4	3	2	2
		8	3	2	1	1
		7	2	1	0	0
		6	1	0	0	—
		5	0	0	—	—
		4	0	—	—	—
	$C + D = 10$	11	6	5	4	4
		10	4	4	3	2
		9	3	3	2	1
		8	2	2	1	0
		7	1	1	0	0
		6	1	0	0	—
		5	0	—	—	—
	$C + D = 9$	11	5	4	4	3
		10	4	3	2	2
		9	3	2	1	1
		8	2	1	1	0
		7	1	1	0	0
		6	0	0	—	—
		5	0	—	—	—

TABLE K—*Continued*

Totals in right margin		B (or A)†	Level of significance			
			.05	.025	.01	.005
$A + B = 11$	$C + D = 8$	11	4	4	3	3
		10	3	3	2	1
		9	2	2	1	1
		8	1	1	0	0
		7	1	0	0	—
		6	0	0	—	—
		5	0	—	—	—
	$C + D = 7$	11	4	3	2	2
		10	3	2	1	1
		9	2	1	1	0
		8	1	1	0	0
		7	0	0	—	—
		6	0	0	—	—
	$C + D = 6$	11	3	2	2	1
		10	2	1	1	0
		9	1	1	0	0
		8	1	0	0	—
		7	0	0	—	—
		6	0	—	—	—
	$C + D = 5$	11	2	2	1	1
		10	1	1	0	0
		9	1	0	0	0
		8	0	0	—	—
		7	0	—	—	—
	$C + D = 4$	11	1	1	1	0
		10	1	0	0	0
		9	0	0	—	—
		8	0	—	—	—
	$C + D = 3$	11	1	0	0	0
		10	0	0	—	—
		9	0	—	—	—
	$C + D = 2$	11	0	0	—	—
		10	0	—	—	—
$A + B = 12$	$C + D = 12$	12	8	7	6	5
		11	6	5	4	4
		10	5	4	3	2
		9	4	3	2	1
		8	3	2	1	1
		7	2	1	0	0
		6	1	0	0	—
		5	0	0	—	—
		4	0	—	—	—

TABLE K—*Continued*

Totals in right margin		B (or A)†	Level of significance			
			.05	.025	.01	.005
A + B = 12	C + D = 11	12	7	6	5	5
		11	5	5	4	3
		10	4	3	2	2
		9	3	2	2	1
		8	2	1	1	0
		7	1	1	0	0
		6	1	0	0	—
		5	0	0	—	—
	C + D = 10	12	6	5	5	4
		11	5	4	3	3
		10	4	3	2	2
		9	3	2	1	1
		8	2	1	0	0
		7	1	0	0	0
		6	0	0		—
		5	0	—	—	—
	C + D = 9	12	5	5	4	3
		11	4	3	3	2
		10	3	2	2	1
		9	2	2	1	0
		8	1	1	0	0
		7	1	0	0	—
		6	0	0	—	—
		5	0	—	—	—
	C + D = 8	12	5	4	3	3
		11	3	3	2	2
		10	2	2	1	1
		9	2	1	1	0
		8	1	1	0	0
		7	0	0	—	—
		6	0	0	—	—
	C + D = 7	12	4	3	3	2
		11	3	2	2	1
		10	2	1	1	0
		9	1	1	0	0
		8	1	0	0	—
		7	0	0	—	—
		6	0	—	—	—

TABLE K—*Continued*

Totals in right margin		B (or A)†	Level of significance			
			.05	.025	.01	.005
$A + B = 12$	$C + D = 6$	12	3	3	2	2
		11	2	2	1	1
		10	1	1	0	0
		9	1	0	0	0
		8	0	0	—	—
		7	0	0	—	—
		6	0	—	—	—
	$C + D = 5$	12	2	2	1	1
		11	1	1	1	0
		10	1	0	0	0
		9	0	0	0	—
		8.	0	0	—	—
		7	0	—	—	—
	$C + D = 4$	12	2	1	1	0
		11	1	0	0	0
		10	0	0	0	—
		9	0	0	—	—
		8	0	—	—	—
	$C + D = 3$	12	1	0	0	0
		11	0	0	0	—
		10	0	0	—	—
		9	0	—	—	—
	$C + D = 2$	12	0	0	—	—
		11	0	—	—	—
$A + B = 13$	$C + D = 13$	13	9	8	7	6
		12	7	6	5	4
		11	6	5	4	3
		10	4	4	3	2
		9	3	3	2	1
		8	2	2	1	0
		7	2	1	0	0
		6	1	0	0	—
		5	0	0	—	—
		4	0	—	—	—
	$C + D = 12$	13	8	7	6	5
		12	6	5	5	4
		11	5	4	3	3
		10	4	3	2	2
		9	3	2	1	1
		8	2	1	1	0
		7	1	1	0	0
		6	1	0	0	—
		5	0	0	—	—

TABLE K—*Continued*

Totals in right margin		B (or A)†	Level of significance			
			.05	.025	.01	.005
$A + B = 13$	$C + D = 11$	13	7	6	5	5
		12	6	5	4	3
		11	4	4	3	2
		10	3	3	2	1
		9	3	2	1	1
		8	2	1	0	0
		7	1	0	0	0
		6	0	0	—	—
		5	0	—	—	—
	$C + D = 10$	13	6	6	5	4
		12	5	4	3	3
		11	4	3	2	2
		10	3	2	1	1
		9	2	1	1	0
		8	1	1	0	0
		7	1	0	0	—
		6	0	0	—	—
		5	0	—	—	—
	$C + D = 9$	13	5	5	4	4
		12	4	4	3	2
		11	3	3	2	1
		10	2	2	1	1
		9	2	1	0	0
		8	1	1	0	0
		7	0	0	—	—
		6	0	0	—	—
		5	0	—	—	—
	$C + D = 8$	13	5	4	3	3
		12	4	3	2	2
		11	3	2	1	1
		10	2	1	1	0
		9	1	1	0	0
		8	1	0	0	—
		7	0	0	—	—
		6	0	—	—	—
	$C + D = 7$	13	4	3	3	2
		12	3	2	2	1
		11	2	2	1	1
		10	1	1	0	0
		9	1	0	0	0
		8	0	0	—	—
		7	0	0	—	—
		6	0	—	—	—

TABLE K—Continued

Totals in right margin		B (or A)†	Level of significance			
			.05	.025	.01	.005
4 + B = 13	C + D = 6	13	3	3	2	2
		12	2	2	1	1
		11	2	1	1	0
		10	1	1	0	0
		9	1	0	0	—
		8	0	0	—	—
		7	0	—	—	—
	C + D = 5	13	2	2	1	1
		12	2	1	1	0
		11	1	1	0	0
		10	1	0	0	—
		9	0	0	—	—
		8	0	—	—	—
	C + D = 4	13	2	1	1	0
		12	1	1	0	0
		11	0	0	0	—
		10	0	0	—	—
		9	0	—	—	—
	C + D = 3	13	1	1	0	0
		12	0	0	0	—
		11	0	0	—	—
		10	0	—	—	—
	C + D = 2	13	0	0	0	—
		12	0	—	—	—
A + B = 14	C + D = 14	14	10	9	8	7
		13	8	7	6	5
		12	6	6	5	4
		11	5	4	3	3
		10	4	3	2	2
		9	3	2	2	1
		8	2	2	1	0
		7	1	1	0	0
		6	1	0	0	—
		5	0	0	—	—
		4	0	—	—	—

TABLE K—*Continued*

Totals in right margin		B (or A)†	Level of significance			
			.05	.025	.01	.005
$A + B = 14$	$C + D = 13$	14	9	8	7	6
		13	7	6	5	5
		12	6	5	4	3
		11	5	4	3	2
		10	4	3	2	2
		9	3	2	1	1
		8	2	1	1	0
		7	1	1	0	0
		6	1	0	—	—
		5	0	0	—	—
	$C + D = 12$	14	8	7	6	6
		13	6	6	5	4
		12	5	4	4	3
		11	4	3	3	2
		10	3	3	2	1
		9	2	2	1	1
		8	2	1	0	0
		7	1	0	0	—
		6	0	0	—	—
		5	0	—	—	—
	$C + D = 11$	14	7	6	6	5
		13	6	5	4	4
		12	5	4	3	3
		11	4	3	2	2
		10	3	2	1	1
		9	2	1	1	0
		8	1	1	0	0
		7	1	0	0	—
		6	0	0	—	—
		5	0	—	—	—
	$C + D = 10$	14	6	6	5	4
		13	5	4	4	3
		12	4	3	3	2
		11	3	3	2	1
		10	2	2	1	1
		9	2	1	0	0
		8	1	1	0	0
		7	0	0	0	—
		6	0	0	—	—
		5	0	—	—	—

TABLE K—*Continued*

Totals in right margin		B (or A)†	Level of significance			
			.05	.025	.01	.005
$A + B = 14$	$C + D = 9$	14	6	5	4	4
		13	4	4	3	3
		12	3	3	2	2
		11	3	2	1	1
		10	2	1	1	0
		9	1	1	0	0
		8	1	0	0	—
		7	0	0	—	—
		6	0	—	—	—
	$C + D = 8$	14	5	4	4	3
		13	4	3	2	2
		12	3	2	2	1
		11	2	2	1	1
		10	2	1	0	0
		9	1	0	0	0
		8	0	0	0	—
		7	0	0	—	—
		6	0	—	—	—
	$C + D = 7$	14	4	3	3	2
		13	3	2	2	1
		12	2	2	1	1
		11	2	1	1	0
		10	1	1	0	0
		9	1	0	0	—
		8	0	0	—	—
		7	0	—	—	—
	$C + D = 6$	14	3	3	2	2
		13	2	2	1	1
		12	2	1	1	0
		11	1	1	0	0
		10	1	0	0	—
		9	0	0	—	—
		8	0	0	—	—
		7	0	—	—	—
	$C + D = 5$	14	2	2	1	1
		13	2	1	1	0
		12	1	1	0	0
		11	1	0	0	0
		10	0	0	—	—
		9	0	0	—	—
		8	0	—	—	—

TABLE K—*Continued*

Totals in right margin		B (or A)†	Level of significance			
			.05	.025	.01	.005
$A + B = 14$	$C + D = 4$	14	2	1	1	1
		13	1	1	0	0
		12	1	0	0	0
		11	0	0	—	—
		10	0	0	—	—
		9	0	—	—	—
	$C + D = 3$	14	1	1	0	0
		13	0	0	0	—
		12	0	0	—	—
		11	0	—	—	—
	$C + D = 2$	14	0	0	0	—
		13	0	0	—	—
		12	0	—	—	—
$A + B = 15$	$C + D = 15$	15	11	10	9	8
		14	9	8	7	6
		13	7	6	5	5
		12	6	5	4	4
		11	5	4	3	3
		10	4	3	2	2
		9	3	2	1	1
		8	2	1	1	0
		7	1	1	0	0
		6	1	0	0	—
		5	0	0	—	—
		4	0	—	—	—
	$C + D = 14$	15	10	9	8	7
		14	8	7	6	6
		13	7	3	5	4
		12	6	5	4	3
		11	5	4	3	2
		10	4	3	2	1
		9	3	2	1	1
		8	2	1	1	0
		7	1	1	0	0
		6	1	0	—	—
		5	0	—	—	—

TABLE K—*Continued*

Totals in right margin		B (or A)†	Level of significance			
			.05	.025	.01	.005
$A + B = 15$	$C + D = 13$	15	9	8	7	7
		14	7	7	6	5
		13	6	5	4	4
		12	5	4	3	3
		11	4	3	2	2
		10	3	2	2	1
		9	2	2	1	0
		8	2	1	0	0
		7	1	0	0	—
		6	0	0	—	—
		5	0	—	—	—
	$C + D = 12$	15	8	7	7	6
		14	7	6	5	4
		13	6	5	4	3
		12	5	4	3	2
		11	4	3	2	2
		10	3	2	1	1
		9	2	1	1	0
		8	1	1	0	0
		7	1	0	0	—
		6	0	0	—	—
		5	0	—	—	—
	$C + D = 11$	15	7	7	6	5
		14	6	5	4	4
		13	5	4	3	3
		12	4	3	2	2
		11	3	2	2	1
		10	2	2	1	1
		9	2	1	0	0
		8	1	1	0	0
		7	1	0	0	—
		6	0	0	—	—
		5	0	—	—	—
	$C + D = 10$	15	6	6	5	5
		14	5	5	4	3
		13	4	4	3	2
		12	3	3	2	2
		11	3	2	1	1
		10	2	1	1	0
		9	1	1	0	0
		8	1	0	0	—
		7	0	0	—	—
		6	0	—	—	—

TABLE K—*Continued*

Totals in right margin		B (or A)†	Level of significance			
			.05	.025	.01	.005
$A + B = 15$	$C + D = 9$	15	6	5	4	4
		14	5	4	3	3
		13	4	3	2	2
		12	3	2	2	1
		11	2	2	1	1
		10	2	1	0	0
		9	1	1	0	0
		8	1	0	0	—
		7	0	0	—	—
		6	0	—	—	—
	$C + D = 8$	15	5	4	4	3
		14	4	3	3	2
		13	3	2	2	1
		12	2	2	1	1
		11	2	1	1	0
		10	1	1	0	0
		9	1	0	0	—
		8	0	0	—	—
		7	0	—	—	—
		6	0	—	—	—
	$C + D = 7$	15	4	4	3	3
		14	3	3	2	2
		13	2	2	1	1
		12	2	1	1	0
		11	1	1	0	0
		10	1	0	0	0
		9	0	0	—	—
		8	0	0	—	—
		7	0	—	—	—
	$C + D = 6$	15	3	3	2	2
		14	2	2	1	1
		13	2	1	1	0
		12	1	1	0	0
		11	1	0	0	0
		10	0	0	0	—
		9	0	0	—	—
		8	0	—	—	—
	$C + D = 5$	15	2	2	2	1
		14	2	1	1	1
		13	1	1	0	0
		12	1	0	0	0
		11	0	0	0	—
		10	0	0	—	—
		9	0	—	—	—

TABLE K—*Concluded*

Totals in right margin	B (or A)†	Level of significance			
		.05	.025	.01	.005
$A + B = 15$ $C + D = 4$	15	2	1	1	1
	14	1	1	0	0
	13	1	0	0	0
	12	0	0	0	—
	11	0	0	—	—
	10	0	—	—	—
$C + D = 3$	15	1	1	0	0
	14	0	0	0	0
	13	0	0	—	—
	12	0	0	—	—
	11	0	—	—	—
$C + D = 2$	15	0	0	0	—
	14	0	0	—	—
	13	0	—	—	—

TABLE L
Probabilities associated with values as small as observed values of U in the Mann-Whitney test (small samples)

$N_2 = 3$

U \ N₁	1	2 ·	3
0	.250	.100	.050
1	.500	.200	.100
2	.750	.400	.200
3		.600	.350
4			.500
5			.650

$N_2 = 4$

U \ N₁	1	2	3	4
0	.200	.067	.028	.014
1	.400	.133	.057	.029
2	.600	.267	.114	.057
3		.400	.200	.100
4		.600	.314	.171
5			.429	.243
6			.571	.343
7				.443
8				.557

$N_2 = 5$

U \ N₁	1	2	3	4	5
0	.167	.047	.018	.008	.004
1	.333	.095	.036	.016	.008
2	.500	.190	.071	.032	.016
3	.667	.286	.125	.056	.028
4		.429	.196	.095	.048
5		.571	.286	.143	.075
6			.393	.206	.111
7			.500	.278	.155
8			.607	.365	.210
9				.452	.274
10				.548	.345
11					.421
12					.500
13					.579

$N_2 = 6$

U \ N₁	1	2	3	4	5	6
0	.143	.036	.012	.005	.002	.001
1	.286	.071	.024	.010	.004	.002
2	.428	.143	.048	.019	.009	.004
3	.571	.214	.083	.033	.015	.008
4		.321	.131	.057	.026	.013
5		.429	.190	.086	.041	.021
6		.571	.274	.129	.063	.032
7			.357	.176	.089	.047
8			.452	.238	.123	.066
9			.548	.305	.165	.090
10				.381	.214	.120
11				.457	.268	.155
12				.545	.331	.197
13					.396	.242
14					.465	.294
15					.535	.350
16						.409
17						.469
18						.531

Source: H. B. Mann and D. R. Whitney, "On a Test of Whether One of Two Random Variables is Stochastically Larger than the Other," *Annals of Mathematical Statistics* 18 (1947): 52–54; with permission of the authors and the publisher.

TABLE L—*Continued*

$$N_2 = 7$$

U \ N₁	1	2	3	4	5	6	7
0	.125	.028	.008	.003	.001	.001	.000
1	.250	.056	.017	.006	.003	.001	.001
2	.375	.111	.033	.012	.005	.002	.001
3	.500	.167	.058	.021	.009	.004	.002
4	.625	.250	.092	.036	.015	.007	.003
5		.333	.133	.055	.024	.011	.006
6		.444	.192	.082	.037	.017	.009
7		.556	.258	.115	.053	.026	.013
8			.333	.158	.074	.037	.019
9			.417	.206	.101	.051	.027
10			.500	.264	.134	.069	.036
11			.583	.324	.172	.090	.049
12				.394	.216	.117	.064
13				.464	.265	.147	.082
14				.538	.319	.183	.104
15					.378	.223	.130
16					.438	.267	.159
17					.500	.314	.191
18					.562	.365	.228
19						.418	.267
20						.473	.310
21						.527	.355
22							.402
23							.451
24							.500
25							.549

TABLE L—*Concluded*

$N_2 = 8$

U \ N_1	1	2	3	4	5	6	7	8	t	Normal
0	.111	.022	.006	.002	.001	.000	.000	.000	3.308	.001
1	.222	.044	.012	.004	.002	.001	.000	.000	3.203	.001
2	.333	.089	.024	.008	.003	.001	.001	.000	3.098	.001
3	.444	.133	.042	.014	.005	.002	.001	.001	2.993	.001
4	.556	.200	.067	.024	.009	.004	.002	.001	2.888	.002
5		.267	.097	.036	.015	.006	.003	.001	2.783	.003
6		.356	.139	.055	.023	.010	.005	.002	2.678	.004
7		.444	.188	.077	.033	.015	.007	.003	2.573	.005
8		.556	.248	.107	.047	.021	.010	.005	2.468	.007
9			.315	.141	.064	.030	.014	.007	2.363	.009
10			.387	.184	.085	.041	.020	.010	2.258	.012
11			.461	.230	.111	.054	.027	.014	2.153	.016
12			.539	.285	.142	.071	.036	.019	2.048	.020
13				.341	.177	.091	.047	.025	1.943	.026
14				.404	.217	.114	.060	.032	1.838	.033
15				.467	.262	.141	.076	.041	1.733	.041
16				.533	.311	.172	.095	.052	1.628	.052
17					.362	.207	.116	.065	1.523	.064
18					.416	.245	.140	.080	1.418	.078
19					.472	.286	.168	.097	1.313	.094
20					.528	.331	.198	.117	1.208	.113
21						.377	.232	.139	1.102	.135
22						.426	.268	.164	.998	.159
23						.475	.306	.191	.893	.185
24						.525	.347	.221	.788	.215
25							.389	.253	.683	.247
26							.433	.287	.578	.282
27							.478	.323	.473	.318
28							.522	.360	.368	.356
29								.399	.263	.396
30								.439	.158	.437
31								.480	.052	.481
32								.520		

TABLE M

Critical values of U in the Mann-Whitney test (critical values of U for a one-tailed test at $\alpha = .025$ or for a two-tailed test at $\alpha = .05$)

N_1 \ N_2	9	10	11	12	13	14	15	16	17	18	19	20
1												
2	0	0	0	1	1	1	1	1	2	2	2	2
3	2	3	3	4	4	5	5	6	6	7	7	8
4	4	5	6	7	8	9	10	11	11	12	13	13
5	7	8	9	11	12	13	14	15	17	18	19	20
6	10	11	13	14	16	17	19	21	22	24	25	27
7	12	14	16	18	20	22	24	26	28	30	32	34
8	15	17	19	22	24	26	29	31	34	36	38	41
9	17	20	23	26	28	31	34	37	39	42	45	48
10	20	23	26	29	33	36	39	42	45	48	52	55
11	23	26	30	33	37	40	44	47	51	55	58	62
12	26	29	33	37	41	45	49	53	57	61	65	69
13	28	33	37	41	45	50	54	59	63	67	72	76
14	31	36	40	45	50	55	59	64	67	74	78	83
15	34	39	44	49	54	59	64	70	75	80	85	90
16	37	42	47	53	59	64	70	75	81	86	92	98
17	39	45	51	57	63	67	75	81	87	93	99	105
18	42	48	55	61	67	74	80	86	93	99	106	112
19	45	52	58	65	72	78	85	92	99	106	113	119
20	48	55	62	69	76	83	90	98	105	112	119	127

Source: D. Auble, "Extended Tables for the Mann-Whitney Statistic," *Bulletin of the Institute of Educational Research at Indiana University* 1, no. 2 1953). Adapted and abridged from tables 1, 3, 5, and 7 D. Auble, with permission of the author and publishers.

TABLE M–1
Critical values of U for a one-tailed test at $\alpha = .05$ or for a two-tailed test at $\alpha = .10$

N_1 \ N_2	9	10	11	12	13	14	15	16	17	18	19	20
1											0	0
2	1	1	1	2	2	2	3	3	3	4	4	4
3	3	4	5	5	6	7	7	8	9	9	10	11
4	6	7	8	9	10	11	12	14	15	16	17	18
5	9	11	12	13	15	16	18	19	20	22	23	25
6	12	14	16	17	19	21	23	25	26	28	30	32
7	15	17	19	21	24	26	28	30	33	35	37	39
8	18	20	23	26	28	31	33	36	39	41	44	47
9	21	24	27	30	33	36	39	42	45	48	51	54
10	24	27	31	34	37	41	44	48	51	55	58	62
11	27	31	34	38	42	46	50	54	57	61	65	69
12	30	34	38	42	47	51	55	60	64	68	72	77
13	33	37	42	47	51	56	61	65	70	75	80	84
14	36	41	46	51	56	61	66	71	77	82	87	92
15	39	44	50	55	61	66	72	77	83	88	94	100
16	42	48	54	60	65	71	77	83	89	95	101	107
17	45	51	57	64	70	77	83	89	96	102	109	115
18	48	55	61	68	75	82	88	95	102	109	116	123
19	51	58	65	72	80	87	94	101	109	116	123	130
20	54	62	69	77	84	92	100	107	115	123	130	138

TABLE M–2
Critical values of U for a one-tailed test at $\alpha = .01$ or for a two-tailed test at $\alpha = .02$

N_1 \\ N_2	9	10	11	12	13	14	15	16	17	18	19	20
1												
2					0	0	0	0	0	0	1	1
3	1	1	1	2	2	2	3	3	4	4	4	5
4	3	3	4	5	5	6	7	7	8	9	9	10
5	5	6	7	8	9	10	11	12	13	14	15	16
6	7	8	9	11	12	13	15	16	18	19	20	22
7	9	11	12	14	16	17	19	21	23	24	26	28
8	11	13	15	17	20	22	24	26	28	30	32	34
9	14	16	18	21	23	26	28	31	33	36	38	40
10	16	19	22	24	27	30	33	36	38	41	44	47
11	18	22	25	28	31	34	37	41	44	47	50	53
12	21	24	28	31	35	38	42	46	49	53	56	60
13	23	27	31	35	39	43	47	51	55	59	63	67
14	26	30	34	38	43	47	51	56	60	65	69	73
15	28	33	37	42	47	51	56	61	66	70	75	80
16	31	36	41	46	51	56	61	66	71	76	82	87
17	33	38	44	49	55	60	66	71	77	82	88	93
18	36	41	47	53	59	65	70	76	82	88	94	100
19	38	44	50	56	63	69	75	82	88	94	101	107
20	40	47	53	60	67	73	80	87	93	100	107	114

TABLE M–3
Critical values of U for a one-tailed test at $\alpha = .001$ or for a two-tailed test at $\alpha = .002$

N_1 \ N_2	9	10	11	12	13	14	15	16	17	18	19	20
1												
2												
3									0	0	0	0
4		0	0	0	1	1	1	2	2	3	3	3
5	1	1	2	2	3	3	4	5	5	6	7	7
6	2	3	4	4	5	6	7	8	9	10	11	12
7	3	5	6	7	8	9	10	11	13	14	15	16
8	5	6	8	9	11	12	14	15	17	18	20	21
9	7	8	10	12	14	15	17	19	21	23	25	26
10	8	10	12	14	17	19	21	23	25	27	29	32
11	10	12	15	17	20	22	24	27	29	32	34	37
12	12	14	17	20	23	25	28	31	34	37	40	42
13	14	17	20	23	26	29	32	35	38	42	45	48
14	15	19	22	25	29	32	36	39	43	46	50	54
15	17	21	24	28	32	36	40	43	47	51	55	59
16	19	23	27	31	35	39	43	48	52	56	60	65
17	21	25	29	34	38	43	47	52	57	61	66	70
18	23	27	32	37	42	46	51	56	61	66	71	76
19	25	29	34	40	45	50	55	60	66	71	77	82
20	26	32	37	42	48	54	59	65	70	76	82	88

TABLE N
Table of random numbers

	1	2	3	4	5	6	7	8	9	10	11	12	13	14	15	16	17	18	19	20
1	03	47	43	73	86	36	96	47	36	61	46	98	63	71	62	33	26	16	80	45
2	97	74	24	67	62	42	81	14	57	20	42	53	32	37	32	27	07	36	07	51
3	16	76	62	27	66	56	50	26	71	07	32	90	79	78	53	13	55	38	58	59
4	12	56	85	99	26	96	96	68	27,	31	05	03	72	93	15	57	12	10	14	21
5	55	59	56	35	64	38	54	82	46	21	31	62	43	90	90	06	18	44	32	53
6	16	22	77	94	39	49	54	43	54	82	17	37	93	23	78	87	35	20	96	43
7	84	42	17	53	31	57	24	55	06	88	77	04	74	47	67	21	76	33	50	25
8	63	01	63	78	59	16	95	55	57	19	98	10	50	71	75	12	86	73	58	07
9	33	21	12	34	29	78	64	56	07	82	52	42	07	44	38	15	51	00	13	42
10	57	60	86	32	44	09	47	27	96	54	49	17	46	09	62	90	52	84	77	27
11	18	18	07	92	46	44	17	16	58	09	79	83	86	19	62	06	76	50	03	10
12	26	62	38	97	75	84	16	07	44	99	83	11	46	32	24	20	14	85	88	45
13	23	42	40	64	74	82	97	77	77	81	07	45	32	14	08	32	98	94	07	72
14	52	36	28	19	95	50	92	26	11	97	00	56	76	31	38	80	22	02	53	53
15	37	85	94	35	12	83	39	50	08	30	42	34	07	96	88	54	42	06	87	98
16	70	29	17	12	13	40	33	20	38	26	13	89	51	03	74	17	76	37	13	04
17	56	62	18	37	35	96	83	50	87	75	97	12	25	93	47	70	33	24	03	54
18	99	49	57	22	77	88	42	95	45	72	16	64	36	16	00	04	43	18	66	79
19	16	08	15	04	72	33	27	14	34	09	45	59	34	68	49	12	72	07	34	45
20	31	16	93	32	43	50	27	89	87	19	20	15	37	00	49	52	85	66	60	44
21	68	34	30	13	70	55	74	30	77	40	44	22	78	84	26	04	33	36	09	52
22	74	57	25	65	76	59	29	97	68	60	71	91	38	67	54	13	58	18	25	27
23	27	42	37	86	53	48	55	90	65	72	96	57	69	36	10	96	46	92	42	45
24	00	39	68	29	61	66	37	32	20	30	77	84	57	03	29	10	45	65	04	26
25	29	94	98	94	24	68	49	69	10	82	53	75	91	93	30	34	25	20	57	27
26	16	90	82	66	59	83	62	64	11	12	67	19	00	71	74	60	47	21	29	68
27	11	27	94	75	06	06	09	19	74	66	02	94	37	34	02	76	70	90	30	86
28	35	24	10	16	20	33	32	51	26	38	79	78	45	04	91	16	92	53	56	16
29	38	23	16	86	38	42	38	97	01	50	87	75	66	81	41	40	01	74	91	62
30	31	96	25	91	47	96	44	33	49	13	34	96	82	53	91	00	52	43	48	85
31	66	67	40	67	14	64	05	71	95	86	11	05	65	09	68	76	83	20	37	90
32	14	90	84	45	11	75	73	88	05	90	52	27	41	14	86	22	98	12	22	08
33	68	05	51	18	00	33	96	02	74	19	07	60	62	93	55	59	33	82	43	90
34	20	46	78	73	90	97	51	40	14	02	04	02	33	31	08	39	54	16	49	36
35	64	19	58	97	79	15	06	15	93	20	01	90	10	75	06	40	78	78	89	62
36	05	26	93	70	60	22	35	85	15	13	92	03	51	59	77	59	56	78	06	83
37	07	97	10	88	23	09	98	42	99	64	61	71	62	99	06	51	29	16	93	15
38	68	71	86	85	85	54	87	66	47	54	73	32	98	11	12	44	95	92	63	16
39	14	65	52	68	74	87	37	78	22	41	26	78	63	06	55	13	08	27	01	50
40	17	53	77	58	71	71	59	36	50	72	12	41	94	96	26	44	95	27	36	99
41	90	26	59	21	19	23	41	61	33	12	96	93	02	18	39	07	02	18	36	07
42	41	23	52	55	99	31	52	23	69	96	10	47	48	45	88	13	41	43	89	20
43	26	99	61	65	53	58	04	49	80	70	42	10	50	67	42	32	17	55	85	74

Source: Table N is taken from Table 33 of Fisher and Yates: *Statistical Tables for Biological, Agricultural and Medical Research*, published by Oliver and Boyd, Edinburgh, and by permission of the authors and publishers.

TABLE N—*Continued*

	1	2	3	4	5	6	7	8	9	10	11	12	13	14	15	16	17	18	19	20
1	53	74	23	99	67	61	32	28	69	84	94	62	67	86	24	98	33	41	19	95
2	63	38	06	86	54	99	00	65	26	94	02	82	90	23	07	79	62	67	80	60
3	35	30	68	21	46	06	72	17	10	94	25	21	31	74	96	49	28	24	00	49
4	63	43	36	92	69	65	51	18	37	88	61	38	44	12	45	32	92	84	88	65
5	98	25	37	55	26	01	91	82	81	46	74	71	12	94	97	24	02	71	37	07
6	02	63	31	17	69	71	50	80	39	56	38	15	40	11	48	43	40	45	06	98
7	64	55	22	21	82	48	22	28	06	00	61	64	13	54	91	82	78	12	23	29
8	85	07	26	13	89	01	10	07	82	04	59	63	69	36	03	69	11	15	83	80
9	58	54	16	24	15	51	54	44	82	00	62	61	65	04	69	38	18	65	18	97
10	34	85	27	84	87	61	48	64	56	26	90	18	48	13	26	37	70	15	42	57
11	03	92	18	27	46	57	99	16	96	56	30	33	72	85	22	84	64	38	56	98
12	95	30	27	59	37	62	75	41	66	48	86	97	80	61	45	23	53	04	01	63
13	08	45	93	15	22	60	21	75	46	91	98	77	27	85	42	28	88	61	08	84
14	07	08	55	18	40	45	44	74	13	90	24	94	96	61	02	57	55	66	83	15
15	01	85	89	95	66	51	10	19	34	88	15	84	97	19	75	12	76	39	43	78
16	72	84	71	14	35	19	11	58	49	26	50	11	17	17	76	86	31	57	20	18
17	88	78	28	16	84	13	52	53	94	53	75	45	69	30	96	73	89	65	70	31
18	45	17	75	65	57	28	40	19	72	12	25	12	74	75	67	60	40	60	81	19
19	96	76	28	12	54	22	01	11	94	25	71	96	16	16	88	68	64	36	74	45
20	43	31	67	72	30	24	02	94	08	63	38	32	36	66	02	69	36	38	25	39
21	50	44	66	44	21	66	06	58	04	62	68	15	54	35	02	42	36	48	96	32
22	22	66	22	15	86	26	63	74	41	99	58	42	36	62	24	58	37	52	18	51
23	96	24	40	14	51	23	22	30	88	57	95	67	47	29	83	94	69	40	06	07
24	31	73	91	61	19	60	20	72	93	48	98	57	07	23	69	65	95	39	69	58
25	78	60	73	99	84	43	89	94	36	34	56	69	47	07	41	90	22	91	07	12
26	84	37	90	61	56	70	10	23	98	05	85	11	34	76	60	76	48	45	34	60
27	36	67	10	08	23	98	93	35	08	86	99	29	76	29	81	33	34	91	58	93
28	07	28	59	07	48	89	64	58	89	75	83	85	62	27	89	30	14	78	56	27
29	10	15	83	87	60	79	24	31	66	56	21	48	24	06	93	91	98	94	05	49
30	55	19	68	97	65	03	73	52	16	56	00	53	55	90	27	33	42	29	38	87
31	53	81	29	13	39	35	01	20	71	34	62	33	74	82	14	53	73	19	09	03
32	51	86	32	68	92	33	98	74	66	99	40	14	71	94	58	35	94	19	38	81
33	35	91	70	29	13	80	03	54	07	27	96	94	78	32	66	50	95	52	74	33
34	37	71	67	95	13	20	02	77	95	94	64	85	04	05	72	01	32	90	76	14
35	93	66	13	83	27	92	79	64	64	72	28	54	96	53	84	48	14	52	98	94
36	02	96	08	45	64	13	05	00	41	84	93	07	54	72	59	21	45	57	09	77
37	49	83	43	48	36	92	88	33	69	96	72	36	04	19	76	47	45	15	18	60
38	84	60	71	62	46	40	80	81	30	37	34	39	23	04	38	25	15	35	71	30
39	18	17	30	88	71	44	91	14	88	47	89	23	30	63	15	56	34	20	47	89
40	79	69	10	61	78	71	32	76	95	62	87	00	22	58	40	92	54	01	75	25
41	75	93	36	47	83	56	20	14	82	11	74	21	97	90	65	96	42	68	63	96
42	38	30	92	29	03	06	28	81	39	38	62	25	06	84	63	61	29	08	93	67
43	51	29	50	10	34	31	57	75	95	80	51	97	02	74	77	76	15	58	49	44
44	21	31	38	86	24	37	79	81	53	74	73	24	16	10	33	52	83	90	94	76
45	29	01	23	87	88	58	02	39	37	67	42	10	14	20	92	16	55	23	42	45
46	95	33	95	22	00	18	74	92	00	18	38	79	58	69	32	81	76	80	26	92
47	90	84	60	79	80	24	36	59	87	38	82	07	53	89	35	96	35	23	79	18
48	46	40	62	98	82	54	97	20	45	95	15	74	80	08	32	16	46	70	50	80
49	20	31	89	03	43	38	36	92	68	72	32	14	82	99	70	80	60	47	18	97
50	71	59	73	05	50	08	22	23	71	77	91	01	93	20	49	82	96	59	26	94

TABLE N—*Concluded*

	1	2	3	4	5	6	7	8	9	10	11	12	13	14	15	16	17	18	19	20
1	22	17	68	65	84	68	94	23	92	35	86	02	22	57	51	61	09	43	95	06
2	19	36	27	69	46	13	79	93	37	55	39	77	32	77	09	85	52	05	30	62
3	16	77	23	02	77	09	61	87	25	21	28	06	25	24	93	16	71	13	59	78
4	03	28	28	26	08	73	37	32	04	05	69	30	16	90	05	88	69	58	29	99
5	78	43	76	71	61	20	44	90	32	64	97	67	63	99	61	46	38	03	93	22
6	93	22	53	64	39	07	10	63	76	35	87	03	04	79	88	08	13	13	85	51
7	78	76	58	54	74	92	38	70	96	92	52	06	79	79	45	82	63	18	27	44
8	23	68	35	26	00	99	53	93	61	28	52	70	05	48	34	56	64	04	61	86
9	15	39	24	70	99	93	86	52	77	64	15	33	59	05	28	22	87	26	07	47
10	58	71	96	30	24	18	46	23	34	27	85	13	99	24	44	49	18	09	79	49
11	57	35	27	33	72	24	53	63	94	09	41	10	76	47	91	44	04	95	49	66
12	48	50	86	54	48	22	06	34	72	52	82	21	15	65	20	33	29	94	71	11
13	61	96	48	95	03	07	16	39	33	66	98	56	10	56	79	77	21	30	27	12
14	36	93	89	41	26	29	70	83	63	51	99	74	20	52	36	87	09	41	15	09
15	18	87	00	42	31	57	90	12	02	07	23	47	37	17	31	54	08	01	88	63
16	88	56	53	27	59	33	35	72	67	47	77	34	55	45	70	08	18	27	38	90
17	09	72	95	84	29	49	41	31	06	70	42	38	06	45	18	64	84	73	31	65
18	12	96	88	17	31	65	19	69	02	83	60	74	86	90	68	24	64	19	35	51
19	85	94	57	24	16	92	09	94	38	76	22	00	27	69	95	29	81	94	78	70
20	38	64	43	59	98	98	77	87	68	07	91	51	78	62	44	40	98	05	93	78
21	53	44	09	42	72	00	41	86	79	79	68	47	22	00	20	35	55	31	51	51
22	40	76	66	26	84	57	99	99	90	37	36	63	32	08	58	37	40	13	68	97
23	02	17	79	18	05	12	59	52	57	02	22	07	90	47	03	28	14	11	30	79
24	95	17	82	06	53	31	51	10	96	46	92	06	88	07	77	56	11	50	81	69
25	35	76	22	42	92	96	11	83	44	80	34	68	35	48	77	33	42	40	90	60
26	26	29	13	46	41	85	47	04	66	08	34	72	47	59	13	82	43	80	46	15
27	77	80	20	75	82	72	82	32	99	90	63	95	73	76	63	89	73	44	99	05
28	46	40	66	44	52	91	36	74	43	53	30	82	13	53	00	78	45	63	98	35
29	37	56	08	18	90	77	53	85	46	47	31	91	18	95	59	24	16	74	11	53
30	61	65	61	68	66	37	27	47	39	19	84	83	70	07	38	53	21	40	06	71
31	93	43	69	96	07	34	18	04	52	35	56	27	09	24	86	61	85	53	83	45
32	21	96	60	12	99	11	20	99	45	18	48	13	93	55	34	18	37	79	49	90
33	95	20	47	97	97	27	37	83	28	71	00	06	41	41	74	45	89	09	39	84
34	97	86	21	78	73	10	64	81	92	59	58	76	17	14	97	04	76	62	16	17
35	69	92	06	34	13	59	71	74	17	32	27	55	10	24	19	23	71	82	13	74
36	04	31	17	21	56	33	73	99	19	87	26	72	39	27	67	53	77	57	68	93
37	61	06	98	03	91	87	14	77	43	96	43	00	65	98	50	45	60	33	01	07
38	85	93	85	86	88	72	87	08	62	40	16	06	10	89	20	23	21	34	74	97
39	21	74	32	47	45	73	96	07	94	52	09	65	90	77	47	25	76	16	19	33
40	15	69	53	92	80	79	96	23	53	10	64	39	07	16	29	45	33	02	43	70
41	02	89	08	04	49	20	21	14	68	86	87	63	93	95	17	11	29	01	95	80
42	87	18	15	89	79	85	43	01	72	73	08	61	74	51	69	89	74	39	82	15
43	98	83	71	94	22	59	97	50	99	52	08	52	85	08	40	87	80	61	65	31
44	10	08	58	21	66	72	68	49	29	31	89	85	84	46	06	59	73	19	85	23
45	47	90	56	10	08	88	02	84	27	83	42	29	72	23	19	66	56	45	65	79
46	22	85	61	68	80	49	64	92	85	44	16	40	12	89	88	50	14	49	81	06
47	67	80	43	79	33	12	83	11	41	16	25	58	19	36	70	77	02	43	00	52
48	27	62	40	96	72	79	44	61	40	15	14	53	40	64	39	27	31	59	50	28
49	33	78	80	87	15	38	30	06	38	21	14	47	47	07	26	54	96	87	53	32
50	13	13	92	66	99	47	24	49	57	74	32	25	43	62	17	10	97	11	69	84

TABLE O

Critical values of T in the Wilcoxon matched-pairs signed-ranks test

N	Level of significance for one-tailed test		
	.025	.01	.005
	Level of significance for two-tailed test		
	.05	.02	.01
6	0	—	—
7	2	0	—
8	4	2	0
9	6	3	2
10	8	5	3
11	11	7	5
12	14	10	7
13	17	13	10
14	21	16	13
15	25	20	16
16	30	24	20
17	35	28	23
18	40	33	28
19	46	38	32
20	52	43	38
21	59	49	43
22	66	56	49
23	73	62	55
24	81	69	61
25	89	77	68

Source: F. Wilcoxon and R. A. Wilcox, *Some Rapid Approximate Statistical Procedures* (New York: Lederle Labs, 1964), adapted from table 2; with permission of the authors and publisher.

TABLE P

Probabilities associated with values as large as observed values of χ_r^2 in the Friedman two-way analysis of variance by ranks ($k = 3$)

$N = 2$		$N = 3$		$N = 4$		$N = 5$	
χ_r^2	p	χ_r^2	p	χ_r^2	p	χ_r^2	p
0	1.000	.000	1.000	.0	1.000	.0	1.000
1	.833	.667	.944	.5	.931	.4	.954
3	.500	2.000	.528	1.5	.653	1.2	.691
4	.167	2.667	.361	2.0	.431	1.6	.522
		4.667	.194	3.5	.273	2.8	.367
		6.000	.028	4.5	.125	3.6	.182
				6.0	.069	4.8	.124
				6.5	.042	5.2	.093
				8.0	.0046	6.4	.039
						7.6	.024
						8.4	.0085
						10.0	.00077

$N = 6$		$N = 7$		$N = 8$		$N = 9$	
χ_r^2	p	χ_r^2	p	χ_r^2	p	χ_r^2	p
.00	1.000	.000	1.000	.00	1.000	.000	1.000
.33	.956	.286	.964	.25	.967	.222	.971
1.00	.740	.857	.768	.75	.794	.667	.814
1.33	.570	1.143	.620	1.00	.654	.889	.865
2.33	.430	2.000	.486	1.75	.531	1.556	.569
3.00	.252	2.571	.305	2.25	.355	2.000	.398
4.00	.184	3.429	.237	3.00	.285	2.667	.328
4.33	.142	3.714	.192	3.25	.236	2.889	.278
5.33	.072	4.571	.112	4.00	.149	3.556	.187
6.33	.052	5.429	.085	4.75	.120	4.222	.154
7.00	.029	6.000	.052	5.25	.079	4.667	.107
8.33	.012	7.143	.027	6.25	.047	5.556	.069
9.00	.0081	7.714	.021	6.75	.038	6.000	.057
9.33	.0055	8.000	.016	7.00	.030	6.222	.048
10.33	.0017	8.857	.0084	7.75	.018	6.889	.031
12.00	.00013	10.286	.0036	9.00	.0099	8.000	.019
		10.571	.0027	9.25	.0080	8.222	.016
		11.143	.0012	9.75	.0048	8.667	.010
		12.286	.00032	10.75	.0024	9.556	.0060
		14.000	.000021	12.00	.0011	10.667	.0035
				12.25	.00086	10.889	.0029
				13.00	.00026	11.556	.0013
				14.25	.000061	12.667	.00066
				16.00	.0000036	13.556	.00035
						14.000	.00020
						14.222	.000097
						14.889	.000054
						16.222	.000011
						18.000	.0000006

Source: M. Friedman, "The Use of Ranks to Avoid the Assumption of Normality Implicit in the Analysis of Variance," *Journal of American Statistical Association* 32 (1937): 688–89; with permission of the author and the publisher.

TABLE P—(*Continued*)
Probabilities associated with values as large as observed values of $\chi_r{}^2$ in the Friedman two-way analysis of varianue by ranks* ($k = 4$)

N = 2		N = 3		N = 4			
$\chi_r{}^2$	p	$\chi_r{}^2$	p	$\chi_r{}^2$	p	$\chi_r{}^2$	p
.0	1.000	.2	1.000	.0	1.000	5.7	.141
.6	.958	.6	.958	.3	.992	6.0	.105
1.2	.834	1.0	.910	.6	.928	6.3	.094
1.8	.792	1.8	.727	.9	.900	6.6	.077
2.4	.625	2.2	.608	1.2	.800	6.9	.068
3.0	.542	2.6	.524	1.5	.754	7.2	.054
3.6	.458	3.4	.446	1.8	.677	7.5	.052
4.2	.375	3.8	.342	2.1	.649	7.8	.036
4.8	.208	4.2	.300	2.4	.524	8.1	.033
5.4	.167	5.0	.207	2.7	.508	8.4	.019
6.0	.042	5.4	.175	3.0	.432	8.7	.014
		5.8	.148	3.3	.389	9.3	.012
		6.6	.075	3.6	.355	9.6	.0069
		7.0	.054	3.9	.324	9.9	.0062
		7.4	.033	4.5	.242	10.2	.0027
		8.2	.017	4.8	.200	10.8	.0016
		9.0	.0017	5.1	.190	11.1	.00094
				5.4	.158	12.0	.000072

appendix: part B
List of computing formulas

The formula number is used throughout the text to refer to the formula. The page number indicates where the formula is first mentioned in the text.

Name and Formula	Formula Number	Page

Percentiles

$$\%\text{tile} = \left[\frac{(N)(\%) - c.f.}{f}\right](i) + l.r.l.$$

2–1   31

Percentile ranks

$$\%\text{tile rank} = \left\{\frac{\left[\dfrac{(X - l.r.l.)}{i}\right](f) + c.f.}{N}\right\}(100)$$

2–2   35

Mean (whole score)

$$\bar{X} = \frac{\Sigma X}{N}$$

4–1   63

Mean (recurring scores)

$$\bar{X} = \frac{\Sigma f X}{N}$$

4–2   65

Mean (grouped data)

$$\bar{X} = \frac{\Sigma f \text{Mid}}{N}$$

4–3   66

Median (50th percentile)

$$\text{Median} = \left[\frac{(N/2 - c.f.)}{f}\right](i) + l.r.l.$$

4–4   70

Standard deviation (deviation score)

$$S = \sqrt{\frac{\Sigma x^2}{N}}$$

5–1   85

Name and Formula	Formula Number	Page

Standard deviation (whole score)

$$S = \sqrt{\frac{\Sigma X^2}{N} - \bar{X}^2}$$

5–2   87

Standard deviation (whole score–calculator)

$$S = 1/N \sqrt{N\Sigma X^2 - (\Sigma X)^2}$$

5–3   88

Standard deviation (grouped data)

$$S = \sqrt{\frac{\Sigma f(\text{Midpoint})^2}{N} - \left[\frac{\Sigma f(\text{Midpoint})}{N}\right]^2}$$

5–4   88

Standard deviation (grouped data–calculator)

$$S = 1/N \sqrt{N\Sigma f(\text{Mid})^2 - [\Sigma f(\text{Mid})]^2}$$

5–5   89

Coefficients of the binomial expansion

$$\binom{n}{x} = \frac{n!}{x!(n-x)!}$$

6–1   123

Mean event of the binomial expansion

$$\mu = np$$

6–2   127

Standard deviation of the binomial expansion

$$\sigma = \sqrt{npq}$$

6–3   127

Standard deviation (estimate of σ)

$$s = \sqrt{\frac{\Sigma x^2}{N-1}}$$

7–2   141

Standard error of mean (estimate of $\sigma_{\bar{x}}$)

$$s_{\bar{x}} = \frac{s}{\sqrt{N}} = \frac{S}{\sqrt{N-1}}$$

7–3   141

Standard error of mean (deviation score)

$$s_{\bar{x}} = \sqrt{\frac{\Sigma x^2}{N(N-1)}}$$

7–4   141

Name and Formula	Formula Number	Page

Standard error of a proportion

$$\sigma_{prop} = \sqrt{\frac{PQ}{N}}$$

7–5   146

Regression equation

$$Y' = r_{xy}\frac{S_y}{S_x}X + \left(\bar{Y} - r_{xy}\frac{S_y}{S_x}\bar{X}\right)$$

8–1   162

Standard error of estimate

$$S_{y \cdot x} = S_y \sqrt{1 - r_{xy}^2}$$

8–2   164

Standard error of the correlation coefficient

$$S_r = \frac{1}{\sqrt{N-1}}$$

8–3   165

Pearson product-moment correlation
(deviation score)

$$r_{xy} = \frac{\Sigma xy}{NS_xS_y}$$

8–4   169

Pearson product-moment correlation
(whole score)

$$r_{xy} = \frac{N\Sigma XY - \Sigma X \Sigma Y}{\sqrt{N\Sigma X^2 - (\Sigma X)^2}\sqrt{N\Sigma Y^2 - (\Sigma Y)^2}}$$

8–5   170

Spearman rank-difference correlation

$$r_S = 1 - \frac{6\Sigma D^2}{N(N^2 - 1)}$$

8–6   174

Contigency coefficient

$$C = \sqrt{\frac{\chi^2}{N + \chi^2}}$$

8–7   177

Maximal contigency coefficient

$$C_{\max} = \sqrt{\frac{k-1}{k}}$$

8–8   178

	Formula	
Name and Formula	*Number*	*Page*

Correction for length of test (split-half reliability)

$$r_{xx} = \frac{2r_{12}}{1 + r_{12}}$$

9–1   189

Standard error of measurement

$$S_e = S_x \sqrt{1 - r_{xx}}$$

9–2   189

Standard error of difference in means

$$s_{D\bar{x}} = \sqrt{s_{\bar{x}_1}^2 + s_{\bar{x}_2}^2} = \sqrt{\frac{s_1^2}{N_1} + \frac{s_2^2}{N_2}}$$

10–1   200

Standard deviation (combined from two samples)

$$s = \sqrt{\frac{\Sigma x_1^2 + \Sigma x_2^2}{(N-1) + (N_2 - 1)}}$$

10–2   201

Standard error of difference in means (whole score)

$$s_{D\bar{x}} = \sqrt{\left(\frac{\Sigma X_1^2 + \Sigma X_2^2 - (N_1 \bar{X}_1^2 + N_2 \bar{X}_2^2)}{N_1 + N_2 - 2}\right)\left(\frac{1}{N_1} + \frac{1}{N_2}\right)}$$

10–5   201

Student's *t*

$$t = \frac{D\bar{x} - \mu}{s_{D\bar{x}}}$$

10–9   217

Standard error of the difference in means for correlated (matched) groups

$$s_{D\bar{x}} = \sqrt{\frac{s^2}{N} + \frac{s^2}{N} - 2r\frac{s^2}{N}}$$

10–10   223

Sandler's Λ

$$A = \frac{\Sigma D^2}{(\Sigma D)^2}$$

10–11   227

Name and Formula	*Formula Number*	*Page*

Chi-square statistic

$$x^2 = \sum \frac{(O - E)^2}{E}$$

11–1   235

Degrees of freedom (two-way classification)

$$df = (r - 1)(c - 1)$$

11–2   239

Chi-square corrected
  for continuity

$$x^2 = \sum \frac{(|O - E| - 0.5)^2}{E}$$

11–3   239

Chi-square for a 2 × 2 table

$$x^2 = \frac{N(|AD - BC| - .5N)^2}{(A + B)(C + D)(A + C)(B + D)}$$

11–4   240

McNemar test of change

$$x^2 = \frac{(|A - D| - 1)^2}{A + D}$$

11–5   241

Z-score conversion for chi-square

$$Z = \sqrt{2x^2} - \sqrt{2df - 1}$$

11–6   242

Mann-Whitney U test (counting)

$$U = N_1 N_2 - U'$$

11–8   249

Mann-Whitney U test (sum of ranks)

$$U = N_1 N_2 + \frac{N_1(N_1 + 1)}{2} - R_1$$

11–9   249

or

$$U = N_1 N_2 + \frac{N_2(N_2 + 1)}{2} - R_2$$

11–10   249

Name and Formula	Formula Number	Page

Total sum of squared deviations

$$SS_t - \sum^{N_t} X^2 - \frac{\left(\sum^{N_t} X\right)^2}{N_t}$$

12–7   ?71

Between-groups sum of squared deviations

$$SS_{bg} = \sum^{k} \frac{\left(\sum^{N_g} X_g\right)^2}{N_g} - \frac{\left(\sum^{N_t} X\right)^2}{N_t}$$

12–8   272

Within-groups sum of squared deviations

$$SS_w = \sum^{k} \left[\sum^{N_g} X_g{}^2 - \frac{\left(\sum^{N_g} X_g\right)^2}{N_g} \right]$$

12–9   272

t test following the analysis of variance

$$t = \frac{\bar{X}_a - \bar{X}_b}{\sqrt{MS_{wg}\left(\frac{1}{N_a} + \frac{1}{N_b}\right)}}$$

12–11   276

Kruskal-Wallis one-way analysis of
   variance for ranks

$$H = \frac{12}{N(N+1)} \sum^{k} \frac{R_j{}^2}{N_j} - 3(N+1)$$

12–12   283

Cochran Q test

$$Q = \frac{(k-1)\left[k\sum^{k} G_j{}^2 - \left(\sum^{k} G_j\right)^2\right]}{k\sum^{N} L_i - \sum^{N} L_i{}^2}$$

12–15   287

Friedman ANV for ranks

$$\chi_r{}^2 = \frac{12}{Nk(k+1)} \sum^{k} R_j{}^2 - 3N(k+1)$$

12–16   289

Name and Formula	Formula Number	Page
Estimate of sample size for *t* test (independent groups)		
$$N = \frac{2s^2 t^2}{D_{\bar{X}}^2}$$	13–1	302
Estimate of sample size for *t* test (correlated groups)		
$$N = \frac{2s^2 t^2 (1 - r)}{D_{\bar{X}}^2}$$	13–2	303

Index

A

Abscissa, 17
Additive theorem of probability,
    116
Algebra, review of, 52–60
Analysis of variance, 259–79
    assumptions underlying, 271
    computational procedure for, 271
    degrees of freedom, 261, 268
        between groups, 268
        total, 268
        within groups, 268
    derivation of, 264–68
    F ratios, 260–70
    mean squares, 269
        between groups, 269
        within groups, 269
    purpose of, 259–62
    ranked information, 282
        Friedman test, 289
        Kruskal-Wallis test, 283
    significance testing, 269
    sum of squared deviations, 265
        between groups, 267
        total, 267
        within groups, 268
    summary table, 274
    t tests in, 275
Anastasi, A., 198
Anastasio, E. J., 317
Areas of normal curve, 106, 332 (Table
    B)
Arithmetic mean, 62

Assumptions
    in analysis of variance, 271
    of chi-square, 242
    Cochran Q test, 286
    of distribution within class interval,
        13, 17, 30, 66
    Friedman test, 289
    Kruskal-Wallis test, 283
    McNemar test of change, 241
    Mann-Whitney U test, 248
    of median test, 246
    of Pearson product-moment correla-
        tion, 171
    Sandler's A test, 226
    of sign test, 252
    of t test, 227
    Wilcoxon test, 254
Average deviation, 84
Averages; see Central tendency

B

Backstrom, C. H., 152
Baggaley, Andrew R., 61
Bar graph, 19
Bashaw, W. L., 61
Binomial expansion, 122–29
    coefficients of, 123, 333 (Table C)
    mean of, 127
    normal curve approximation, 125
    probability, 122
    standard deviation of, 127
Binomial test, 233, 340 (Table I)

Blommers, P., 38, 103, 231
Boneau, C. A., 231

C

Cartesian coordinates, 17
Cattel, J. M., 2
Central limit theorem, 144
Central tendency, measures of, 62–78
  comparison of measures of, 72
  mean
    arithmetic, 62
    geometric, 71
    harmonic, 71
  median, 70
  mode, 70
  when to use each measure, 72
Chi-square, 234–42
  assumptions, 242
  computation of, 237, 238, 240
  degrees of freedom for, 236, 239
  distribution of, 236
  formula for, 235, 239
  McNemar test of change, 241
  median test, 246, 282
  one-way classification, 235
  restrictions on use of, 242
  sign test, 252
  table J, 341
  as test of independence, 237
  two-way classification, 237
  with more than 30 df, 242
  Yates' correction, 239
Class interval, 11–16
  limits of, 15
  midpoint of, 13, 17, 30, 66
  size of, 13
Cochran, W. G., 258, 286, 293, 336,
    (Table F)
Cochran Q test, 286
Combinations, in probability, 123
Confidence intervals, 142
  correlation coefficient, 165
  difference in means, 199
  proportions, 146
  regression prediction, 164
  true score, 189

Continuous numbers, 15
Correlation, 153–81
  accuracy of prediction, 164
  assumptions, 171
  and causation, 154
  contingency coefficient, 176
  degree and direction, 154
  nonparametric, table of, 174
  Pearson product-moment (r), 153–73
    assumptions for, 171
    computational procedure, 168
    in prediction, 157
    summary of, 156, 157
  r; *see* Pearson product-moment
  regression equation, 162
  scatter diagram, 159, 160
  Spearman rank difference correlation,
    174
  testing, significance of, 165, 176
Cox, D. R., 152, 317
Cronbach, L., 198
Cumulative frequency distribution, 27

D

Deciles, 37
Degrees of freedom, 219
  analysis of variance, 261, 268
  chi-square, 236, 239
  matched samples, 223
  t test, 219
Dependent variable, 17, 295, 306
Derived score, 184
Descriptive statistics, 3
Discrete numbers, 16
Dispersion, measures of, 79–95
  average deviation, 84
  interquartile range, 81
  range, 8, 13, 80
  semi-interquartile range, 81
  standard deviation, 85
  variance, 86
Distributions
  chi-square, 236, 341 (Table J)
  cumulative, 27
  frequency, 11, 24

Distributions—*Cont.*
  normal, 104, 332 (Table B)
  Poisson, 130
Dixon, W. J., 130
Durkheim, E., 2

E

Edwards, A., 155, 293
Error, 105
  sampling, 140
  Type I, 206
  Type II, 209
    factors which affect, 214
Evans, S. H., 317
Experimentation, 295–317
  matched samples in, 299
  noncomparable groups, 300
  planning experiments, 301
Extraneous variable, 297
Ezekiel, M., 180

F

F ratio, 259–79, 336; *see also* Analysis
    of variance
  assumptions, 271
  computation, 271
  degrees of freedom, 261, 268
  derivation, 265
  mean squares, 269
  multiple internal comparisons, 277
  proportion of variance, 314
  relationship to *t* test, 275
  sum of squares, 265
  summary table, 274
  table F, 336
Fechner, G. T., 2
Festinger, L., 317
Fisher, Sir R. A., 259
  table D, 334
  table G, 338
  table J., 341
Fisher-Yates test, 243
  exact probabilities, 243
  significance levels, 243, 342 (Table
      K)

Fox, K. A., 180
Freeman, Linton C., 181
Frequency distribution, 11
  description of, 24
  grouped, 11
  interval size, 13
  simple, 11
Freund, J. E., 103
Friedman analysis of variance for ranks,
    289

G

Galton, Sir Francis, 2
Gauss, C. F., 25
Geometric mean, 71
Gosset, W. S., 216
Graphing, 16–27
  Cartesian coordinates, 17
  histogram, 18
  labeling, 24
  ogive, 29
  polygon, 17
  rules, 21
Grouping scores, 7–15

H

Harmonic mean, 71
Helmstadter, G. C., 198
Histogram, 18
Homogeneity of variance, 203, 270, 271,
    275
Homoscedasticity, 172
Hursh, G. D., 152
Hypothesis, null, 204

I

Independent samples
  interval or ratio data
    analysis of variance, 262
    t test, 216
  nominal data
    binomial test, 233
    chi square test, 234–42
    k-sample, 279

Independent samples—*Cont.*
  nominal data—*Cont.*
    chi square test—*Cont.*
      one-way classification, 235
      two-way classification, 237
    Fisher-Yates test, 243
  ordinal data
    Kruskal-Wallis test, 283
    Mann-Whitney *U* test, 248
    median test, 246, 282
Independent variable, 17
Inferential statistics, 3, 75, 95
Interval scale, 46

J–K

Jenkins, W. L., 181
k-sample tests
  interval data, 262
  nominal data, 280, 286
  ordinal data, 282, 289
Katz, D., 317
Kish, L., 152
Kruskal-Wallis test, 283
Kurtosis, 26

L

Lev, J., 340 (Table I)
Limits of numbers, 15
Lindquist, E. F., 38, 103, 231
Linearity of regression, 162, 172

M

McNemar, Q., 38, 78, 103, 130, 181,
    239, 317
McNemar test of change, 241
Mann-Whitney *U* test, 248
  moderate samples, 252
    table M, 360
  small samples, 250
    table L, 357
Massey, F. J., Jr., 130
Matched samples, 299
  interval data
    Sandler's *A* test, 226

Matched samples—*Cont.*
  interval data—*Cont.*
    t test, 221
  nominal data
    Cochran *Q* test, 286
    McNemar test, 241
  ordinal data
    Friedman test, 289
    sign test, 252
    Wilcoxon test, 254
Maxwell, A. E., 258
Mean, 62–69
  combining means, 68
  computation of, 65
    with grouped data, 65
    with ungrouped data, 65
  definition of, 63
  estimation of μ, 76
    bias, 76
    consistent, 77
    efficient, 76
  geometric, 71
  harmonic, 71
  sampling distribution of, 137
Mean squares, 269
Measurement, scales of, 40–50
  interval, 46
  nominal, 40
  ordinal, 42
  ratio, 49
Median, 70
Median test, 246, 282
Merrington, M., 336 (Table F)
Messick, David M., 130
Mode, 70
Mosteller, F., 58, 294
Multiple internal comparisons, 277

N

Nominal scale, 40
Nonparametric, 232–58, 279–93
  binomial test, 233
  chi-square, 234
  Cochran *Q* test, 286
  contingency coefficient, 176
  Fisher-Yates test, 241

Nonparametric—*Cont.*
  Friedman test, 289
  Kraskal-Wallis test, 283
  McNemar test of change, 241
  Mann-Whitney *U* test, 248
  median test, 246, 282
  sign test, 252
  Spearman Rank correlation, 174
  tests compared, 257, 308
  Wilcoxon test, 254
Normal curve, 104–14
  and the binomial expansion, 125
  approximation for chi-square, 242
  area of, 106, 332 (Table B)
  definition of, 105
  estimating probabilities with, 126
  formula, 105
Null hypothesis, 204
  one-tailed and two-tailed, 212
  tests of, 308 (Table 13–1)
Nunnally, J. C., Jr., 198

O

Ogive, 29
Olds, E. G., 335 (Table E)
One-tailed versus two-tailed tests, 212,
    271
  binomial, 233
  F ratio, 261
  Fisher-Yates test, 243
  sign test, 252
  t tests, 212
Ordinal scale, 42
Ordinate, 17

P

Pearson, K., 157
Percentile ranks, 33–35
  computation of, 34
Percentiles, 27–33
  computation of, 29
Poisson distribution, 130
Polygon, frequency, 17
Population, 132
Power of a statistic, 209
Probability, 113–25
  actuarial, 113

Probability—*Cont.*
  additive theorem, 116
  binomial expansion, 122
  contingent, 120
  definition of, 114
  multiplicative theorem, 118
  normal curve approximation, 125
  subjective, 113
  theoretical, 114
Product-moment correlation coefficient;
    see Correlation
Proportion, 146–50
  confidence interval, 146
  sampling error, 146

Q–R

Quartiles, 38, 81
r; *see* Correlation
Random numbers, 136, 364 (Table N)
Random sampling, 133, 309
Range, 8, 13, 80
  semi-interquartile, 81
Rank-difference (rank-order) correlation
    coefficient, 174
  formula for, 174
  interpretation of, 175
  testing the significance of, 176, 335
    (Table E)
Ratio scale, 49
Ray, W. S., 293, 317
Regression, 153–68
  equation of, 162
Reliability, 186–91
  comparable form, 187
  split-half, 188
  test-retest, 187
Research results, interpreting, 309
Rho, 174
Robustness of *t*, 228
Rourke, R. E. K., 258, 294

S

Sampling, 131–52
  central limit theorem, 144
  confidence intervals, 142
  defining the population, **132**

Sampling—*Cont.*
  distribution of sample means, 137
  finite population, 135
    with replacement, 135
    without replacement, 135, 312
  multi-stage, 134
  random numbers, 136, 364 (Table N)
  random selection, 133, 309
  selecting the sample, 131
  standard error of the mean, 140
  stratified sampling, 132
Sandler's *A* test, 226; *see also t* test
Scaling, 39–52
  characteristics, summary of, 50
  interval, 46
  nominal, 40
  ordinal, 42
  ratio, 49
Semi-interquartile range, 81
Senders, V., 61
Siegel, S., 61, 181, 231, 258, 294
Sign test, 252
Significance
  choice of level, 149, 206, 212
  of correlation
    Pearson product-moment, 165, 334
      (Table D)
    Spearman rank-difference, 176, 335
      (Table E)
  of deviation from expected
    binomial test, 233
    chi-square test, 234
    Fisher-Yates test, 241
    McNemar test of change, 241
    Mann-Whitney *U* test, 248
    median test, 246
    sign test, 252
  of difference
    correlated samples, 221
    independent samples, 200, 209
    for means, 203
  of variance
    among several means, 262
    between two samples, 259
Skewness, 25
Spearman rank-difference correlation,
  174

Standard deviation, 85
  estimate of σ, 95
    bias, 96
    consistent, 96
    efficient, 96
  deviation formula, 85
  grouped data computation, 88
  relationship to range, 93
  whole score computation, 86
Standard error of difference between
  means, 200
Standard error of mean, 140
Standard error of proportion, 146
Stevens, S. S., 40, 61
Sum of square, analysis of variance,
  265

T

t test, 216
  assumptions, 227
  computation, 221, 224, 226
  definition, 217
  degrees of freedom, 219, 223
  distribution of, 217
  with matched samples, 221
  relationship to *F* test, 275
  robustness of, 228
  Sandler's *A* test, 226, 339 (Table H)
  table G, 338
Tchebysheff's inequality, 91
Tests, one- and two-tailed, 212
Thompson, C. M., 336 (Table F)
Thurstone, L. L., 82
Tippet, L. H. C., 93

U

U test, 248
Underwood, B. J., 317

V

Validity, 191–96
  concurrent, 192
  criterion related, 191
  effects of unreliability, 195
  predictive, 192

Variability, 79–103
  average deviation, 84
  range, 80
  semi-interquartile range, 81
  standard deviation, 85
  variance, 86, 259
Variables
  continuous, 15
  dependent, 17, 295
  discrete, 16
  extraneous, 297
  independent, 17, 95
Variance; *see also* Analysis of variance
  proportion of, 313
    with analysis of variance, 314
    with correlation, 213

W

Walker, H. M., 61, 78, 340 (Table I)
Wilcoxon matched pairs signed-ranks
  test, 254, 367 (Table O)
Winer, B. J., 293
Wundt, W. M., 2

Y–Z

Yates, F., 239, 243
  table D, 334
  table G, 338
  table J, 341
Z scores, 97–101
  formula for, 97
  uses of, 98

*This book has been set in 11 and 10 point
Garamond, leaded 2 points. Chapter numbers
are in 30 point Scotch Roman italic and chapter
titles are in 18 point Scotch Roman italic. The
size of the type area is 27 × 45 picas.*